THE PRACTICE OF
SILVICULTURE

EIGHTH EDITION

THE PRACTICE OF
SILVICULTURE

DAVID M. SMITH

Morris Jesup Professor of Silviculture
Yale University School of Forestry
and Environmental Studies

JOHN WILEY & SONS
New York • Chichester • Brisbane • Toronto • Singapore

Library of Congress Cataloging in Publication Data:

Smith, David M. (David Martyn)
  The practice of silviculture.

  Rev. ed. of: The practice of silviculture / Ralph C.
Hawley. 7th ed. 1962.
  1. Forests and forestry.
The practice of silviculture.   III. Title.

SD391.S57 1986   634.9   85-26410
ISBN 0-471-80020-1

Printed in the United States of America

10 9 8 7 6 5

To Ralph C. Hawley, pioneer American silviculturist, who published the first edition of this work in March 1921

P R E F A C E

The first five editions of this book were written by the late Professor Ralph C. Hawley; the first was published during the month when the current author was born. It was Hawley's view that silvicultural practices should be prescribed by foresters who had their feet on the ground, literally and figuratively, with their minds at work.

In subsequent editions, I have attempted to carry Professor Hawley's philosophy into the present. Even more emphasis has been put on applying scientifically analytical reasoning to the formulation of silvicultural solutions to the diversity of management problems that foresters face. These problems are still construed as including both biological and socioeconomic considerations.

Previous editions of this book were intended entirely for use in North America but have come to be used on other continents. This edition has, therefore, been given a broader geographical scope although it is still intended primarily for North America. However, the continent has so many different kinds of forests that it is not a long step from them to the rest of the world. The principles of silviculture are independent of geography, even though geographical diversity makes application of the principles highly variable.

In view of the transition from the English to the metric system of measurements in the United States both systems are used interchangeably. The metric system is the preferred one but not to the extent of pedantic references to saplings 2.54 cm in diameter. The useful concept of the board foot is not abandoned.

The list of references at the end of each chapter is designed mainly to lead the reader to publications that bring together as much other literature as possible. The references chosen are usually those likely to be accessible to students in North America. Heavy reliance is placed on the recent phenomenon of the symposium with published proceedings; these often have the desirable feature of including the ideas of practicing foresters as well as researchers. However, the volume of literature has become so great that it is no longer possible to attempt the kind of reviews of American writings that appeared in earlier editions. The original sources of many ideas have gone uncited.

I am indebted to Professor Hawley for the wealth of ideas and the material expressive of them that has been inherited from the long history of this book. Thanks are due to Professors Craig Lorimer and Bruce Larson, as well as Gro Flatebo, Betsy Hale, and Mir Javed Hussain, for critical editorial assistance.

The opportunity to make selections of illustrations from the collections of the U. S. Forest Service, American Forest Institute, American Forestry Association, and other organizations credited in the text has been invaluable.

This book is a collection of ideas about silviculture. I am humbly and respectfully grateful to the host of foresters and others concerned about forests from whose thoughts, experience, and studies these ideas have been gathered.

David M. Smith

New Haven, Connecticut

CONTENTS

PART TWO
Regeneration   **189**

THE PRACTICE OF
SILVICULTURE

CHAPTER 1

Silviculture and Its Place in Forestry

Silviculture has been variously defined as: the art of producing and tending a forest; the application of knowledge of silvics in the treatment of a forest; the theory and practice of controlling forest establishment, composition, structure, and growth (Spurr, 1979). Silvicultural practice consists of the various treatments that may be applied to forest stands to maintain and enhance their utility for any purpose. The duties of the forester are to analyze the natural and social factors bearing on each stand and then devise and conduct the treatments most appropriate to the objective of management.

Silviculture is to forestry as agronomy is to agriculture in that it is concerned with the technology of crop production. Like forestry itself, silviculture is an applied science that rests ultimately upon the more fundamental natural and social sciences. The immediate foundation of silviculture in the natural sciences is **silvics**, which deals with the principles underlying the growth and development of single trees and of the forest as a biological unit. Among the sources of information about silvics are books by Spurr and Barnes (1980), Daniel, Helms, and Baker (1979), Hocker (1979), Kramer and Kozlowski (1979), Kozlowski (1971), and the U. S. Forest Service (1965). This textbook is written primarily for students who have studied silvics and understand its main concepts and terminology.

The competent practice of silviculture, whether it be crude or elaborate, demands as much knowledge of such fields as ecology, plant physiology, entomology, and soil science as a forester can acquire. It is through silviculture that a major part of the growing store of knowledge about trees and forests is applied. These things are not learned once for a lifetime. The forestry practitioner must continually keep abreast of new knowledge and ideas through communication with other members of the profession and familiarity with the results of formal research. While formal research is indispensible, it does not lead to total knowledge nor does it relieve the forester from responsibility for additional thought. In applied science, one

1

is always condemned to act, on the basis of thoughtful judgment, in the absence of total knowledge. Skillful practice itself is a continuing, informal kind of research in which understanding is sought, new ideas applied, and old ideas tested for validity. The observant forester, who is wise to seek to explain what is observed, will find answers to many silvicultural questions in the woods by examining the results of accidents of nature and earlier treatments of the forest.

The practice of silviculture is concerned with the social as well as the biological aspects of forestry. The implicit objective of forestry is to make forests useful to society. Since all management is, therefore, aimed at economic objectives, it is almost impossible to separate the biological aspects from the economic. The mere fact that the silviculturist grows trees so that they will be useful rather than merely vigorous from the physiological standpoint automatically introduces an economic purpose.

The Purpose of Silviculture

Silviculture is normally directed at the creation and maintenance of the kind of forest that will best fulfill the objectives of the owner. The growing of timber and other forest commodities is the most common objective, but it is not always the primary one and there are instances where it is not an objective at all. Sometimes it is just a way of helping to finance or otherwise facilitate the attainment of other purposes of forest management. However, silviculture for timber production is usually the most intricate because both the kind and quality of trees are of greater concern than with other forest uses. In this book greatest emphasis is placed on timber production only because silviculture aimed at managing forests for water, wildlife, grazing, recreation, or aesthetics is usually less complicated and more simply perceived.

The important point is that the objectives should be clearly defined and silviculture directed to their attainment. There is also the implicit objective of working for the good of the forest as an entity, not as an end in itself but as a means of ensuring that it will be a permanently productive source of goods and benefits.

Improving on Nature through Silviculture

The most magnificent forests that are ever likely to develop were present before civilization and grew without human assistance. Furthermore, under reasonably favorable conditions, forests may remain productive even after long periods of mistreatment. Therefore, it is logical to consider why foresters should attempt to direct any of the powerful natural forces at work in the forest.

The purely natural forest is governed by no purpose unless it be the unceasing struggle of all the component plant and animal species to perpetuate themselves. Human purpose is introduced by preference for certain kinds of trees or stand structures that have desirable characteristics.

Where desirable forests have developed in nature they are usually found to have been the result of fortuitous events followed by long periods of growth. In silviculture, natural processes are deliberately guided to produce forests that are more useful than those of nature, and to do so in less time.

The time required for growth was not a factor in utilizing virgin forests because no one had to pay any costs of holding land while the trees were growing. In managed forests the rate at which value is produced, and not the final value, is the important consideration. Unmanaged or mismanaged forests, like poorly treated farm lands, do not yield products and benefits of the kind or value usually desired.

The productivity of the managed forest is improved through attainment of the objectives listed in the following sections of this chapter.

Control of Stand Structure and Process

Silviculture is forest architecture aimed at the design of stands with outward shape and internal construction that will serve the intended purposes, be in harmony with the environment, and withstand the loads imposed by environmental influences. Since the stands grow and change with time, the designing is more sophisticated and difficult to envision than that of static buildings. In this sense it is akin to the kinds of engineering in which dynamic processes and not just structures are designed. Furthermore, the stands alter their own environment enough that the forester is partly creating a new ecosystem and partly adapting to the one that already exists.

The possible variations in stand structure and process are almost infinite. The shapes and sizes of stands can be altered for many purposes. A few among these are expediting silvicultural treatments and harvesting, creating attractive scenery, governing animal or pest populations, trapping snow, and reducing wind damage. The shapes of stands should be fitted to the immutable patterns dictated by soils and terrain. While arrangement of stands in checkerboard patterns has a certain administrative appeal, the natural characteristics of land are not arranged in such ways.

The internal structure of a stand is determined by considerations such as variation in species and age classes (or lack of it), the arrangement of different layers or stories of vegetation (usually differing as to species), and the distribution of diameter classes. Much of this book is concerned with the purpose and means of achieving these kinds of structural variations.

Control of Composition

Undesirable species or inferior individuals of otherwise desirable species appear in almost any forest. One objective of silviculture is to restrict the composition of stands to that most suited to the location from economic and biological standpoints. This frequently means that the total number of species in a managed forest is less than that which could grow there under purely natural conditions. Unwanted plants commonly flourish at the expense of the desirable, so every reasonable effort should be made to keep them in check.

The basic means by which species composition is controlled is through regulating the kind and degree of disturbance during periods when new stands are being established. In this way environmental conditions can be adjusted to favor stages of natural succession most nearly dominated by the desired species.

Regulation of natural succession by itself is not always enough to provide adequate control over stand composition. It is often necessary to supplement this approach by direct attack on the undesirable vegetation during or after periods of stand establishment. Cutting, poisoning, controlled burning, or regulated feeding by various forms of animal life may be used to restrict the competition and regeneration capacity of undesirable vegetation.

Desirable species and genotypes can be favored in more positive fashion by planting or artificial seeding. By this means it is also possible to improve upon nature through introduction of species that do not occur in the native vegetation, provided they are adequately adapted to the environment.

The goal of controlling stand composition is to achieve the best fit between the purposes of management and the natural constraints imposed by the site.

Control of Stand Density

Inadequately managed forests are commonly too densely or too sparsely stocked with trees. If stand density is too low, the trees may be too branchy or otherwise malformed and the unoccupied spaces are likely to be filled with unwanted vegetation. This condition arises from inadequate provision for regeneration; it is most common in the early life of a stand but its consequences may linger after the trees have grown to occupy all of the space available. Excessively high stand density causes the production to be distributed over so many individual trees that none can grow at an optimum rate and too many may decline in vigor. Unless stand density is carefully controlled at the time a stand is established or during its development, it is almost sure to depart from optimum density at some stages of its life.

Restocking of Unproductive Areas

Without proper management many areas of land potentially suited to growth of forests tend to remain unstocked with trees (Fig. 1-1). Fires, destructive logging, animal browsing, and ill-advised clearing of land for agriculture have already created many large, open areas that can be put back into immediate timber production only by planting.

Protection and Reduction of Losses

In unmanaged stands, severe losses are commonly caused by damaging agencies such as insects, fungi, fire, and wind, as well as by the loss of merchantable trees through competition. Substantial increases in production may be achieved merely by salvaging material that might otherwise be lost. Proper control of damaging agencies can result in further increases in production. Forest protection often involves modification of silvicultural techniques. Adequate protection should be extended to all forests, even the poorest, because fire and pest outbreaks developing in stands where they may be of little concern can sometimes spread to those deemed worthy of protecting. Those areas set aside for wilderness, scenery, or scientific study require some protection; sound policies about their care and use inevitably involve something other than leaving them entirely alone.

Control of Rotation Length

Stands of trees are not immortal. In most situations there is an optimum size or age to which trees should be grown. The period of years required to grow a stand to this specified condition of either economic or natural maturity is known as the rotation. Premature cutting is a common type of mismanagement. Trees allowed to grow beyond the optimum do not continue to increase in value at rates sufficient to provide acceptable return on either the costs of growing them or the investment represented by their own value. The risk of decay or other damage may increase the possibility that they will decline in value, be lost, or become some sort of hazard. Proper regulation of stand density can shorten rotations by making the final-harvest trees grow to the desired sizes at earlier ages.

Facilitating Harvesting

In the unmanaged forest, timber, like gold in the hills, is where one finds it; the greater the amount extracted, the more difficult and expensive it becomes to find and extract more. In the managed forest it is possible to plan the growth of stands so that any use of them is on a more efficient, economical, and predictable basis. It becomes possible to produce good stands so located that the cost of transporting timber from them is kept under control.

Conservation of Site Quality

Paramount among the objectives of forestry in general and of silviculture in particular is the maintenance of the productivity of the forest. The **site** is the total combination of the factors, living and inanimate, of a place that determine this productivity of life.

The site factors that are most subject to long-lasting harm are those of the soil, which is the most nearly nonrenewable of the resources used in

FIGURE 1-1   Much of silviculture has always consisted of rehabilitation efforts and of knowing what will happen as a result of treatments of the forest. This sequence of pictures from 1938, 1949, and 1969 shows a planted stand in the St. Joe National Forest in northern Idaho in three stages of development. The tract had been logged over from a logging railroad in 1930–1931 and both burned and acquired just before the first picture was taken. Planting of western white pine and Engelmann spruce was done in 1939 and 1940. The subsequent pictures show the development, to age 30, of the mixture of planted trees and other conifers that seeded-in naturally.

silviculture. The basic supply of solar energy, the most vital site factor, is beyond human influence. While silviculture rests heavily upon manipulation of the *microclimate* of a site, its effects on the *macroclimate* are minute and any benefits probably outweigh any harm.

The living organisms of a place are site factors themselves; however, they can reproduce themselves and are thus the epitome of the renewable resource. If none are rendered extinct, damage to these living components of the site is not likely to be permanent even though it can be serious and long-lasting. There are always uncertainties over the extent to which silviculture should discriminate against "undesirable" forms of life.

The most obvious and least reparable kind of damage to the soil is physical erosion. It is entirely possible for careless treatments to cause several years of accelerated erosion that may negate the soil formation processes of a thousand years. A more subtle kind of chemical erosion can result if the remarkable capacity of forest vegetation to recycle nutrients in place is so impaired that large amounts of vital chemicals are irretrievably lost to surface runoff or leaching. These two kinds of erosion cause harm twice because they reduce the productivity of the soil and the quality of the water that flows from it. Soil damage impairs the capacity of the site to yield all of the primary tangible benefits of the forests—vegetation, animals feeding on plants, and good water.

It is entirely within the realm of possibility to conduct forestry permanently without the degradation that is almost inevitable in agriculture and other "higher" uses of land. However, realization of this potentiality is not automatic.

Silviculture as Applied Ecology

Silviculture is the most ancient conscious application of the science of ecology; the association arose before the word "ecology" was coined. The reliance on ecological knowledge is all the firmer for resting on the virtue of necessity and not only on philosophical principle. The economic returns from forestry are not high enough to make it feasible to shield forests from all the vicissitudes of nature. Therefore, silviculture is usually far more of an imitation of natural processes than of substitution for them. Those of forest growth may be improved upon, channelized, and limited; however, excessive disruption leads to severe losses, high costs, and other evil consequences, immediate or delayed.

The necessity that nature should be understood and emulated does not mean that silviculture should slavishly follow either the reality of natural processes or abstract theories about them. Most forests live longer than people. It is, therefore, not easy to detect that the natural disturbances that renew forests, often after intervals of centuries, are usually catastrophes such as fires, windstorms, and insect outbreaks (Oliver, 1981; Means, 1982). There are also forests that are slowly and continuously renewed by minor disturbances, but these are far from being any universal or typical form of nature. It is important to know the successional developments of the vegetation and to recognize that there may be a final climax stage in which changes slow and cease. However, this does not mean that there would necessarily be enough freedom from disturbance to permit such a stage to be reached or that it would be ideal if achieved. Major portions of the original forest of North America, including some of the most magnificent and productive, were disturbed too frequently to allow stable climax vegetation to develop.

The continuity of life ultimately depends upon completion of ecological cycles of vital materials. Some are closed in seconds within the space of a few millimeters whereas those which involve the natural loss of certain gases to the atmosphere will not be closed within a single continent or year. Some purely natural cycles are not completed, if ever, until nutrients lost to streams are heaved up from the ocean floor millions of years later.

It is desirable that these cycles be closed as nearly as possible within distances of time and space that are not needlessly large. However, this does not mean that they need to be closed and contained within single hectares on an annual basis; they were not in nature and it is not feasible to do so in practice. In other words, the concept of the mixed, all-aged stand of stable climax vegetation with some ancient trees and no net gain or loss of materials is more of an abstraction than a reality. Diagrams depicting such forests are useful for conveying ideas succinctly but the concept that they depict may be to forest ecology and silviculture what the perfect blackbody is to physics and engineering.

Ecological Technology

Silviculture is not conducted in a pure state of nature and this state ceased to be pure when society advanced beyond the food-gathering stage. To the extent that civilization is partly artificial it can be argued that forestry must also be partly artificial in ways calculated to keep the world ecosystem in balance. However, the wider any step departs from nature the more perilous it is if only because it becomes less possible to predict the results.

The web of life is so intricate that it is easy to argue that one should do nothing to the forest for fear of doing something wrong. This is the indictment continually brought by pure science against the technology of all applied science. The charge can never be entirely refuted but society requires practitioners of applied science to act in the absence of full knowledge.

Silviculture is conducted on the basis of ecological principles and not in spite of them. There is virtually no way of using land resources so heavily with so little risk of depleting their productive capacity. The goods and benefits that flow from forests under sound, long-term management depend on living processes and are thus renewable to the extent that basic productive site factors are maintained; they can even be increased if these factors are permanently improved. The wood produced by forests is by far the most important structural substance in human use; unlike alternative mineral or agricultural materials, its production requires little that would damage or pollute. In fact, the growing of it increases the stock of resources even as it cleans both air and water. If forest vegetation were more efficient in yielding human food and in concentrating sources of fuel, the future of the world ecosystem would be much brighter for mankind.

Economic and other social factors decide the silvicultural policy for any given area; the objective is to operate so that the value of benefits derived from a forest should exceed the value of efforts expended. The most profitable forest type is not necessarily the one with the greatest potential growth or that which can be harvested at lowest cost. One must also consider the silvicultural costs of growing the crop and the prospective losses to the damaging agencies. In fact, it is usually the inroads of insects, fungi, and atmospheric agencies that ultimately show where silvicultural choices have run afoul of the laws of nature. The majority of the best choices are imitations of those natural communities, not necessarily or even commonly climax communities, that have grown well in nature.

It is not entirely safe to accept the success of modern agriculture as justification for highly artificial kinds of silviculture. The environment of a cultivated field is much more thoroughly modified and readily controlled than that of a forest stand. Furthermore, forest crops must survive winter and summer over a long period of years while most agricultural crops need survive only through a single growing season. One disastrous year harms the production of but one year with an annual crop but can destroy the accumulated production of many years in a stand of trees. Neither economic nor ecological principles permit the forester to engage in the wholesale, routine use of pesticides and fertilizer on which intensive agriculture often rests. Any silvicultural application of refinements borrowed from agriculture must be combined with *all* the kinds of measures appropriate to the intensity of agriculture imitated. Forestry can profitably borrow much more than it ever has from the science on which modern agriculture is based, but there is little place for uncritical imitation.

Some silvicultural measures depart very far from natural precedent. These usually involve the introduction of exotic species of the creation of communities of native species unlike anything that might come into existence naturally. Departures of this sort cannot be condemned out of hand but should be viewed with reservations until they have been tested over long periods. Otherwise, most of the choices can be thought of in terms of the degree to which natural succession is accepted or arrested, pursued or reversed. Any departure from nature that still looks good at the end of a rotation is probably sound.

Choice of Successional Stages

The best approach to the matter is to determine which stage of successional development is most desirable in a given situation. In the Pacific Northwest, for example, the forester must often decide whether to perpetuate pure stands of Douglas-fir or allow them to be succeeded by western hemlock. In the Lake Region, he may have to choose between the pioneer aspen association and later stages such as spruce–fir association. In the South, he must determine whether to let old–field stands of loblolly pine revert to hardwoods. Several generalizations of wide, but not universal, application may be introduced at this point.

In the first place, the most valuable commercial species in any region tend to be relatively intolerant trees representative of the early or intermediate stages in natural succession. Species such as pines, Pacific Coast Douglas-fir, yellow-poplar, and ash definitely fall in this category. It is no coincidence that intolerant species are important commercially because they are the ones most likely to lose their lower branches through natural pruning. It is of significance that some of them are adapted to reproduce mostly after major disturbances that happen seldom. If they are to survive from one major disturbance to another they must be long-lived; as a result they are likely to develop the economically desirable attributes of large size and resistance to decay or other relatively minor sources of damage. Late successional forest types, characterized by species such as hemlock, true firs, and beech, are frequently composed of branchy trees that produce less valuable wood. Because of their shade tolerance they can reproduce almost continuously so the ability of individuals to endure for long periods is not so crucial in the survival of the species. Of course, many pioneer species have even less capacity for individual survival; however, they usually exhibit good natural pruning and the necessity that they grow rapidly to seed-bearing age is an economically desirable attribute.

Natural succession proceeds most rapidly and vigorously on the better sites, that is, on soils that are both moist and well aerated. Here it is sometimes impossible to resist natural succession without expensive silvicultural treatments. Furthermore, good sites are hospitable to the growth of such a large assortment of species that silvicultural treatment becomes complicated and difficult. These considerations have the paradoxical effect of making silviculture most profitable on sites of intermediate quality where uncomplicated stands can be maintained without strenuous effort. In fact, on poor sites it may occasionally be virtually impossible for succession to proceed beyond an intermediate stage, which is sometimes referred to as a **physiographic climax.** For example, stands of both jack and red pine occasionally represent valuable physiographic climaxes on certain dry, sandy soils in the Lake Region.

It has also been claimed that late successional types may be more resistant to, and more cheaply protected from, fire, insects, fungi, wind, and weather than earlier stages. This advantage results more from the diversity of species and age classes in a climax type than it does from position in the successional scale. Similar advantages can prevail in mixed stands with a variety of age classes that are still typical of earlier successional stages.

The point is often advanced that natural communities are in a stable and favorable equilibrium with the physical and biological environment. Perfect stability and complete favorability do not exist, so one must think in terms of relative degrees of each quality. For example, the balance achieved by long-continued natural processes, operating more or less at random, is not necessarily more favorable to the trees than to the organisms that feed upon them. The more artificial equilibrium produced by prudent silviculture may be less *stable* but ought to be more *favorable* from the standpoint of the integrated effect of all socio-economic factors. If the dynamic equilibrium created by treatment ultimately balances at a condition of economic disaster, the silviculture was hardly prudent.

The naturalistic doctrine of silviculture did not arise from any clearly demonstrated disadvantages of early stages of natural forest succession. It developed largely from disappointments with attempts to create unnatural types, particularly with exotic species or those not indigenous to the sites involved. The most extreme manifestation of this viewpoint is, in some respects, merely an unwarranted extension of a sound observation.

Difficulties on Specific Sites

It is better to think of the applicability of naturalistic doctrines in terms of specific cases and causes rather than just in generalizations. For example, there has long been suspicion that repeated planting of pure conifers might cause soil deterioration and "second-rotation declines" in productivity. The classic case involved Norway spruce in northern Germany (Rehfuess, 1981), where it was ultimately found that the particular problem was confined to certain heavy clay soils of lowlands. The spruce had been brought down from the cooler, moister climate of neighboring mountains and planted where deeper-rooted hardwoods had been cleared away. As the old root canals formed by the hardwoods closed, internal soil drainage became poorer. The soils were then so saturated and poorly aerated in the spring that the spruces became even more shallow-rooted than they were under normal conditions. Therefore, during the summer, when the low rainfall was further reduced in effectiveness by interception, the spruces suffered from drought. This might in turn reduce resistance to root-rotting fungi and other damaging agencies (possibly including acid fog). The basic difficulty was not mysterious and it was confined to clay soils.

A more recent case of second-rotation decline involves planting the Californian *Pinus radiata* on old sand dunes in South Australia. While the matter remains under examination (Pritchett, 1979), it seems most likely that supplies of nitrogen compounds that could not be used by the previous wild vegetation had become available to the pines and were depleted during the first rotation. It is also suspected that too much organic matter had been destroyed in preparing the sites for the second rotation. However, the important point is that the problem is confined to deep sands poor in water, nutrients, and organic matter. There is no universal "second-rotation decline" in radiata pine in the Southern Hemisphere.

Both of these cases are important lessons for foresters everywhere but not to the extent that one may conclude that repeated crops of pure conifers always cause soil deterioration. The spruce case shows that proper soil aeration is necessary for trees to grow well and that shallow-rooted trees may allow internal drainage to deteriorate on soils with a high clay content. The little-leaf disease of shortleaf pine on badly eroded clay soils in the Piedmont Plateau of the southeastern United States is a more spectacular example of the same thing, although planting loblolly pine is one of the best solutions. The South Australia case shows that the supply of nutrients and organic matter on old sand dunes is small and must be carefully husbanded. However, it is well to note that neither these cases nor any others will ever be absolutely and fully understood.

Scope and Terminology of Silvicultural Practice

Silvicultural practice encompasses all treatments applied to forest vegetation. While there is much more to the understanding of these treatments than their definitions and nomenclature, the terminology must be understood and *used carefully and precisely.* Sloppy use of the terms has caused all manner of misunderstanding within the forestry profession and in dealings with the general public. For example, some categorize all cutting as either "clearcutting" or "selective cutting." This not only stunts the development of their own understanding of forestry practice and causes blunders but also keeps everyone else confused. The terminology in this book generally adheres to that promulgated by the Society of American Foresters and the Commonwealth Forestry Bureau (Ford-Robertson, 1978); it departs only where further improvement in clarity or precision seems imperative.

The scope of silvicultural treatment and some of its terms may be defined under two categories. (1) **Methods of reproduction** refer to treatments of stand and site during the period of regeneration or establishment while (2) **tending** or **intermediate cutting** refers to treatments at other times during the rotation (Fig. 1-2). A program for the treatment of a stand during a whole rotation is called a **silvicultural system.** Although the silvicultural system includes both the method of reproduction and any tending operations, it is customarily given the name of the reproduction method employed. This confusing custom arose because the reproduction method has a decisive effect on the structure and treatment of the stand throughout its life. The clearcutting system, for example, leads to reproduction by the

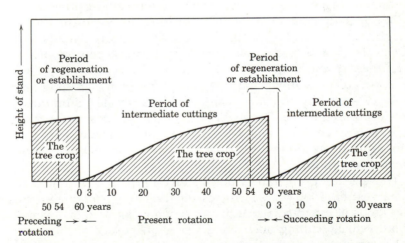

FIGURE 1-2 The relationship between the period of regeneration and the period of intermediate cuttings for a sequence of even-aged stands managed on a 60-year rotation according to the shelterwood system.

clearcutting method but may also include thinning during the period of tending.

The act of replacing old trees, either naturally or artificially, is called **regeneration** or **reproduction** and these two words, which are synonymous in this usage, also refer to the new growth that develops. **Advance reproduction** or **regeneration** is that which appears spontaneously or is induced under existing stands. **Reproduction** or **regeneration cuttings** are made with the twin purposes or removing the old trees and creating environments favorable for establishment of regeneration. The period over which such regeneration treatments extend is the **reproduction** or **regeneration period.** Regeneration cuttings range from one to several in number and the regeneration period may extend from several years to several decades. In truly uneven-aged stands, handled under the selection system, regeneration is almost always under way in some part of the stand. The regeneration period begins when preparatory measures start and ends when young trees, free to grow, are dependably established in acceptable numbers. The **rotation** is the period during which a single crop or generation is allowed to grow.

The names of the various methods of regeneration (see Chapter 12) denote the patterns of cutting in time and space that determine the structure of the stands created or maintained by the process. However, the natural and artificial sources of regeneration are numerous and the different measures that can be taken to foster it are even more so (Chapter 7). The name of the regeneration method tells something about the cutting pattern usually only begins to describe the details of treatment.

After a new stand is established, a long period ensues during which the crop grows through various stages until it is mature and ready for replacement by a succeeding generation. Various **intermediate cuttings** or **tending operations** may be conducted during development from the regeneration stage to maturity. These are done to improve the existing stand, regulate its growth, and provide for early financial returns, *without any effort directed at regeneration.*

Intermediate cuttings that are aimed primarily at controlling the growth of stands by adjusting stand density are called **thinnings**. Treatments conducted to regulate species composition and improve the quality of very young stands are **release operations;** cuttings made in older stands for these purposes are **improvement cuttings;** those that involve only the branches are **pruning.**

Those tending operations that involve outright investment without any harvest of wood are sometimes referred to as **stand improvement.** Many kinds of intermediate cutting or tending can now be accomplished without actually cutting down trees. Therefore, usage of the word "cutting" is now often correct only if viewed in a traditional and figurative sense.

Protection of the stand against injury is as much a part of silviculture as harvesting, regenerating, and tending to the crop. It is, in fact, so important that it has led to fields of academic specialization in forestry such as entomology, pathology, and fire control and is no longer included as a separate section of this book. The details of almost any successful silvicultural system include significant modifications designed to reduce injuries. Where such measures fail or are inadequate it is sometimes desirable to conduct **salvage cuttings** to recover the values represented by damaged trees or stands.

Role of Cutting in Silviculture

The techniques of silviculture proceed on the basic assumption that the vegetation on any site tends to extend itself aggressively to occupy the available growing space. The limit on growing space is usually set by the availability of light, water, inorganic nutrients, or carbon dioxide. In a general way the available amount of growing space is set by the most limiting of these factors, although an abundant supply of one factor can partially offset deficiency of another.

If the vegetation nearly fills the growing space, the only way that the forest can be altered or controlled is by killing trees and other plants. In reproduction cutting this is done to provide room for the establishment of new trees; in intermediate cutting, to promote the growth of desirable trees already in existence. Paradoxical as it may seem, and repugnant as it may be to certain influential segments of public opinion, useful forests are created and maintained chiefly by the destruction of judiciously chosen parts of them. One of the characteristics of life is death; if there were no death, there would be no space for new life.

The axe and other means of killing trees can be used for the construction as well as the destruction of the forest. The importance of cutting as a means of harvesting wood for human use should not obscure its role as the major means by which forests are established and tended.

Preoccupation with the trees should not cause one to overlook the lesser vegetation and the animals that are a part of the forest community. The animals ultimately depend on the vegetation for food and thus do not compete directly for the growing space. However, whether they be defoliating insects or carnivores that feed on herbivorous mammals, they can exert a major influence on the nature of the vegetation even as they are, in turn, controlled by it. The fauna and nonarborescent vegetation of the forest are as likely to respond to cutting as the trees.

Effect of Cutting on Growing Stock

Cutting controls not only the composition and form of forest stands but also the relationship between trees reserved for growth and the amount of growth available for cutting. It is important to understand the long-term, cumulative effect of cutting operations in building, or degrading, a forest.

Timber growing is one of the few kinds of creative processes in which both product and productive machinery are the same thing (Heiberg, 1945). The wood of the stem cannot be removed without destroying the machinery that produced it. A clear distinction must, therefore, be drawn between the trees that must be left to produce more wood and the surplus trees that can be regarded as product and harvested.

The trees that must be reserved somewhere in the forest to continue production are the **growing stock** or **forest capital.** The volume of wood that is grown in the future depends on the quantity and condition of growing stock that is maintained. Cuttings regulate the amount of this growing stock and its distribution within individual stands or among the various stands that comprise the forest. The regulation of growing stock is of most crucial importance in silviculture when partial cuttings are applied within stands.

Structure of Forest Stands

A **stand** is a contiguous group of trees sufficiently uniform in species composition, arrangement of age classes, and condition to be a distinguishable unit. The internal structure of stands varies mainly with respect to the degree that different species and age classes are intermingled. The simplest kind of structure is exemplified by that of the pure, even-aged plantation consisting of trees of a single species. The range of complexity can extend to a wide variety of combinations of age classes and species in various vertical and horizontal arrangements.

Arrangement of Age Classes

True regeneration cuttings and natural lethal disturbances of similar magnitude determine the times when new trees appear or start active development on any given unit of ground area. Each new aggregation of trees so produced is an **age class** of trees all of essentially the same age. Differences in the timing of regenerative events create various spatial patterns of age classes. The area occupied by a given age class can be of any size, provided that it is large enough that some new trees can continue to grow in height without being arrested by expansion of the crowns of older adjacent trees. Only those truly regenerative events that leave new or small trees free to grow really affect the arrangement of age classes. Intermediate cuttings such as thinnings do not leave new trees free to grow and thus have no effect on age-class arrangement.

There are three general types of age-class structure within stands: even-aged, stands with two age classes, and uneven-aged. They are most easily distinguished in pure stands, so these will be considered first and the complexities of mixed stands will be taken up later.

In an **even-aged stand** (Fig. 1-3) all trees are the same age or at least of the same age class; a stand is considered even-aged if the difference in age

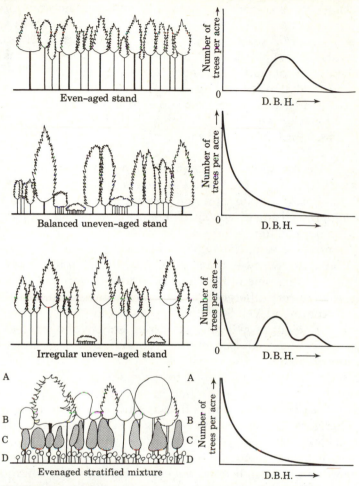

FIGURE 1-3   Typical examples of four different kinds of stand structure, showing appearance of stands in vertical cross section and corresponding graphs of diameter distribution in terms of numbers of trees per unit of area. The trees of the first three stands are all of the same species but the fourth consists of several species but all of the same age.

between the oldest and youngest trees does not exceed 20 percent of the length of rotation. An **uneven-aged stand** contains at least three age classes intermingled intimately on the same area. **Stands with two age classes** represent an intermediate category. All gradations of age distribution may be found in nature or created by cuttings designed to make new way for new age classes.

Distinction should be made between balanced and irregular uneven-aged stands. **A balanced uneven-aged stand** consists of three or more different age classes, each of which occupy an approximately equal area;

the age classes are also spaced at uniform intervals all the way from newly established reproduction to trees near rotation age. Such stands, once created, may function as self-contained, sustained yield units. **Irregular uneven-aged stands** do not contain all the age classes necessary to ensure that trees will arrive at rotation age at short intervals indefinitely. Uneven-aged virgin stands and stands that have been culled over without plan are almost always irregular in age distribution. In fact, irregular uneven-aged stands are common and may be highly desirable, so long as they are recognized and treated for what they are.

Identification of Age Classes.   The profile of a stand is a good criterion of age distribution because trees of the same age grow in height at roughly the same rate, provided site conditions are uniform; those that do not keep pace are suppressed and disappear. Therefore an even-aged stand tends to be almost smooth on top. An uneven-aged stand is usually distinctly irregular in height; the greater the number of age classes, the more uneven the canopy. There are two exceptional cases in which stands with more than one age class can become rather smooth on top. In very old stands, all of the trees, even of very different age classes, may have culminated in height growth at a common level. Sometimes isolated older trees are sufficiently decelerated in height growth that more numerous younger trees around them catch up and both age classes continue growing slowly in one smooth-topped stand.

Although it might seem that fat trees are always older than thin ones, diameter is not a very good criterion of age and must be used as such cautiously. The diameter growth of trees is much more variable than that in height. Therefore, the trees in an even-aged stand are not as uniform in diameter as they are in height. If one plots the number of trees in each diameter class over diameter for a given even-aged stand, the distribution approximates the normal, bell-shaped curve (Fig. 1-3). The continuing loss of small trees through suppression accounts for the typically abrupt slope of the left-hand side of the curve. It should be borne in mind that even-aged stands can have a wide range of diameter classes; age distribution of a stand cannot be determined merely from the *range* of diameter classes present.

Uneven-aged stands are composed essentially of small even-aged groups of different ages; the distribution of diameters in each group also fits a bell-shaped curve. However, as each little even-aged group grows older the number of trees in it declines, rapidly at first and more slowly later on, until the point may even be reached where but 1 tree remains from 100 or more. Therefore, the composite diameter distribution curve for a balanced uneven-aged stand (Fig. 1-3) is "J-shaped," if each age class occupies the same area.

If the age classes of an irregular uneven-aged stand differ widely in age, they are revealed as humps on the diameter distribution curve. The diameter distribution of each even-aged component broadens with age and will also be modified if the age class is composed of different species that grow at varying rates.

The most accurate assessment of the age-class structure of a stand comes from actual counts of annual rings. It is seldom reliable to depend on the criteria illustrated in Fig. 1-3 until direct age determinations have been made in representative stands typical of a locality. When such counts are made, consideration should be given to the fact that many species start as suppressed advance regeneration beneath older trees. In such instances, the **effective age** is more important than the **chronological age** and equals the period since the trees were released. In other words, any core of fine growth rings around the pith is best discounted in assigning a tree to its proper class. This interpretation will often reveal that seemingly complicated stands are, for practical purposes, even-aged.

Mixture of Species.   Differences in age distribution are most easily recognized in pure stands and in mixed stands composed of species with rates of height growth so nearly identical that the trees of a single generation are aggregated into a single stratum in the crown canopy. However, even-aged mixtures of tree species usually segregate into different canopy strata and exist as stratified mixtures (Fig. 1-3). It is rare that any two associated species grow in height at precisely the same rate throughout life even when they are not actually competing with each other. If they are intimately intermingled, the species with the most rapid rate of juvenile growth in height will gain ascendancy over the slower growing species, which will lag even farther behind because of lack of light. If the slower growing species is not sufficiently tolerant of shade few will survive and a nearly pure stand of the faster growing species will remain. However, if the slower growing species is sufficiently tolerant, it will persist as a lower story beneath the main canopy formed by the other species. In the simplest kind of **stratified mixture** the different species ultimately tend to occupy different strata of the total crown canopy.

In general, there will be as many strata as there are groups of species that differ from one another in height growth and tolerance. The stratification is not always perfect or readily apparent. Even when the stratification is well differentiated a few individuals of species that go with one stratum may by chance have grown upward into a higher stratum or have been left behind in a lower one. Furthermore, the observer standing beneath a stratified mixture will have to look closely and exercise some imagination to perceive the different strata.

The different strata are designated *A*, *B*, and *C* downward, a terminology first applied in the forests of the moist tropics, where the concept of the stratified mixture originated. Stratum *A* may be continuous but is more often composed of scattered and isolated **emergents** that either grow faster or continue growing longer than their associates.

There are many variations in the pattern of complex stands. Differences in age class in the horizontal can be combined with differences in species in the vertical. Furthermore, different strata over the same point can be of different age.

While the lowermost strata of shrubs or herbaceous vegetation are usually not counted among the arborescent strata, they may play an important role in the total structure and function of the forest ecosystem. Subordinate strata are very common even beneath stands of single tree species that are casually thought of as absolutely pure.

In mixed stands, diameters tell almost nothing about ages when different species are compared. A slender tree of a *C*-stratum species may be as old or even older than a fat emergent of another species above it. Large differences in diameter *within* a species normally denote truly different age classes. In fact, it is common for *even-aged*, stratified mixtures to have varying degrees of semblance of the "J-shaped" diameter distribution that would be a plausible indicator of the balanced, *uneven-aged* structure if the trees were all of the same species. In such cases, the different diameter classes generally represent species with different schedules of height growth rather than different age classes. If there are truly different age classes in mixed stands, there are likely to be real variations in the height of the top of the main canopy in different parts of the stand. If shade-intolerant species normally found only in the upper strata exist in very different diameter classes within a stand, one may suspect that the stand is not even-aged. However, in such complex stands, there may be no substitute for direct observations of age in determining what the age-class structure really is.

Relationship with Wood Utilization

Wood for timber products, including paper and fuel, is, in terms of tonnage, the most important raw material of civilization. With proper soil protection and wise management forests are an infinitely renewable resource. Silviculture directed at wood production must, therefore, be aimed at growing useful trees, setting the stage for efficient harvesting, and simultaneously maintaining the capacity of the site and vegetation to produce more.

The utilization and harvesting of wood are generally more costly than growing trees; they also have a certain immediacy in satisfying human wants. Consequently it is not easy to keep them from becoming such dominating considerations that they defeat long-term objectives. Many of the solutions devised to strike balances between long-term goals and immediate wants are put into effect by choices among different silvicultural procedures. Because silvicultural practices determine what kinds of stands and trees will exist in the future, much, perhaps too much, of the ethical concern of forestry for future productivity has traditionally been viewed as a silvicultural consideration.

Mainly because it can truly control some remote future, silviculture for wood production is both blessed and cursed with the most distant planning horizons that actually govern present actions. Most foresters harvest trees that they did not grow and grow trees that they are not likely to harvest. They must attempt to predict what kinds of trees will be useful in the future and then start to grow them now.

Predictions of future wants are always fraught with uncertainty. Fortunately the technology of wood utilization seems to develop faster than trees can be grown. The trend is toward enabling use of poorer and smaller trees of an increasing variety of species. This makes the silviculture easier but can also cause harm and generate complacency. If more kinds of trees can be harvested economically, the greater will be the opportunity for silvicultural manipulation of stands. However, if everything can be harvested, the greater is the temptation to damage forests by harvesting everything.

If everything can be harvested, it is usually because markets have been developed for small or poor trees for some single product such as pulp or fuel. This can lead to the temptation to grow trees only for such purposes. This policy may be justified in places where there is overriding local demand for pulp or fuel or where the sites are too poor to produce anything else. However, the only important cases in which timber stands have become technologically unemployed because of changes in demand have been ones in which small or poor trees were grown for some single product. It is sometimes argued that one could grow fine trees of traditionally valuable species and then find that no one wanted them. The instances in which this has actually happened are rare and, even then, involved such specialized products as curved oak timbers for ribs of wooden ships.

Trees that are large enough to saw or slice into solid or laminated wood products command higher prices than those that will be chipped or burned. Furthermore, because of the high energy requirements of substitute materials (mostly from nonrenewable resources), solid-wood products also represent the best way to use wood for energy conservation. Ideally wood fuel and products from chipped wood would come only from mill or logging residues and the large numbers of trees that never become large or good enough for any other use. There are plenty of circumstances in which this ideal is unattainable, so the silviculture is modified accordingly.

It is generally prudent to grow trees of species, sizes, and qualities that will make them usable for a variety of products. Because of market uncertainties and pathological risks, it is better to grow the species best adapted to a given site than to try to force the site to produce the one that sells best at the beginning of the rotation. Even though logging machinery changes, it is usually well to assume that comparatively simple and uniform stands will be easier to log than those with intricate spatial variations. It is generally best to start by formulating silvicultural procedures that will create good, productive, healthy stands and then modify the actual practices to meet nonsilvicultural problems. Difficulties often arise when planning starts from some goal of utilization that does not fit the natural circumstances.

Relationship with Forest Management and the Social Sciences

The decisions made in silvicultural practice are based fully as much on economic constraints and social objectives as on the natural factors that govern the forest. Recognition of objectives and limitations set by society in any given case reduce the silvicultural alternatives that need be considered. Even though intelligent application of silviculture can make a positive contribution to the tactics, the strategy for solving problems associated with social sciences is not academically part of silviculture. Such matters are dealt with in the field of forest management, which is concerned with both the organization of collections of stands into integrated forests and the administrative aspects of forestry (Davis, 1966; Leuschner, 1984). The social sciences impinge at all levels of application from the political aspects of the world forest economy down through the public relations psychology of the single forestry enterprise to silvicultural maximation of the net financial returns from single trees.

It is important to note that silviculture and forest management are interdependent and not parallel, alternative approaches to the same problem. Silviculture, because of its dominant concern for efficient application of the natural sciences, is not any less "practical" than forest management, with its tendency toward preoccupation with economic considerations. No management plan is better than the silviculture it stipulates or any silvicultural treatment better than the usefulness of the results it produces.

The Stand and the Forest

The essential unit of silviculture is the little world of the stand. Forest management is primarily concerned with the **forest**, in a special sense meaning a collection of stands administered as an integrated unit. The distinction between stand and forest is important in regulating the cut from the forest. The objective of this type of planning usually is achievement of a sustained, annual yield of products. The forest, and not the stand, is the unit from which this sustained yield is sought. Management determines the volume of timber that should be removed from the whole forest in a given period; silvicultural principles should govern the sequence and manner in which individual stands are cut to produce the required volume. The tendency to treat large groups of dissimilar stands as if they conformed to a uniform, hypothetical average should be studiously avoided. However, an arbitrary decision must always be made regarding the minimum size of stand that can be regarded as a separate entity.

The size and number of stands recognized depend on the intensity of practice, the value of the stands, and the diversity of site conditions. Where intensive forestry is feasible, stands as small as one acre may be recognized; under crude, extensive practice the same forest might be divided into units no smaller than several hundred acres. The best policy is to recognize the smallest stands that can be conveniently delineated on the maps of forest types and age classes used in administration.

Even after type maps have been put on paper the forester must still deal with variations that actually exist within each stand. From the technical standpoint, each portion is best treated separately, although acceptance of too many variations would eventually create a mosaic of conditions that would be awkward for both logging and administration.

Silviculture and the Long-Term Economic Viewpoint

It is said that money does not grow on trees; the bane of forestry is the popular view that trees exist but do not grow. Conventional economic theory holds, in effect, that the silviculturist cannot win while certain naturalistic ecological theories warn against trying. The economic time scale of forestry is so vast and unique that to many investors and in terms of ordinary economic theory it really is not economic at all. There is scarcely any part of forestry in which this issue must be faced more squarely than in silviculture, especially when investments in establishing or treating young stands are considered. It takes a certain kind of ambivalence to keep the economics of forestry in perspective. The decision to practice forestry is usually a matter of ethics, politics, and social concern for posterity but not basically one of conventional economics. However, once the decision is made, it become logical to apply economic analysis to determine how best to execute the details. Any conflict is not between "silviculture" and "economics" but between the long-term economic viewpoint of forestry itself and customary short-term outlooks on financial matters.

The holding of land for future production of wood crops or other benefits involves silviculture even if nothing more is done than to let nature take its course and to harvest trees occasionally. Ownership incurs costs and these constitute investments in the future even if nothing is invested in treatments to increase future production. The question is not whether an owner is investing in silviculture because he or she already is, even if unintentionally; it is whether the capacity for long-term investments in silviculture and other parts of the forest enterprise is being used to optimum advantage.

Funds are rarely available for all the silvicultural work likely to prove profitable, so the forester must ensure that those that are invested are expended on the most remunerative lines of work possible. In any situation it is logical to apply first those treatments that will yield the greatest increase in value of benefits per dollar of investment. One then adds additional treatments of successively lower ratio of benefit to cost until the cumulative margin between the value of all benefits and all costs reaches a maximum (Fig. 1-4). This procedure is more easily described than done, and it requires successful intuition in addition to careful factual analysis. It is rare that this sequence of successive investments has actually been pursued to a point even close to that of maximum long-term profit.

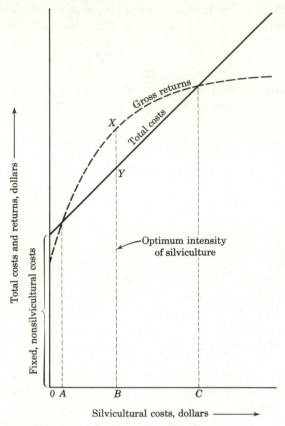

FIGURE 1-4   Relationship between costs, returns, and profit from silviculture in a single hypothetical situation. In this example, the returns from holding the land without any silvicultural investment are exceeded by the fixed costs of ownership. If *A* dollars are invested in appropriate silvicultural operations, the returns balance costs without profit or loss. With further increases in intensity of silviculture, profit increases until, when *B* dollars are invested, the vertical distance *XY* reaches a maximum, indicating the highest possible profit. Beyond that point, the law of diminishing returns operates and the profit margin declines even though gross return continues to increase slowly. In fact, silvicultural investments greater than *C* dollars result in a loss.

The first stage in the evolution of silvicultural practice (Spurr, 1979) is that in which continued production is actively sought but not with any cash investments in it. This "no-investment" silviculture places emphasis on treatments that can be accomplished by removing merchantable timber without significant increase in harvesting costs. Some forests are sufficiently easy to control that treatments made on this basis give reasonably good results. This kind of silviculture is practiced over wide areas and will doubtless continue to be for a long time. It rarely makes the ownership of forests very attractive financially. The idea of taking values out of the forest without really reinvesting anything in future production has a powerful appeal. It almost completely dominated American silviculture for decades and there are still many instances in which it is consciously or unconsciously regarded as the only feasible economic viewpoint.

Orderly policies of long-term investment in silviculture emerge if economic conditions and natural productivity are favorable, provided that adequate experience has accumulated. The kind and amount of investment is limited only by the economic law of diminishing returns. The actual amount expended on this kind of silviculture varies widely but is not likely to be trivial.

The application of different policies regarding silvicultural investments is something that varies in time and place in complicated fashion. If the existence, cause, and logic of this pattern are not recognized both the application and analysis of silvicultural practice become hopelessly confused. If one becomes too much accustomed to one policy it becomes difficult to understand any other or to change to another when and where circumstances warrant.

Variations in Intensity of Practice

The amount of effort expended on the treatment and care of stands, that is, the **intensity** of silviculture, varies widely, depending chiefly on economic circumstances. The converse of **intensive** silviculture is **extensive** silviculture. The degree of intensity is usually estimated in terms of such things as the amount of money invested in cultural treatment, the frequency and severity of cuttings during the rotation, and the amount of concern accorded to future returns relative to immediate returns.

The appropriate intensity of silviculture varies with accessibility, markets, site quality, objective of management, and nature of ownership. The proper level often must be chosen for each stand because the application of a single degree of intensity will not give optimum results throughout a given forest, unless it is exceedingly small and uniform. The more favorable is the combined economic effect of all factors, the higher is the appropriate level of intensity of silviculture. The place for extensive silviculture is found in remote areas, on poor sites, or where owners are not willing or able to make more than minimum investments. It often plays a role where timber production is secondary to other purposes of forest management.

In the past, American forests have been exploited in such a manner that the poorest and most ill-treated stands are often found on the best sites and in the most accessible areas, such as those along permanent roads. This situation, which is the reverse of that which should apply, results from the fact that the best and most conveniently located stands have been exploited first, most heavily, and most frequently. Ultimately silviculture of the highest intensity should be practiced in many of these situations rather than in the remote areas where the best forests are now often found. Permanent roads and good markets for a diversity of forest products do not automatically ensure optimum practice, but they are essential to intensive management.

The intensity of silviculture depends in large measure on the nature and objectives of ownership. Variations in the species and sizes of trees desired may necessitate different procedures on adjoining lands that are fundamentally similar. Stability or longevity of ownership also controls intensity of silviculture. Large corporations and public agencies, which are relatively immortal, are in a far better position to practice intensive silviculture than individuals or small corporations of uncertain stability.

The intensity of silviculture often depends on the extent to which the owner processes the wood grown in his forest. The closer the product is carried to the ultimate consumer, the greater is the ability to capture the values added by increases in intensity of practice in the woods. Prices for **stumpage**, that is, standing trees, do not necessarily reflect all the values added to the product by silvicultural measures to improve the quality of wood. Therefore, the owner who is not in a position to do more than sell stumpage may not be able to practice silviculture as intensive as that suitable for owners who also harvest, manufacture, or sell the final product. This relationship is, however, modified by ability and willingness to make long-term investments. For example, public forestry agencies usually confine their operations to producing stumpage; they may, however, practice intensive silviculture without great concern for profit on their investments in order to discharge their long-term responsibilities to the national economy.

Financial Analysis of Silviculture

The financial techniques employed in measuring and predicting silvicultural accomplishment are part of forest management and economics (Leuschner, 1984). A preliminary understanding of them is necessary because, both as detailed methods and as concrete expressions of economic viewpoints, they are indispensable to sound choices among silvicultural measures.

The returns from silvicultural treatment aimed at wood production are most objectively assessed in terms of the increase in stumpage values that is attributable to treatment. Where the silviculture is directed at production of benefits such as wildlife, recreation, and control of watersheds, the values are usually so intangible that the whole analysis tends to be intuitive.

One of the most peculiar and important problems of the financial analysis of silviculture results from the fact that many of the benefits are not reaped until many years after the investments in treatment are made. The act of waiting has a certain kind of cost so decisions between alternatives of treatment must be based on some means of comparing not only the values created by various treatments but the costs of waiting for varying periods to obtain the benefits. For example, it may be necessary to decide whether to invest $1.00 in a planting operation that will yield $20.00 in stumpage 50 years hence or in a precommercial thinning that will increase the value of wood harvested 10 years from now by $3.00.

The methods of comparing such differing alternatives employ the device of compound interest in the same sense used in connection with accounts in savings banks (Gunter and Haney, 1984). If the appropriate rate of compound interest is known, it is possible to estimate both present values of future benefits and the future costs represented by present expenditure. Usually the procedure is to reduce all estimates of future costs and benefits to the present time so that various alternatives of procedure can be compared on the same basis. The rate of interest used depends on the circumstances, including the estimated risks and the compound-interest return available to the owner from alternative investments.

The rates should not include the allowances for inflation that apply to interest paid on bonds and bank deposits. This is because the forestry investments are in tangible, inflation-proof assets and not in promises to pay in paper money. However, there should be some component of the expected rate to quantify the risk that what is being produced may be worth less, in real money, some years hence than it is now. That risk may be offset in some degree by the historically plausible possibility that it will be worth more. Experience with investments in general shows that it is unrealistic to anticipate that compound-interest returns of more than 5 percent in uninflated money can be sustained for long periods. For example, if the Indians who sold Manhattan Island in 1626 had invested their $24 at 6 percent compound interest, they would have been wiser than those who bought it, even if the modern dollar were still worth as much.

The following example illustrates the increase in the value of a $1.00 investment with time at 4 and 8 percent compound interest:

Time, years	20	40	60	80
Value at 4%	$2.19	$4.80	$10.52	$23.05
Value at 8%	$4.66	$21.72	$101.26	$471.96

In other words, if one demands a 4 percent return on the investment of $1.00 in planting made now, a gross return of $10.52 is expected on it 60 years hence. If the ultimate return *was* $10.52, it would be worth $1.00 to the investor now. To take a purely hypothetical case, if the $1.00 were invested in pruning rather than planting and yielded an increase in stumpage value of $101.26 at the end of 60 years, this would represent a return of 8 percent and would clearly be a better investment. If $1.00 were invested in release cutting and produced an increase in value of $21.72 at 40 years, it would also represent an 8 percent return and would have the same **present worth** as the pruning investment. However, it would be the better of the two alternatives because an investor would demand an indefinably larger interest rate on a 60-year investment than on a 40-year investment.

These analytical techniques find their real utility in guiding choices between different long-term silvicultural programs aimed at the same general objectives. They are not used in conventional accounting. In practice, silvicultural costs are usually charged against current income, although the matter is endlessly complicated by the favorable and unfavorable details of tax laws.

The delay between silvicultural treatments and their financial fruition is such that anything that can shorten the delay is beneficial. The inevitable existence of such delay forces diligent search for the modest inputs of effort that yield large benefits. Fortunately the natural productivity of the forest is plastic enough to respond to gentle guidance. However, it would be most remarkable if silviculture were the unique form of economic activity that yielded maximum profits with little or no outright investment. Owners of some forests may indeed find that the minimum costs of protection and other carrying charges exceed the value of benefits derived from such management (Fig.1-4).

The valuation of benefits from silviculture directed at production of benefits such as those from wildlife, recreation, and control of watersheds are more difficult to assess because they are so intangible. On the other hand, they are usually more quickly obtainable than those from timber production.

Silviculture in Application

The practice of silviculture does not consist of rigid adherence to any set of simple or detailed rules of procedure. For example, there is no part of this book that could be safely used as a manual of operations. Many of the techniques of cutting are described in simplified form shorn of the myriad of refinements and modifications necessary to accommodate the special circumstances and local variations encountered in practice. Each procedure described is merely an illustration designed to demonstrate the application of a set of treatments designed to meet a uniform set of circumstances. Even though uniform stands have important advantages that make them worthy of creation, the stands encountered in the field may lack uniformity and thus call for variation in treatment.

Any consideration of silviculture covers a variety of treatments wider than is likely to be practiced in any locality at a particular time. In times when the forests of a locality are all immature, silvicultural practice may be limited to intermediate cuttings and anything connected with regeneration may be limited to the reforestation of vacant areas. In areas where it is customary to secure regeneration by planting, the forester may regard methods of natural regeneration only as matters of intellectual exercise. Where attention is for the time being concentrated on replacing old-growth forests with regeneration, consideration of intermediate cuttings may seem quite unessential. At times and places where economic conditions support only the crudest kind of extensive silviculture, intensive treatments may seem visionary indeed.

The subject matter of this book reflects a wide variation in intensity of silvicultural practice because an attempt is made to describe all known techniques that seem applicable in any significant forest area of North America within the foreseeable future. The procedures characteristic of the more intensive kinds of silviculture cannot be described as briefly as those associated with extensive silviculture and so get more attention. This does not mean that a management program must include a long series of different treatments to be silviculture. Some of the most astute silviculture is the kind conducted at low intensity in which much is accomplished with a very limited amount of treatment.

The student forester interested in only one particular region cannot safely restrict attention to the kind of silviculture currently practiced there. Foresters move, times change, and ideas from other places are often as fruitful as the indigenous ones. Scientific knowledge and technology also grow at an accelerating pace.

There are many places where the impractical of twenty years ago is now the routine, and may prove to be the naive and inadequate of a decade hence. Because of cutting and growth, the forests of a locality often change and such change calls forth new methods of treatment. This is especially true in North America, where the forests of many localities tend to be uniform either because they have been scarcely touched or because they were all cut over in a short space of time. This book may seem to contain more techniques and ideas than a forester might need in a professional lifetime. While some may go unused or go quickly out of date, there are really only enough to provide a starter.

It is not enough for the forester to know what to do and how to do it. The important questions in silviculture, whether they involve matters of natural or social science, start with the word "why." As in other applied sciences, action proceeds from the knowledge represented by the answer, or sometimes the merest inkling of an answer. The forester can find as many solutions in the woods as in the printed word. However, it is necessary to ask oneself the questions that generate the solutions and also to be ready to take the time to observe how forest vegetation develops.

Silvicultural Literature

Each chapter of this book ends with a listing not only of the references cited in that chapter but also of others chosen to lead the way to as much literature as possible on topics covered in the chapter. It often helps to find a recent article on the chosen topic and then follow it backwards in time through use of the references in that article. Any sort of literature review is useful.

Abstracting journals can be very valuable, especially *Forestry Abstracts,* a British publication that covers significant forestry literature on a worldwide basis. The professional and scientific journals of forestry of one's locality can also be helpful, especially if they have annual or cumulative indexes used with appropriate imagination. The use of computerized information retrieval systems is in its promising beginning stages.

More detailed information and many additional literature references about silviculture in the United States can be obtained from at least three consolidated publications. In *Regional Silviculture in the United States* (Barrett, 1980), various silviculture professors have written about their localities. Research scientists of the U. S. Forest Service (1965, 1983) have summarized information about the ecological characteristics of tree species and about the silviculture of the important forest types. One advantage of these three sources is that they help with the location of many of the large number of publications issued by research and extension agencies of governments and universities.

The *Forestry Handbook* of the Society of American Foresters (Wenger, 1984) has much information about silviculture and closely associated topics as do similar compendia designed to help the practicing foresters of a locality.

The written word can bring the forester ideas from distant places. Not all of the problems of growing loblolly or ponderosa pine have to be solved entirely by study of these individual species. Much has also been learned about the silviculture of pines in Finland or Australia and knowing about teak in Asia may also help. In fact, new and useful insights may come even faster from distant sources. Most of the world literature of forestry is in English, although English-speaking forestry students should be more ambitious about mastering other languages.

One does not read about silviculture just to absorb information. What is written is not Revealed Truth anyhow. Reading should be a stimulus to thought, a way of synthesizing new patterns of understanding, and of both expanding and testing one's ideas. It can make comprehension of what one sees in the woods surer and more serviceable.

BIBLIOGRAPHY

Barrett, J. W. (Ed.) 1980. *Regional silviculture of the United States.* 2nd ed. Wiley, New York. 551 pp.

Champion, H. G., and S. K. Seth. 1968. *General silviculture for India.* Govt. of India Press, Delhi. 511 pp.

Daniel, T. W., J. A. Helms, and F. S. Baker. 1979. *Principles of silviculture.* 2nd ed. McGraw–Hill, New York. 500 pp.

Davis, K. P. 1966. *Forest management: regulation and valuation.* 2nd ed. McGraw–Hill, New York. 519 pp.

Ford-Robertson, F. C. (Ed.) 1978. *Terminology of forest science, technology, practice and products.* Society of American Foresters, Washington, D. C. 349 pp.

Gunter, J. E., and H. L. Haney, Jr. 1984. *Essentials of forest investment analysis.* Oreg. State Univ. Bookstores, Corvallis. 333 pp.

Heiberg, S. O. 1945. "Is forestry unique?" *J. For.* **43**:294–295.

Hocker, H. W., Jr. 1979. *Introduction to forest biology.* Wiley, New York. 467 pp.

Kozlowski, T. T. 1971. *Growth and development of trees.* 2 vols. Academic, New York. 957 pp.

Kramer, P. J., and T. T. Kozlowski. 1979. *Physiology of woody plants.* Academic, New York. 811 pp.

Leuschner, W. A. 1984. *Introduction to forest resource management.* Wiley, New York. 298 pp.

McIntosh, R. P. 1978. *Phytosociology.* Academic, New York. 416 pp.

Means, J. E. (Ed.) 1982. *Forest succession and stand development research in the Northwest.* Oreg. State Univ., For. Res. Lab., Corvallis. 171 pp.

Oliver, C. D. 1981. Forest development in North America following major disturbances. *For. Ecol. and Mgmt.* **3**:169–182.

Packham, J. R., and D. J. L. Harding. 1982. *Ecology of woodland processes.* Edward Arnold, London. 262 pp.

Pritchett, W. L. 1979. *Properties and management of forest soils.* Wiley, New York. 500 pp.

Rehfuess, K. E. 1981. *Waldboden: Entwicklung, Eigenschaften und Nutzung.* Parey, Hamburg. 193 pp.

Shugart, N. H. 1984. *A theory of forest dynamics, the ecological implications of forest succession models.* Springer, New York. 304 pp.

Spurr, S. H. 1979. Silviculture. *Scientific American* **240**:76–82, 87–91.

Spurr, S. H., and B. V. Barnes. 1980. *Forest ecology.* 3rd ed. Wiley, New York. 687 pp.

U. S. Forest Service. 1965. Silvics of forest trees of the United States. *USDA, Agr. Hbk.* 271. 762 pp.

U. S. Forest Service. 1983. Silvicultural systems for the major forest types of the United States. *USDA, Agr. Hbk.* 445. 191 pp.

Wenger, K. F. (Ed.) 1984. *Forestry handbook.* 2nd ed. Wiley, New York. 1335 pp.

West, D. C., H. H. Shugart, and D. B. Botkin (Eds.) 1982. *Forest succession: concepts and application.* Springer, New York. 516 pp.

PART ONE

Tending and Intermediate Cutting

CHAPTER 2

Thinnings and Their Effect on Growth and Yield

The next five chapters deal primarily with the tending of stands that have already been established or regenerated. The most important kind of tending or intermediate cutting is thinning, which is a kind of partial cutting designed to guide the production of stands and trees along desirable channels. Before considering the practice of thinning itself it is desirable to consider some of the fundamentals of production of wood and other organic matter by stands.

The populations of most organisms and the production represented by their growth tend to increase to the limit of food available to them. Trees make their own food by using solar energy to combine water from the soil and carbon dioxide from the air into glucose, the simple sugar that contributes most of the basic building materials for all the compounds that provide the trees with energy and structural material. The ultimate limits on production are set by the ability of the soil to supply water and chemical nutrients, the amount of carbon dioxide and light available, and the time during which temperature and other environmental conditions will permit photosynthesis and other essential processes.

PRODUCTION OF DRY MATTER

Forests are among the most productive plant communities, that is, in capacity for photosynthetic carbohydrate production as measured in terms of oven-dry organic substance, "dry matter," or "biomass." The high primary productivity of closed stands of tall, woody, perennial plants results from their deep canopies of sugar-producing foliage. The perennial woody stem is an efficient device both for the display of foliage in vertical depth and for conducting the upward movement that brings water and nutrients from the soil to the foliage. The vegetation of shallow waters and tidal marshes can be equally productive; while it does not have to divert as much production on powerful supportive stems, it depends heavily on special vertical currents or the tides to deliver nutrients. Certain tall grasses, such as sugarcane or corn (maize) of agriculture, can be more productive but mainly because of fertilization, good water supply, or comparatively low losses to respiration.

The biological productivity of forests is so great that it is estimated that woody plant cover, occupying about a third of the land surface and a fifth of the total surface of the earth, accomplishes almost half of the world annual photosynthetic fixation of carbon (Woodwell *et al.* 1978). Cultivated agricultural lands account for about a twentieth. Furthermore, roughly 85 percent of all the carbon fixed in the world's living plants is stored in communities of woody plants.

It is much more difficult to comprehend the production of material by woody perennials than that of typical agricultural annuals in which the production of a single year can be easily seen and assessed. Most of the comparatively greater production of a typical forest is laid down in forms such as the thin annual sheath of wood or new shoots that are not readily visible and can only be assessed with detailed measurements. At the very age when a stand is producing useful material most rapidly the trees appear to change little in a year or even in a decade unless one measures their change in stem diameter.

Occupancy of Growing Space

In the initial stages of development, a new stand of trees does not fully occupy the site. Herbaceous annuals and perennials or shrubs develop more rapidly and temporarily use most of the growth factors. The trees can overcome them only because they can grow taller. Gradually the roots of woody plants assert occupancy of almost the whole stratum of soil in which the supply of water and oxygen is sufficiently ample and stable to allow them to persist.

Perhaps because they do not have to build tissue to support themselves, tree roots ordinarily approach full occupancy of available space more rapidly than the foliar crowns. The foliar surface likewise tends to expand to form a complete crown canopy, unless severely limited by damaging agencies, water shortages, or other unfavorable conditions. The available crown space if fully occupied when the canopy has not only closed horizontally but also developed enough in the vertical so that the lower branches die and the insides of the crowns become vacant of foliage (Assmann, 1970). The stand then has as much foliage as it ever will have; the amount may remain stable or decline somewhat thereafter, but it is unlikely to increase.

After the crowns close there may be a period in which the branches of adjoining trees interlace with one another. When they grow taller, however, the wind load on the crowns increases and there is enough twig breakage that the crowns become separated again and the gaps between them widen as trees sway and crown friction increases. This effect and more subtle ones gradually diminish the extent to which the original trees occupy the site and new woody plants may appear in the understory to claim the unused growing space.

The natural, competitive suppression of trees by the shading effect of more aggressive ones does not create vacancies in the growing space; the death of such trees is merely a sign that their growing space has already been usurped by the others. At least until a stand is middle-aged, most of any silvicultural treatment of it can be viewed as modifying a plant community that almost fully uses the available site factors and growing space. If scattered trees are removed, the remaining trees expand their crowns and root systems; the amount of foliage and effective root surface soon returns to that which existed before treatment. Intraspecific root grafts may enable the residual trees to effect immediate capture of part of the root systems of cut trees.

Operation of the Foliar Sugar Factory

The amount of sugar-producing foliage that a stand can maintain depends upon species and, to a lesser extent, the ability of the soil factors to supply the foliage. Evergreen species maintain more foliar surface than deciduous species, if only because the leaves persist longer than one growing season. Shade-tolerant species have more than the intolerants because their leaves can function at lower light intensity and thus form deeper crowns. Within a species, the total amount of foliage in a closed stand is somewhat less on a very poor site than on the best.

Poor sites produce less dry matter than good sites not so much because of major differences in the amount of foliage but because the foliage cannot function as efficiently (Assmann, 1970). Deficiencies in the supply of nutrients and water from the soil can slow photosynthetic rates and the creation of new tissues. If water become totally unavailable because of depletion, low temperature, or soil-oxygen deficiency, photosynthesis must cease. Excessively high temperature can cause stomatal closure and cessation of photosynthesis. In other words, the unfavorable environmental factors of a poor site can substantially reduce both the rate and duration of photosynthetic activity even if the amount of foliage is the same as on a more favorable site.

Not all the carbohydrate produced by a stand is converted to permanent stem and root tissue. Some must be used for temporary tissues such as leaves, small twigs, rootlets, and fruits or cones. A major portion is expended in respiration to supply energy to the leaves, roots, meristematic tissues, and other living parts of the trees.

The production of new tissues by a tree or stand of trees may be viewed as depending upon the relationship between the **photosynthetic** or foliar surface and the **aphotosynthetic** surface (Satoo and Madgwick, 1982). The aphotosynthetic surface is composed of the meristematic tissues of boles, branches, and roots; the growth of permanent tissues and respiration other than that of leaves and rootlets takes place at or near this surface. If the amounts of foliage and carbohydrate produced by a stand are fixed quantities and the rootlets completely occupy the available soil space, the amount of new stem and root tissue laid down will be determined by the amount of carbohydrate left for the aphotosynthetic surface after the requirements of respiration are met. From this standpoint, the goal of removing part of the trees in thinning could be viewed as reducing the aphotosynthetic surface substantially but in ways such that the foliage or photosynthetic surface quickly grows back to nearly its maximum extent. If the amount of foliar (and root) surface available to nourish each square centimeter of cambium is thus increased, the amount of carbohydrate left after respiration for wood formation should theoretically be greater and increased diameter growth should result.

Quantification of Production

Estimates of the distribution of gross dry-matter production among the component uses have been prepared for whole stands by Mar:Møller, Müller, and Nielsen (1954) for European beech on good sites in Denmark and are shown in Fig. 2-1. These indicate that, at least with beech, respiration consumes more of the gross dry-matter production than is channeled into the permanent stem tissues. Respiration, being immediately essential

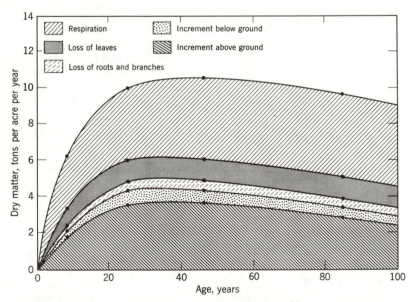

FIGURE 2-1 An estimate, adapted from Mar:Møller, Müller, and Nielsen (1954), of the allocation of carbohydrates in an even-aged stand of European beech during the course of about one rotation. The vertical distances between the curves define the amount of carbohydrate expended for various functions. The total area below the lowest curve, for example, also defines the total weight of stems and permanent branches produced during 100 years.

to continued life, commands the highest priority in the carbohydrate metabolism of the trees.

The data in Fig. 2-1 also show that the gross production is not constant with age. It does not approach its maximum value until the stand achieves full occupancy of the site and the foliage weight becomes constant or at about the twenty-fifth year in this example. With further increase in age, however, the annual production of dry matter commences to dwindle even though foliage weight remains constant with age. Ovington (1957) observed a similar pattern of variation of dry-matter production with age in Scotch pine in England. The ultimate decrease in production with age takes place in all organisms and remains largely inexplicable. The decline is less with trees than with most other organisms. It is possible that some of it may result from actual decreases in foliage and increases in respiration as stands grow taller. One may also reflect on the likelihood that organisms seem more fundamentally adapted to perpetuating their own kind than to producing substance.

One of the most precise studies of energy transformation and dry-matter production by forest vegetation was conducted by Whittaker and Woodwell (1969) in a poor, 45-year-old stand of oak and pitch pine on a droughty site on Long Island, New York. While it would be easy to find stands that would be more productive, especially in economic terms, the study provides an example of the details of basic forest production.

Total solar radiation was 130,300 cal per cm^2, of which 56,030 were in photosynthetically active wavelengths. The total amount of sugar (gross primary production) manufactured yearly was estimated at 2647 g per m^2 or the energy equivalent of 1244 cal per cm^2 which is only 0.91 percent of the energy annually received in the photosynthetically active spectrum. Forests on better sites have somewhat greater efficiency, but probably not much over 1.2 percent, even though they represent one of the most efficient ways of capturing energy from the sun. In this case, the respiration of the plants themselves consumed 1452 g per m^2 leaving an annual net primary production of 1195. Nearly half of this, including a major part of that which went into such temporary tissues as leaves, was consumed by insects, fungi, and other dependent organisms. This left a net ecosystem production (of new material remaining in the stand) of 542 g per m^2 per year.

All of this production was the result of photosynthesis in only 384 g per m^2 of leaf tissue. The net annual primary production (g per m^2) was distributed as follows:

	Above Ground	Below Ground	Total
Trees	796	260	1056
Shrubs	60	73	133
Herbs	2	4	6
Total	858	337	1195

The proportion of new root tissue on this dry site is probably greater than it would be in most forests; it is noteworthy that the shrubs and herbs grew more below than above ground, but the trees did not. Only 149 g per m^2 of the annual tree production was in the form of main-stem wood; the efficiency of conventional wood utilization is low in relation to the massive productivity of forests. Acorns, cones, and other reproductive structures accounted for 22 g per m^2 annually. As is the case with most forests, the tree stratum produced much more substance than the subordinate vegetation.

Analysis of Increment

It is important to distinguish between two different ways of expressing the annual rates at which stands or individual trees add substance to themselves. Because they are perennial plants some special bookkeeping is necessary to understand their production rates.

Current annual increment (C.A.I.) is the amount (in any unit of measure) that is actually added in a given year. Problems with measurement errors make it difficult to estimate reliably. The **periodic annual increment (P.A.I.),** which is the increment of a short period (normally 5 or 10 years) divided by the number of years in that period, is usually taken as nearly equivalent to C.A.I. The effects of climatic variations and of errors in estimates of beginning and ending values are diluted or smoothed over by using a period longer than one year.

A basically different mode of expression is **mean annual increment (M.A.I.),** which is the amount of substance accumulated to date divided by age. Because age must be known, M.A.I. can be determined only for even-aged stands or individual trees of known age.

As shown in Fig. 2-2, these two different kinds of annual increment bear a definite relationship to each other. In an even-aged stand, C.A.I. or P.A.I. is equal to M.A.I. when M.A.I. culminates at its maximum. C.A.I. or P.A.I. can be spectacularly high early in the life of a stand but represents production rates that cannot be maintained. The M.A.I. provides the best estimate of the maximum production rate that can be continuously sustained by a given combination of species and site quality, provided that stands are not replaced much before or after the age of maximum M.A.I. The peak M.A.I. is a key value for determining how much can be harvested under sustained-yield management and how long the rotations should be.

It is not possible to determine M.A.I. for uneven-aged stands since they have no single age. However, the theoretical, balanced all-aged forest would have, during every year, a C.A.I. virtually equal to the peak M.A.I. of an even-aged stand *of the same species grown to the same rotation age on a similar site.* This conclusion that age-class distribution does not alter average production rates in the long run is one that emerged from long-term studies of forest production by Burger (1919–1953) in Switzerland.

The basic relationships between P.A.I. and M.A.I. curves shown in

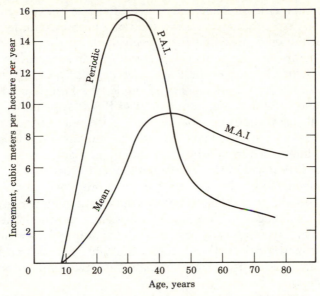

FIGURE 2-2 An example of the forms and interrelationships of curves of periodic and mean annual increment in an even-aged stand showing how the two rates are equal when M.A.I. is at its peak.

Fig. 2-2 are the same regardless of the units of measurement if each curve is for the same unit. Figure 2-2 is based on the merchantable cubic volume of stemwood; the curves do not originate at age zero because there is no merchantable volume until one tree reaches merchantable size. The curves would originate from zero and be shifted to the left if based on total dry-matter production. If they were based on board-foot volume with some large minimum top diameter for logs, the origins of the curves and their peaks would shift far to the right. If the measurement units define product objectives, the age of maximum M.A.I. is the optimum rotation length where the economic objective is to maximize production from a limited land base.

If the P.A.I. of a given stand exceeds the simultaneous M.A.I., this means that the stand has not yet reached the culmination of M.A.I.; one may also make an educated guess about the amount of M.A.I. and the time of its culmination. P.A.I. becomes negative after the age when decay and mortality become equal to growth, as in very old stands. M.A.I. would decrease to zero only when the very last tree of the initial stand was gone.

Distinction should also be made between gross and net increment. **Gross increment** includes material that formed earlier in the life of the stand or measurement period but was no longer present at the end; **net increment** includes only that present at the end. However, the values of any of these units can be quite different depending upon whether the production of temporary tissues such as leaves and rootlets during the final year is or is not included. This is the chief difference between net primary and net ecosystem productivity as in the Long Island example.

Most of the estimates that have been made of the total dry-matter production of forests have been of P.A.I. or C.A.I. in immature stands. The rates of production by such stands are often impressively high but cannot be sustained over long periods. For example, the Long Island study involved a stand in which the culmination of P.A.I. appears to have been reached 5–10 years earlier but that of M.A.I. had not yet been attained.

Studies of total production elucidate the fundamental biology of the effects that differences in species, site factors, and silvicultural treatment have on production in forests. The much longer history of studies involving measurements of production of stemwood, often only in merchantable units, provides a large and useful body of information. However, efforts to analyze both biological and economic considerations simultaneously can be detrimental to either category of understanding. Unfortunately the study of dry-matter production exhibits the same inconsistency in techniques and ambiguous use of terminology that have confused more traditional analyses.

Further reference in this chapter to values of dry-matter production will, unless otherwise specified, be in terms of net M.A.I. at or near culmination age, for aboveground materials, including leaves of the current year, in metric tons per hectare per year. A metric ton is 1000 kg and one metric ton per hectare is equal to 100 g per m^2 or 0.446 English short tons of 2000 pounds each per acre.

Interspecific Variations in Production

It is very clear that some species produce more than others even when the site factors are the same. Evergreen species are basically more efficient than deciduous ones because they use their investment in a year's production of foliage for more than one year. Their foliage can also start photosynthesis whenever light, water, and temperature are favorable without any delay while new leaves form.

While conifers are often thought of as more productive than angiosperms they are not inherently more efficient. Broad-leaved evergreens, such as eucalypts, can be just as productive as evergreen conifers. However, many coniferous species have lighter wood and longer lengths of straight central stems than angiosperms. These attributes may make a conifer more productive than some hardwoods in terms of volume of usable stemwood even if its biomass production is no greater.

The arrangements of branches within crowns and of leaves on twigs appear to cause some of the interspecific differences of photosynthetic production (Hallé, Oldemann, and Tomlinson, 1978). For example, one highly efficient form of crown structure is that akin to a candelabrum in which leaves are displayed in a single flat plane by wide-spreading branches. Although this form, typified by some understory palms, has high photosynthetic efficiency it is so weak structurally that the trees must remain small. Some early successional species, such as poplars, exhibit rapid growth in youth because they have branch arrangements that allow them to develop closed sheaths of foliage in balloon-shaped crowns that cover the ground quickly. However, it is postulated that certain structural inefficiencies make it difficult for these species to sustain the early high rates of production.

Another highly efficient crown form is that in which there are substantial gaps between foliated branches. However, this structure is not conducive to swift occupancy of growing space. Stands of trees with this crown structure attain high, long-sustained values of C.A.I. only after the early stages of development.

The mode of arrangement of leaves on twigs that is photosynthetically most efficient is the two-ranked one in which the leaves do not shade one another. However, this arrangement is the least efficient in preventing the leaves from losing water and becoming overheated. Since there is no way that the open stomata of land plants can admit carbon dioxide for photosynthesis without letting water escape, high rates of photosynthesis have to be paid for with transpirational water loss.

Dry-site species are often less productive than those of moist sites because they close their stomata and halt photosynthesis earlier during episodes of moisture stress. This adaptation to drought is one of several that reduce production but increase survival.

Shade-tolerant species, whether evergreen or deciduous, are usually more productive than the intolerants. Actually it is more accurate to think of shade tolerance as the result rather than the cause of this difference in photosynthetic efficiency. The organs that confer efficiency are leaves that can manufacture much sugar and also live at low light intensity. Species that have such leaves have more of them and also deeper crowns than the less efficient species. Perhaps because they have more leaves in the lower and cooler parts of the crown canopy, shade-tolerant species do not use such a high proportion of their gross photosynthate in respiration (Assmann, 1970).

Species with dark-green leaves denoting high chlorophyll content tend to be relatively efficient. Another useful criterion is the **leaf-area index,** which is the number of layers of leaf surface above each point on the ground. Values of this index vary from 3 to 12 in closed stands but are ordinarily about 5 (Waring, Newman, and Bell, 1981).

Interspecific variation among species is well illustrated by estimates of gross M.A.I. (exclusive of roots but including current leaf production and material previously removed in thinnings) for stands of 40–47 years, at or near peak M.A.I., on comparable soils in the favorable climate of England (Ovington and Pearsall, 1956; Ovington, 1956). *Castanea sativa* and *Quercus robur*, deciduous species of moderate shade tolerance, had M.A.I.'s of 3.8 and 3.9 tons per hectare per year, respectively and that of *Fagus sylvatica*, a very tolerant hardwood, was 4.9 tons. The evergreen conifers were more productive: *Pinus sylvestris*, 8.0; *Pinus nigra*, 9.7; *Pseudotsuga menziesii*, 9.9; and *Picea abies*, 9.9 tons per hectare per year. *Larix decidua*, a deciduous conifer, had a rate of 6.7 tons, intermediate between the hardwoods and evergreen conifers. Data from other parts of Britain indicated that even higher rates than those of spruce and Douglas-fir are achieved by very tolerant conifers such as redwood, the hemlocks, and true firs. Intolerant pioneer hardwoods such as *Betula alba* produced at lower rates than those of chestnut and oak. The yield of moderately intensive mixed agriculture in the same locality was 4.5 tons per hectare annually.

Intuition and some data suggest that the highest production rates are in the evergreen tropical rain forest, where site factors are unfavorable to photosynthesis only at night. Annual production rates are at least 8 and possibly as much as 20 or 30 tons per hectare. Actually the high temperatures also induce such profligate respiration that the production rates are not clearly higher than those of the most favorable parts of the temperate zone.

The highest proven rates of production are actually for evergreen species in limited areas of wet, mild, strongly maritime climates at middle latitudes. One of the highest known values of C.A.I. is 30.7 tons per hectare in a stand of western hemlock and Sitka spruce at an effective age of 21 years on the Oregon coast (Fujimori, 1971). The M.A.I. was 9.2 tons per hectare and might culminate at about 20 some decades hence. Such favored midlatitude areas include the coastal fringe of western North America, parts of western Europe, southern Japan, and certain coastal areas in the strongly oceanic climate of the Southern Hemisphere. If the rainfall is sufficient and the soil favorable, the equable maritime climates of these areas provide very long growing seasons without the high temperatures that stimulate rapid respiration. Production rates approximating 15 tons per hectare have been observed with shade-tolerant conifers, evergreen oaks of southern Japan, and *Pinus radiata* in New Zealand (Cannell, 1982).

Most of the world's forests grow under less favored conditions. In widely varying degrees, production is reduced by seasonal moisture restriction (which can exist in some regions at any latitude), periods when low temperature immobilizes water, deficiencies of soil oxygen induced by excess water, low light intensity at high latitudes, or excessive respiration from high temperatures. However, the production rates of the poorest vegetation that might plausibly be called forest are not less than one ton per hectare per year.

Observed production rates vary so much within species and genera because of site variations that it is perilous to assign numerical values to any of them. The total production of a given species on a poor site is commonly only a half or a third of that on the best. In general the greater the tolerance for shade the higher is the total production, although the trees may be slow to achieve such rates. The true firs are among the most productive of the conifers, yet the peak rates of aboveground M.A.I. can range from 17 tons per hectare in Britain down to a tenth of that on the wettest sites where balsam fir will grow in eastern North America.

The 2- and 3-needled or hard pines, which are so often grown for timber, are generally not as productive of dry matter as more shade-tolerant conifers. Many observations suggest maximum values of M.A.I. around 6 tons per hectare. However, these species are most commonly grown on sites where there are significant seasonal moisture deficiencies. On better sites in parts of the Southern Hemisphere production rates of Monterey and other pines can often be 12 tons per hectare or as high as 17 in the case of New Zealand. This shows that site factors can be as important as foliar efficiency.

It has sometimes seemed plausible that very fast growing pioneer species or stands of coppice shoots grown on short rotations might sustain remarkably high values of M.A.I. in terms of dry tonnage. The culmination of C.A.I. for these species does come very early. However, such values of peak M.A.I. as can be found are generally less than those of species with slower rates of juvenile growth in height on the same sites. The high respiration rates of these species may offset the rapid gross photosynthesis associated with quick juvenile growth.

The culmination of M.A.I. in terms of aboveground dry-matter production, for a given species and situation, does not appear to come at ages much less than those of the peak values of M.A.I. for total cubic volume of stemwood. This should not be surprising because stemwood makes up a high proportion of the dry matter that accumulates in stands of trees.

Studies of total production tell much about the fundamental biological factors controlling forest production and the tremendous role that it plays in matters such as world budgets of energy and carbon. They also show the tantalizing possibilities of any technology that might enable closer utilization of the massive production of organic matter by forests. However, most use of wood is confined to those parts of tree stems that are large enough to be handled economically. One key problem of timber-production silviculture is manipulating this production so that as much of it as possible is channeled into utilizable stem sizes commonly defined in terms of merchantable cubic and board-foot volume.

THINNING AND ITS OBJECTIVES

The yield of merchantable timber volume by stands can be optimized by judicious, temporary reductions in stand density that enhance diameter growth. This process can be thought of as a controlled and accelerated reduction of the number of trees or the amount of photosynthetic area left to regain full occupancy of the growing space. Artificial reductions in stand density cause temporary reductions in gross total production but enable the remaining trees to accelerate their occupancy of growing space and their diameter growth. Some of this same effect can also be achieved by control of initial stand density in the spacing of planted trees (see Chapter 11).

Surplus trees are removed in thinning to concentrate the potential wood production of the stand on a limited number of selected trees. Total yield is augmented largely by harvesting trees that would ultimately die of suppression. However, the value, utility, and health of the remaining trees are increased because of their more rapid growth.

It is usually well to keep thinnings light enough to restrict the growth of any shrubs, vines, or undesirable trees that will cause problems at the time of regeneration. The gaps created in the crown canopy by most thinnings are so small that they close before any understory trees can start the kind of rapid, sustained growth in height of the sort necessary to count as new age classes of regeneration. Thinning is done to regulate the distribution of growing space for the benefit of the existing crop and not to vacate enough space to start a new one.

Attempts, either ambitious or casual, to obtain regeneration and secure the benefits of thinning simultaneously often wind up doing neither thing well. Desirable regeneration may appear under stands as a result of thinning or even without any treatment. If it is persistent enough or is deliberately sought to be a new crop of trees, then it is advance growth and is best regarded as shelterwood regeneration rather than as a result of what was intended as thinning. This semantic switch helps focus attention on the source and nature of the regeneration process, which is a more crucial step in the silvicultural system than thinning.

The general principles of thinning are most expeditiously considered in connection with aggregations of trees that are even-aged, of a single species, and thus all in the same crown-canopy stratum. While it is simplest to think of such single-canopied aggregations as being whole stands, they could also be the even-aged groups that are immature components of uneven-aged stands. Thinnings can also be done in mixtures of species. This includes those with differing rates of height growth that segregate into stratified mixtures with several canopy strata; these will be considered in Chapter 17.

Natural Development of the Stand

The basis of the theory followed in thinning is found in the natural development of the stand. The typical stand starts life with a relatively large number of small trees, usually thousands or tens of thousands per acre. The number of trees decreases as they grow larger, at first rapidly but more slowly with each passing decade. When the stand is ready for reproduction cuttings, it has been reduced to a few hundred trees per acre or to even less than one hundred.

This continual diminution in numbers is the result of competition and rigorous natural selection and is the expression of one of the most fundamental biological laws of silviculture. Those trees that are most vigorous or best adapted to the environment are most likely to survive the intense competition for light, moisture, and nutrients. However, the process is not entirely a steady and progressive selection of the fittest because it may be interrupted or temporarily reversed by natural accidents that eliminate trees purely at random. Furthermore, the individuals that are the fittest from the standpoint of natural selection are not necessarily the best from the forester's viewpoint.

Growth in height is the most critical factor in competition, although those trees that increase most rapidly in height are almost invariably the largest in all dimensions, especially in size of crown. As the weaker trees are crowded by their taller associates their crowns become increasingly misshapen and restricted in size; unless freed by the random accidents mentioned in the previous paragraph, such trees gradually become overtopped and ultimately die. In this constant attrition the weaker trees are progressively submerged and the strongest forge ahead. The process is known as **differentiation into crown classes;** four standard crown classes are recognized (Figs. 2-3 and 2-4).

Very few trees ever recover a dominant position after they have fallen behind in the race for the sky. Once the crown of a tree has been reduced in size by the competition of its more vigorous neighbors it cannot always be restored to a dominant position by cultural treatment. Therefore, the most common policy followed in thinnings is to encourage the growth of the leading trees rather than to resuscitate those that have fallen behind. Nevertheless, there is enough risk of accidents and inherent fluctuations in rates of development that it is unwise to proceed with plans of stand treatment that depend upon precise predictions of exactly which will be the leading trees decades hence.

The intraspecific competition between trees for places in the crown canopy is readily visible. The corresponding competition for growing space

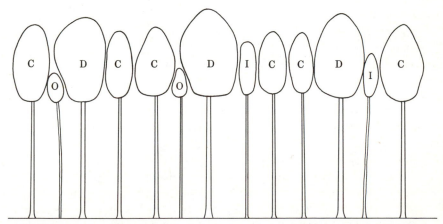

FIGURE 2-3 The relative positions of trees in the different crown classes in an absolutely even-aged, pure stand that has not been thinned. The letters, D, C, I, and O, stand for dominant, codominant, intermediate, and overtopped crown classes.

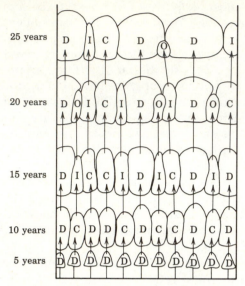

FIGURE 2-4   The relative spatial relationships of trees in the canopy of a the same pure even-aged stand at successive intervals of age showing the differentiation of trees into crown classes. This figure illustrates the suppression, as a result of competition, of some trees that were initially dominants.

in the soil is also important, but more intricate and exceedingly difficult to observe. The roots can wander in quite devious patterns and form intra-specific root grafts. Through these root grafts adjacent trees can either nourish each other mutually or engage in subtle kinds of parasitism, the patterns and results of which are obscure.

The struggle for existence in dense, unthinned stands is so fierce that it reduces the growth and vigor of all the trees in the stand. Even those that express dominance and go on to become the largest and most vigorous survivors usually develop live crowns that are too short and narrow in proportion to the total height of the trees. The resultant reduction in diameter growth is not entirely disadvantageous because such trees are likely to be well-pruned or small-branched. The severe competition also causes the stems to be straighter and more nearly cylindrical than they might be had the growing space been less restricted.

In the practice of thinning these advantages of competition between trees are at least partly preserved by refraining from thinnings so heavy that orchard-like stands of widely spaced trees result. The primary objective is, however, to keep the more promising trees growing steadily by removing less desirable, neighboring trees before their competition becomes injurious. In accomplishing this purpose the number of trees per acre is reduced as it would be under purely natural conditions, but at a substantially more rapid rate. An ideal program of thinning might, just for example, gradually reduce the number of trees in a stand so that it would have at 60 years the same number of trees and about the same average D.B.H. (diameter at breast height) that it would normally attain at an age of 90 years.

Classification into Crown Classes

The outward signs of the struggle for existence are the relative position and condition of the tree crowns. Vigorous trees that have outstripped their neighbors gain superior positions in the crown canopy and normally have the best chance of surviving competition in the future. The less vigorous lapse into successively lower positions in the crown canopy until they succumb. Since thinnings are conducted to accelerate or modify the course of the struggle, the position of the crown is an important and convenient criterion in deciding which trees to cut and which to favor.

The crown classification (Fig. 2-3) most widely used in the practice of thinning in North America is as follows:

Dominant: Trees with crowns extending above the general level of the crown cover and receiving full light from above and partly from the side; larger than the average trees in the stands, and with crowns well developed but possibly somewhat crowded on the sides.

Codominant: Trees with crowns forming the general level of the crown cover and receiving full light from above but comparatively little from the sides; usually with medium-sized crowns more or less crowded on the sides.

Intermediate: Trees shorter than those in the two preceding classes but with crowns extending into the crown cover formed by codominant and dominant trees; receiving a little direct light from above but none from the sides; usually with small crowns considerably crowded on the sides.

Overtopped: Trees with crowns entirely below the general level of the crown cover, receiving no direct light either from above or from the sides. *Synonym:* "suppressed."

The crown classification is simple and fits the ordinary requirements of thinning practice. It has the drawback of being essentially qualitative. More precise subdivision can be useful but it depends on measurements of relative heights or other crown dimensions.

In the classification given here, the intermediate and overtopped classes are well-defined and easily distinguished from the superior categories. However, it is difficult to draw a sharp line of distinction between the dominant and codominant classes. It would be desirable if the codominant class, which usually includes the majority of the main canopy trees, could be objectively subdivided into thrifty and less thrifty categories. Special recognition is sometimes given to a category of **super-dominant** trees that have so far outstripped their contemporaries as to develop excessively large branches or poor bole form. Anything growing in a stratum definitely below the main crown canopy is better considered part of the **understory** rather than as part of the overtopped crown class.

Further Classification

The selection of the trees to be favored and of those to be cut in thinnings is based not only on (1) the relative position and condition of the crown but also on (2) the health of the tree and (3) the condition and quality of the bole. In mixed stands choices between species and their relative rates of height growth also affect the selection.

The correlation between crown class and health of trees within a pure, even-aged group is very high, except when some of the trees have recently been injured by agencies capable of reducing growth. The relationship between crown class and condition of bole is less definite, particularly if the stand tends to depart from the even-aged condition. Trees that have grown into exceptionally vigorous dominants often develop coarse branches and poor form. Furthermore, injuries sufficient to deform the bole of a tree frequently have surprisingly little effect on its growth. The extent of natural pruning is also greater the lower the crown class.

It is difficult to generalize about the relationship between crown class and straightness of bole. There is, however, a tendency for trees with excurrent branching, like conifers, to have straighter stems when grown in a subordinate position than when they are strong dominants. Most hardwoods, on the other hand, develop deformed or shortened boles when grown in the lower crown classes. This difference is the result of the fact that most hardwoods, being relatively phototropic, tend to lean toward the light, while conifers, being almost exclusively geotropic, tend to grow vertically. In hardwood silviculture this situation presents a problem that is best attacked by seeking the prompt and dense reproduction required to produce uniform stands with dominants of good form.

In selecting trees to be removed in thinnings consideration should always be given to injuries affecting the condition of the bole and the health of the tree, even though subdivisions are not specifically set up in the classification for this purpose. The diminution of number of trees with advancing age is so great that there is rarely any necessity to formalize distinctions between relative degrees of imperfection among trees that ought to be removed in the earlier thinnings. In practical application of thinnings, it is best to focus attention on the relatively small number of trees that are the most promising and not let all the others obscure them.

Quantifying Objectives

A program of thinnings is essentially a series of temporary reductions made in stand density (measured in terms of basal area or some similar parameter) to maximize the net value of products removed during the whole rotation. Among the factors determining this net value are the quantity, quality, utility, and size of the products as well as the costs of harvesting and manufacture. Production in terms of quantity or volume of wood is usually considered to be the factor of greatest importance. In this connection, clear distinctions should be drawn between all the different ways in which the growth and yield of stands and trees are measured. Thinning practice cannot be adequately understood without a thorough comprehension of the relationships between the very different mensurational units in which growth can be evaluated.

It must be kept in mind that thinning is primarily intended to optimize economic returns for timber production. This means that the mensurational units used to assess various treatment alternatives, including no thinning, are meaningful only to the degree that such units of measure indicate the true net value of the material grown and removed. Unfortunately conventional mensurational parameters are, at best, only preliminary approximations of this, mainly because they do not automatically reflect the variations in value associated with tree diameter and wood quality. It should not be inferred in any situation that each unit of cubic volume, weight, or board measure is as valuable as all others.

Even the board-measure log rules do not perfectly indicate true net value. All that they are intended to show is that the sawing waste that comes from converting round logs into square-edged boards is less in large logs than in small ones. They do not otherwise reflect the higher handling costs of small trees, except perhaps to the extent that inventions such as the Doyle log-rule, which discounts small logs, may happen to do so accidentally. Even after wood is converted into flat boards, it is necessary to recognize that a board foot of a certain grade is usually worth more in a wide board than in a narrow one.

Effect of Thinning on Wood Production

It is very important to make a clear distinction between the yield of forests and their production. **Yield** is the amount actually harvested. **Production,** which is harder to determine, is the amount deposited by growth whether harvested or not.

Foresters reliably observed several centuries ago that thinning could increase the yield of usable wood from stands. It has taken much longer to deduce that the total production is ordinarily decreased or left unchanged. This statement about production has to be equivocal because much contradictory evidence still obscures the matter.

Some of the data about production after thinning are subject to large variances because of inherent variability in productivity within tree species and the plot of ground on which their growth is measured. It is also difficult to measure production precisely. A very large amount of confusion has arisen from attempts to measure production in terms of merchantable or quasi-merchantable cubic volume. These approximate yield better than they do production and the diameter limits used in their definition are seldom standardized enough to compare results effectively.

If total production is measured in terms of tonnage of dry matter there are extremes of stand density at which production would suffer. If the trees being measured did not fully occupy the growing space, their production would clearly be less than if they did. Some other vegetation that was not counted would tend to fill any unused space; its production, if counted, might or might not offset the production deficiency of the trees.

If the stand density is very great, there are some instances in which total production is diminished. Where this happens, it may be because a large proportion of the fixed amount of stand foliage goes to support the respiration of such a large amount of living and consuming tissue that the surplus available to form new tissues, including wood, is reduced. There are also cases, in climates conducive to heavy accumulation of undecomposed organic matter, in which thinning can speed decomposition and nutrient cycling enough to cause real increases in production. Some trees, even of a given species, are inherently more productive than others; thus it is theoretically possible that stand production could be increased if such trees were favored in thinnings.

Otherwise it appears that the thinning of ordinary stands decreases rather than increases the kind of gross, total production that counts all organic materials in the stand including the trees that died during stand development. If yield and production were indeed the same thing and everything that the stand produced could be harvested, it would be logical to confine thinning to the salvage of suppressed tress just before they died. The vacancies in the growing space left by heavier thinning, regardless of how small or temporary they may be, usually seem to cause uncompensated decreases in true total production.

It is economically impossible to turn all production into yield. Therefore, it is practical to consider how variations in stand density induced by thinning affect production in terms of total cubic volume of stemwood or of potentially merchantable wood. This departure from scientific measurement units such as total dry matter opens the Pandora's box full of measurement units that differ in so many large or subtle ways that they have probably caused much of the confusion.

Almost any mode of measuring cubic volume of wood involves some restrictive choice of the smallest stem diameter or of the kind of stem material that will be counted. **Total cubic volume** is customarily taken to be all the wood in the central main stem from tip to ground level; this parameter is one for which bark is specifically included or excluded. It is reasonably well-defined for most conifers and some hardwood species that have excurrent branching and single central stems. However, it involves a concept that is difficult to apply to species with deliquescent branching and single stems that fork far below the tip of the crown. Even in excurrently branched trees, so-called total volume does not include the wood of the branches or roots.

Merchantable cubic volume can be defined only in terms of some specification of the smallest diameter or other characteristic of stem components regarded as merchantable or, often, as potentially merchantable. Usually the restriction is some minimum diameter but sometimes it is the lowest point on the main stem at which forks or the presence of large branches prevent utilization. Bark may or may not be included. Most European observations involve a diameter limit of 7 cm including bark and branches, but American specifications vary. Usually whole trees that are of less than a certain D.B.H. are excluded.

Practical reasons for setting such restrictions are obvious but they are probably also the source of part of the conflicting evidence about the relationship between stand density and production. The greater the density of a stand the slower is the diameter growth of all stem components, including branches. Therefore, the denser a stand the smaller is the proportion of the total amount of wood that is included in merchantable or even "total" cubic volume. When production is assessed in such terms, it often appears to be higher at moderate levels of stand density, such as those induced by thinning, than at higher stand densities. The larger the restrictive minimum diameter set on measured cubic volume the more accentuated is the effect; it becomes even more pronounced if production is measured in the American board-foot unit. While such practical mensurational adjustments have clouded the scientific study of controlling stand production, they have advantages in considering the technology of thinning, especially if the nature of their effects is recognized.

Alternative Hypotheses

Fig. 2-5 shows three alternative interpretations that have been proposed about the relationship between stand production in gross merchantable cubic volume after thinning and density to which stands were reduced in thinning. Stand density is expressed in basal area after thinning and cubic volume includes stemwood to minimum diameters of 10 cm. Production is expressed as gross periodic annual increment during periods of 5–15 years between thinnings. Gross production includes that laid down on trees that died during the period; net production would ordinarily deduct the entire volume of such trees and will be discussed later. In Fig. 2-5, the differentially thinned stands being compared are all of the same single species, all of the same age, and on the same site. In other words, stand density after thinning is the only independent variable.

The first interpretation (*A*) is that production increases right up to the highest level of stand density that can be maintained in nature. Any vacancy in the growing space is counted as reducing total volume production.

It may be well to digress to point out that there is a long-standing procedure under which it is presumed that Alternative *A* is represented by a straight-line relationship. It has been common in North America to predict yield from adjustments of so-called "normal" yield tables. These tables are based on stand densities equal to or only slightly less than the

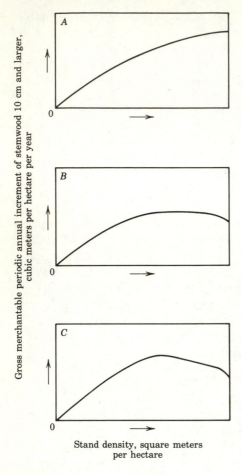

FIGURE 2-5   Graphs depicting three hypotheses about the effect of changes in stand density, induced by thinning, on the production of merchantable stemwood in pure, even-aged stands, all of the same species, site quality, and age. See text for discussion.

highest ones attainable; such densities are actually so uncommon as to be *abnormal* in the ordinary sense of the word. Predictions of yield are often made by the simple assumption that a stand that has 80 percent of the "normal" stand density will give 80 percent of the "normal" production. This assumption would be correct if 20 percent of the growing space of the stand was vacant and remained so throughout the rotation. Although most stands have some permanent vacancies, these tend to refill to some extent so the assumption of a straight-line relationship between stand density and production would generally lead to underestimates. Thinnings are supposed to be light enough that all vacancies soon fill up again.

Alternative *A* would probably fit the relationship between total production of dry matter and ·stand density, except that production might decline at very high densities in grossly overcrowded stands. However, it is clear that when production is assesed in merchantable cubic volume, Alternative *A* fits only part of the cases.

The second alternative (*B*) is that production remains constant and optimum over a wide range of stand density from some lowest level at which there is full occupancy of growing space up to those at which excessive competition is postulated to restrict production. This interpretation received its greatest impetus from studies by Mar:Møller (1947, 1954) and others in Scandinavia. Alternatives *B* and *C* rest on the assumption that full occupancy of the growing space can exist at comparatively low levels of stand density.

The data graphically presented in Fig. 2-6 indicate an effect of differences in site quality that may partly clarify the situation. The data are from observations of many thinned stands of loblolly pine in the southeastern United States (Nelson *et al.*, 1961). The data were fitted by statistical techniques not obviously biased in favor of any of the three alternatives. On the better sites, production increased to nearly the highest levels of basal area, about as Alternative *A* would suggest. However, on the poorer sites, Alternative *B* best fits the results.

It is here postulated that, on the poorer sites, seasonal moisture deficiencies or other soil factors limit production more at high stand densities than at intermediate levels consistent with full or nearly full occupancy. On the better sites, production increases almost up to the limit set by the

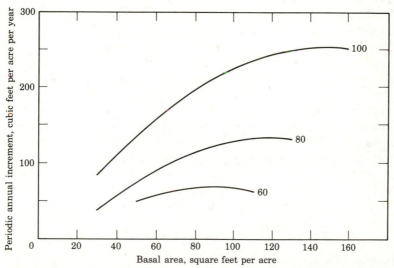

FIGURE 2-6   The 5-year periodic annual increment, of merchantable cubic volume of stands of loblolly pine, averaging 40 years old, thinned to various levels of basal area. The stands were on sites of different quality in the southeastern United States and these curves represent regression equations fitted to the experimental data by Nelson *et al.* (1961). These relationships suggest that production in thinned stands increases with increasing stand density on the best sites where water is not limiting but it is optimum and nearly the same over a wide range of stand density on more restrictive soils.

amount of light because soil factors do not set much restriction on the amount or efficiency of foliage.

It is tempting, but not necessarily valid, to conclude that on excellent, nonrestrictive soils, the best cubic-volume production comes with light thinning or none (Alternative *A*). However, on the more ordinary and common soils, which are restrictive, optimum basal area gradually increases with age according to Alternative *B* or *C* in youth and middle age, but Alternative *A* best fits the final stages of a rotation.

It should be noted that the foregoing applies to production of merchantable cubic volume with quite small diameter limits. If the diameter limits were even smaller, Alternative *A* would probably approximate the truth in more cases. However, if the minimum diameter limits were increased, the logic of Alternatives *B* and *C* and of thinning to ranges or points of moderate basal area would be strengthened. For example, if one used the American board-foot unit, which weights production in large stem-wood more than that in small, there would presumably be many cases in which Alternative *C* and rather heavy thinning would become most logical.

Part of the difference in interpretation between Alternatives *A*, *B*, and *C*, results merely from the bewildering variability that would become obvious in the scatter of points representing the actual observations on which graphs such as those of Fig. 2-5 are based. Some of this variance is due to factors such as genetic differences, variation in site quality or stand density *within* experimental thinning plots, and observational errors, especially those involving the effect of taper on stem volume. In fact, the hypothesis of *B* may rest on nothing more than the fact that in many instances there is no universally demonstrable upward or downward trend in total periodic annual increment with respect to moderate artificial decreases in stand density.

Even if Alternative *B* were incorrect, its tacit acceptance would have the thought-provoking effect of focusing attention on those considerations about thinning that can be important regardless of what its effect on production of total cubic volume might be. If decades of study have produced so much contradictory evidence, one is certainly entitled to conclude that, so long as stands remain nearly closed, the effect of stand density on production is not large enough to be any sort of exclusive consideration in determining how to thin.

Differences between Species

There is ample evidence to indicate that levels of optimum basal area tend to be greater for shade-tolerant species than for intolerant ones and greater for evergreens than for deciduous species. The more efficient the foliage of a species the greater is not only the production but also the amount of growing meristematic tissue that can be efficiently nourished.

The approximate ranges of appropriate residual basal area, after thinning, for middle-aged stands of these four categories of species may be tentatively stated as follows:

	ft²/acre	m²/hectare
Tolerant evergreens	130–230	9–16
Intolerant evergreens	80–130	6–9
Tolerant deciduous	70–160	5–11
Intolerant deciduous	50–80	4–6

It should not be inferred that the limits stated here are precise. The basis for the implication that the logical ranges are greater for tolerant than for intolerant species is weak, although stands of shade-tolerant species do seem to be more flexible in their response to thinning than those of intolerant species.

Where differences in site quality are substantial, the weight of evidence indicates that the logical basal area for stands of a given species would be less on a poor site than on a good one. If the site is capable of supporting a stand with a closed canopy, the difference may be small. However, it is worth noting that in arid forest regions, such as those of parts of the interior ponderosa pine region, there can be full closure of the root systems with as little as 10 percent canopy cover. Ideas about levels of basal area appropriate to more humid forest regions would scarcely be applicable there.

Effect of Thinning on Economic Yield of Stands

Practical understanding of thinning procedures depends on knowing how they can be applied to increase the economic yield even though they do not substantially increase, and may even slightly decrease, gross production. The general approach is to allocate the production to some optimum number of trees of highest potentiality to increase in value; the other trees are systematically removed in such a sequence as to obtain maximum economic advantage from them. The various advantages that can be gained are summarized as follows:

1. Salvage of anticipated losses of merchantable volume.
2. Increase in value from improved diameter growth.
3. Yield of income and control of growing stock during the rotation.
4. Improvement of product quality.
5. Opportunity to improve stand composition, to prepare for establishment of new crops, and to reduce risk of damage.

Salvage of Anticipated Losses

The gross production of wood by a stand should not be confused with the actual yield in terms of usable volume removed in cuttings. Not all the cubic feet of wood produced by the growing stand remain stored on the stump until the end of the rotation. In fact, a high proportion of the total production will be lost from death and decay of the large numbers of trees that fail to survive the struggle for existence. From the economic standpoint, any part of this perishable volume that can be salvaged by removing doomed trees in thinnings represents an increase in the quantitative yield of the stand.

If a typical stand is grown on a rotation long enough to produce trees of conventional sawtimber size without thinning, as much as one-third or even half of the potentially merchantable cubic volume may be lost to suppression. This kind of loss can be reduced by ending rotations when such mortality becomes serious, but this makes it hard to get trees of the diameters often desired. Diameter growth can be increased by widening the initial spacing of the stand, but this maneuver prevents suppression loss partly by leaving growing space vacant and thus merely preventing the production that would become loss. Thinning is the best solution to this dilemma and is therefore practiced whenever other considerations do not preclude it.

Thinnings designed to anticipate and salvage losses from natural suppression are the only proven means, other than site improvements, of increasing yield of total cubic volume or tonnage from a stand. Only part of the prospective losses can be recovered unless stands are annually gleaned for dead and dying materials no matter how small.

Not all of the material removed in anticipation of loss need come from small trees of the suppressed and intermediate crown classes. In certain methods of thinning, it is possible to take out part of this volume in larger trees thus forestalling the death and stimulating the growth of their subordinates. This particular technique must be used carefully and then only in situations where the trees of the lower crown classes are still healthy enough to respond to release.

Increase in Value from Improved Diameter Growth

Up to this point, the discussion of the effect of thinning on the quantitative yield of a stand has dealt mainly with production of cubic volume. However, not all units of cubic volume of roundwood are equally valuable; those that come from large trees are generally of greater value than those from small. One of the most important objectives of most, but not all, kinds of thinning is to reduce the stocking of a stand so that it eventually has fewer trees, but of larger average diameter, than it would without thinning.

The most general reason for the greater unit value of trees of large diameter is that the cost, per unit value of ultimate product, is less for

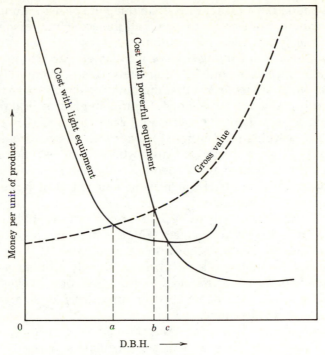

FIGURE 2-7 The effect of the diameter of a tree on the cost of processing it and the value of the product, *per unit of product,* when the harvesting and processing are done with two different sets of equipment. The set of light equipment is low in capacity and cost; the powerful equipment is high in initial investment and hourly operating cost. Points *a* and *b* are the smallest diameters at which it becomes economical to use each set of equipment; for trees larger than diameter *c* it would be logical to use the more powerful set. Net value at any diameter is the vertical distance between the appropriate curve of cost and that of gross value. Note that certain costs, such as those of roads or of handling trees collectively in bundles, are not affected by tree size. The curve of gross value would be nearly horizontal for such products as fuelwood or that chipped for pulp or other reconstituted products. Gross value can also decline with increasing tree size if the incidence of rot or similar defect increases with age or size. Note that the ideal thinning program would be one in which trees are caused to move as swiftly as possible into diameter classes that can be harvested most profitably and also with equipment appropriate to those sizes.

processing them than for small ones. (Fig. 2-7) This relationship applies to the cost of handling each tree or log individually during harvesting and subsequent processes. The monetary return for growing a tree is usually the difference between the value of the ultimate unit of product and the cost of harvesting and processing it. The forest owner who grows a hundred units of cubic volume in fat trees saves handling costs for the buyer and is thus entitled to a higher price than one who grows a hundred units in a greater number of slender trees. Of all the economic reasons to thin, this one is usually the most important.

A second reason for trying to improve diameter growth is that wood quality usually increases with tree diameter. This is mainly because the outer rind of wood has fewer knots and grain irregularities than the central core; ordinarily it also has superior anatomical and mechanical properties. The outer wood is usually stronger, more easily worked, longer-fibered, less subject to warping, and in most ways much easier to use than the inner wood. With some species the inner wood improves when it turns to heartwood but this change also depends on diameter growth. The more outer wood that can be added by increasing diameter growth the greater is the inherent value of the wood.

One of the most common and valuable ways of using trees is as sawn lumber or other solid-wood products. There is probably no other way of producing structural materials that is more economical of the world's energy supply. Unfortunately it is not correspondingly economical either of the wood itself or of the time required to grow trees of suitable sizes. Both of these problems can be mitigated by thinnings designed to grow trees of larger diameter in shorter time.

The waste of wood in lumber production is in the sawdust, slabs, and edgings generated when round stems are sawn into square-edged boards. Given the ordinary requirement that boards be of certain minimum thicknesses, the proportion of wood that is wasted is greater in trees of small diameter than large ones. In fact, there is usually some minimum diameter of stemwood below which the material is simply left in the woods; the proportion thus abandoned in severed tops is also inversely related to diameter.

The board-measure log rules that are used in North America are, in effect, computational devices designed to estimate the volume of lumber that will remain after round logs are sawn into square boards. It would be possible to convert a cubic foot of wood into 12 board feet only if a tree were a square timber throughout its length and could be sliced into boards without sawdust or other waste. Cubic volume is, in other words, round and not square. The ratio of board feet to cubic feet ranges from roughly 5 in trees about 8 inches DBH to a nearly stable value of 7 or slightly more in trees that are more than 20 inches DBH. If trees are grown for lumber and the cubic volume growth is fixed, this difference in proportion of waste is another reason to seek improved diameter growth. Even if the waste is used for reconstituted wood products, loss of potential value still takes place.

Not all units of board measure are equally valuable. Those that are in wide boards are worth considerably more than those in narrow ones. This is yet another advantage that accrues to improved diameter growth.

Merely because the calculations are easier, the results of thinning trials are often interpreted in terms of common mensurational units, especially cubic volume, without any consideration of the distribution of diameters in the stands. The purpose of most regimes of stand management is to maximize the production of net value and not just volume.

Rough comparisons can be made between possible treatments by observing the distribution of numbers of trees or of volume by diameter classes. Another simple test parameter is the average diameter of a fixed number of the largest trees per unit area, such as the 100 largest per acre. It must be borne in mind that the average diameter of a whole stand is automatically increased by removing some of the smaller trees.

To reap the rewards of cutting large trees late in the rotation it is necessary to cope with the high cost per unit volume of harvesting and utilizing small trees in earlier thinnings. This difficulty can be mitigated partly by using different equipment and logging methods for the small trees than are later used for the large. The kind of equipment that is most efficient for handling final-crop trees is not necessarily the best for the thinnings, ordinarily being heavier, more cumbersome, more costly, and more expensive to operate for a given length of time.

The problem can also be evaded by delaying the thinnings until the trees to be removed grow large enough to handle economically. Another evasion is to confine the first removals to any large or medium-sized trees of poor potentiality and ignore the small ones until they become larger. Trees that are destined never to be worth harvesting can be either left to die of suppression or killed by the cheapest means available.

The ideal solution is to manipulate the equipment and sequence of thinnings such that the equipment is matched with the sizes of trees to be removed and as many trees as possible are cut when they have grown into the range of diameter that can be harvested efficiently. The objective should be to maximize the return from harvesting the entire crop of one rotation rather than to minimize the cost of each separate operation in the whole sequence of cuttings.

Increase in stem diameter is so advantageous that it is generally best to thin stands down to the lowest density that will not cause poor tree form or other unacceptable side effects. Deliberate sacrifice of production of gross tonnage or total cubic volume is often desirable; that of merchantable volume may or may not be depending on the relative merits of growing large trees and securing a large total yield of wood of any sizes. Some countries have sufficient concern for maintaining high wood production that they prohibit thinning so heavily that yield of merchantable cubic volume is sacrificed.

Yield of Income and Control of Growing Stock during Rotation

An unthinned crop of trees is an asset that increases in value throughout the rotation but remains frozen under passive management. No cash income is realized, yet carrying charges that must be paid for by the crop accumulate. After some trees have become merchantable thinning can become a form of continual asset management.

In whole forests that are young and immature because of earlier heavy cutting in a locality, thinnings and other intermediate cuttings may provide the *only* source of income for long periods.

Thinning helps meet two rather cruel financial tests often applied to long-term investments in timber production. Both tests involve compound interest and affect decisions about rotation length. Consequently the two different tests often get confused with each other.

The first test requires an adequate rate of compound-interest return on all out-of-pocket costs of establishing the crop and carrying it to maturity. The second test requires that the rate of interest earned on the stumpage value of the trees or stands not fall below some desired rate. Neither of these tests is easy to meet but they are at least not additive. The rate demanded under the first test, which involves whole rotations and hard cash, may be higher than that of the second, which operates over shorter periods and merely involves income temporarily forgone. Their status as separate and distinct tests is a manifestation of the fact that there is no single, universal test of financial wisdom.

The first test is that in which the cost of establishing a stand and all subsequent costs are carried at compound interest to the time when the major income is recouped at the end of the rotation. Thinning helps meet this test by providing income during the rotation and thus earlier in time. By the reckoning involved, the money received at age 20 may be worth twice that at age 35. It is usually advantageous to pay off the compounded costs of stand establishment early in the rotation rather than, in effect, letting them mushroom until the end. If the thinning enables trees to grow fatter and more valuable or increases the total yield, the net income against which the charges can be made is increased. If these improvements make it financially prudent to have longer rotations, the average annual expense for regeneration in the whole forest is reduced, so the overall result can be still further improved.

The second test involves the crucial task of improving the return on the large investment represented by the liquidation value of merchantable growing stock in managed forests. If an owner requires a rate of return of 8 percent, each dollar left in growing stock at the beginning of the decade must grow to $2.16 in value during the decade to justify having left the tree or stand. In this test, unlike the first, any money invested in growing the tree does not count; only that money for which it could have been sold at the beginning of the period is considered. Thinning can be applied in such a manner as to reduce the investment in growing stock and increase the value of gross growth by removing trees of low earning potentiality and increasing the growth of those of high potentiality. Since it is possible to reduce growing stock without greatly reducing volume growth, this becomes a way of eating one's cake and having it too.

The second test embodies one of the most useful ways of determining which trees to cut and which to leave in thinning. Those that cannot be made to increase enough in value to yield an acceptable rate of return on their own value are removed. Those that can are not only left but they are also granted enough additional growing space that the compound interest that they earn is increased. The trees most likely to continue yielding the desired rate of return for the longest time are those of highest quality and rate of growth. In a pure application of this approach, the first trees to be cut would be those of low quality or slowest growth, although they would theoretically not be cut until they had acquired a positive value.

In applying the idea it is best to allocate growing space to the trees that earn the most. This requires estimates of the earning capacity of trees of different classes of diameter, crown class, and stem quality. Once a few calculations are made and one has the basic principle in mind, it is then necessary to continue with intuitive judgement in determining which trees to cut or leave because the detailed computations are complicated. It is especially difficult to take differences in tree quality into account. The easy assumption that all trees of the same diameter in a stand are of the same value can impair the application of the principle.

The periodic removal of the poor earners and encouragement of the good is a means of decelerating the rate at which the overall rate of compound-interest return on the growing stock of a stand inevitably declines. The rate declines between thinnings and can be elevated again by each thinning. When it ceases to be possible to restore the rate to the desired level it is, according to this mode of analysis, logical to replace the stand with a new one and thus end the rotation. However, if rotation length is being set on the basis of return on capital investments, the first test of return on monies actually spent is a criterion that must be considered also. Since people tend to want high rates of compound interest, rotations set on the basis of return on capital may be shorter than those of maximum mean annual increment of volume of product, especially if the product is sawtimber. The latter approach, sometimes called that of maximized forest rent, is usually followed where the amount of forest land rather than of capital is viewed as the limiting factor and the object is to get as much wood as possible from the land.

Regardless of how their length is determined, thinning provides ways of extending rotations and also of lending some flexibility to their control. The rotation necessary to grow trees of given size can be shortened; larger and more valuable trees can be grown on rotations of fixed duration. Short rotations do not necessarily have the financial values so often ascribed to them, especially if all of the financial factors involved are really considered. If thinnings have produced enough revenue to amortize establishment costs, are yielding an adequate income, and are maintaining good rates of growth, there may be little need to rush into the risk and expense of starting a new crop that will be some years in regaining full occupancy of the site. The forest manager who refrains from thinning usually is condemned to short rotations, frequent regeneration problems, and limited opportunity to maneuver to meet all sorts of emergencies and changes in markets or management objectives.

Improvement of Product Quality

The value of a fixed total production can be enhanced simply by favoring the trees of best potential quality and discriminating against the poor. *This effect of thinning on wood quality is vastly more important than any other.* The superior diameter growth induced by thinning usually improves wood quality because the large trees tend to be of better quality than small ones. The general effect of thinning on wood quality is not harmful, in spite of much opinion and some evidence to the contrary. This matter can, however, be dealt with more effectively after considering the effect of thinning on the growth of individual trees.

Stand Composition, Regeneration, and Protection

The opportunity to intervene in the development of the stand during the rotation is convertible into a number of other economic benefits. It enables continuing control of undesirable species not eliminated during the period of regeneration. If mixed stands of dissimilar species are being deliberately maintained, thinning may represent the only means of taking adequate advantage of those species that tend to mature earlier than the major components.

The long period of control over stand composition enables one to prepare for natural regeneration by reducing the seed source of undesirable species and fostering that of the good. Thinning also builds up the physiological vigor, mechanical strength, and seed production of the individual trees of the final crop so that there is wider choice among methods of regeneration cutting. Unthinned stands are likely to have to be replaced during an abbreviated period of establishment even when more gradual replacement might enable greater use of the financial potentialities of the crop.

Consideration will be given later to the ways, in addition to salvage, by which thinning increases the strength and health of stands and thus reduces the total monetary loss to damaging agencies.

BIBLIOGRAPHY

The bibliography for Chapters 2 to 4 is at the end of Chapter 4.

C H A P T E R   3

The Individual Tree and Its Treatment

Whether one grows trees by thousands in forest stands or as shade trees on city streets one must know how they develop as individuals. Among the attributes of trees that can be regulated are the size, shape, and structure of their stems as well as their branching characteristics. This can be done partly by governing the amount of growing space allocated to them, as in thinning, and partly by artificial pruning or other surgery.

Response of Trees to Increased Growing Space

The amount of carbohydrates produced by a tree depends mainly on the size of the crown or leaf surface and the ability of the roots to supply the foliage. When a tree is released in thinning, any prompt acceleration of growth is largely from increase in water and nutrients supplied by the roots. The amount of foliage does not increase until there has been time for the crown to enlarge, although this delayed effect is ultimately the most important. Not all units of leaf surface are equally efficient in photosynthesis. Leaves that are severely exposed to sun and wind produce somewhat less than slightly protected ones; heavily shaded leaves do little more than supply themselves. The part of the crown above the point of horizontal crown closure, especially the upper middle portion, produces much more than anything below.

The roots extend more widely and swiftly than the crown. Since they do not have to support themselves like aerial branches their extension is not limited by structural necessities. They may extend into any unoccupied soil space even if this means intermingling with or going around the root systems of other trees. The root systems of healthy trees are much wider than the crowns. Many species, but not all, form intraspecific root grafts so it is possible for a tree to incorporate part of the root system of an adjacent cut tree. In fact, in some pure stands the trees that are individuals above ground share some semblance of a common root system.

Priorities in Allocation of Carbohydrates

It is logical to assume that carbohydrates are used in ways that increase the ability of the tree to survive. Some functions are more crucial than others so a priority schedule appears to exist.

Respiration comes first because the living tissues would perish immediately without it. The greater the amount of crown surface per unit of living tissue the greater will be the proportion of carbohydrate left for formation of new tissues. The amounts then used for the indispensible renewal of foliage and the extension of crown and root system depend mainly on the room for expansion that is made available by thinning or can be captured from less vigorous competitors. Part of this process is growth in height, which is so important to the survival of competing trees that it commands high priority; it varies surprisingly little with tree vigor even though it is strongly controlled by site quality and differs greatly between species. Seed production is so vital to species survival that whenever it takes place it does so at the expense of other functions.

Another vital function that must be fulfilled is conduction of materials between crown and root. Phloem must be renewed annually and so ordinarily must the xylem. However, most trees lay down far more conductive tissue than is absolutely necessary. This can be verified by observing how small a "bridge" of tissues is necessary to maintain an imperfectly girdled tree in good vigor.

Most of the wood laid down in the stem and larger roots serves to support the crown. Although this is a crucial function it does not appear to have enough immediate survival value to command high priority. In any event, the extent to which provision is made for mechanical support seems to depend on the vigor of the tree and the amount of carbohydrate that remains after provision for more vital functions. The part of the growth that is of greatest economic concern is thus low enough on the scale of biological priority that it is subject to great variation. It is for this reason that growth in diameter of stems is so readily controllable by thinning or other means of regulating stand density and tree vigor.

The tree stem as a whole tends to develop as a mechanical structure designed to bear the vertically acting load of its own weight and the horizontally acting load of wind. It seems possible that the better the carbohydrate supply the closer the stem approaches the ideally engineered form of a cantilever beam of uniform resistance to bending throughout its length. If tree weight were the only load, the stem would be cone-shaped above the base of the crown. Since there is always some effect of the horizontally acting wind, the stem assumes the tapered form of the modified paraboloid recognized in mensurational work. At least in the lower portions of the stem, the amount of wood formed appears to be controlled by the bending stress that is maximal at the ground line because that is the fulcrum of a long moment-arm of bending. This is the cause of the butt swell at the base of the stem. However, if the tree is not able to grow well, the butt swell ceases to continue to develop but may remain obvious as an inheritance from an early period of more rapid growth.

The annual layer of wood deposited on the stem is not uniform in thickness, density, strength, or anatomical structure from top to bottom. One of the thickest parts is slightly above the base of the live crown (Fig. 3-1). If the tree is very vigorous there will be an even thicker portion at the base of the tree in the region of the butt swell. However, if the tree is not well nourished with self-generated building material, the very thinnest part of the wood layer, that is, the slowest diameter growth, will be at ground line. The pattern of diameter growth at various points in the upper part of the stem gives credence to the hypothesis that stem form is controlled by relative proximity to the carbohydrate source in the crown. That in the lower bole accords with the view that stem growth is a mechanically controlled response to bending. Both kinds of response seem to be involved to some degree and are, in either case, governed by complex hormonal systems (Larson, 1963).

The fact that the thickness of the annual layer decreases for some distance below the crown base is partly the result of the spreading of similar amounts of substance over increasing circumference. If the cross-

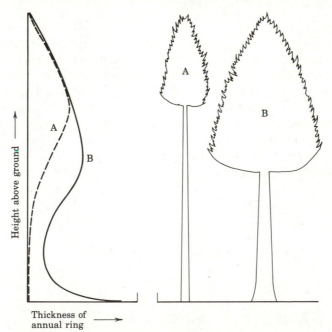

FIGURE 3-1 The variation in thickness of the sheath of wood annually laid down on the central stem for (A) a small-crowned tree that is barely surviving and (B) a vigorous one with well-developed crown and butt-swell. Note that there are peaks in ring thickness near the base of the crown of each tree.

sectional area or mass of material in the annual layer is measured rather than the radial thickness, it usually appears to be nearly uniform from the base of the crown to the top of any butt swelling that is still developing. Furthermore, it may be noted that an annual ring of given thickness laid down on a 15-inch stem adds 50 percent more basal area than on a 10-inch one.

The general effect of thinning is to make the lower branches live longer, become thicker in diameter, and also produce more carbohydrate. This in turn provides the structural material that enables the tree to grow faster in diameter. However, the diameter growth is increased much more in the lower than in the upper parts of the stem so that it becomes more tapering and the whole structure becomes stronger.

It is important to note that the wide variations that can be induced in diameter and basal area growth in the zone of butt swell can give exaggerated impressions of increases or decreases in growth of stem volume. Except in small trees this zone definitely includes the breast-height level. It is unfortunate that trees are tall and people are short, because it is more reliable to assess diameter at some higher point.

Tree Growth as a Guide to Thinning

Patterns of annual-ring thickness are handy indicators of tree development but they need to be interpreted with some mathematical sophistication. An ideal stem might have wide rings that were uniformly thick but this can be produced only if the crown surface expands exceedingly rapidly and growth in basal area accelerates (Fig. 3-2). This may happen with an isolated shade tree but is seldom possible with trees in closed stands. The ring pattern of a timber crop tree that has been released in a series of timely thinnings is one in which ring thicknesses decrease slowly but steadily outward from the pith. A tree often actually accelerates in volume growth even while ring width slowly declines.

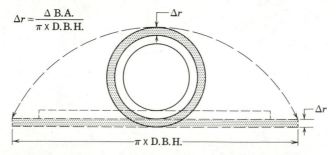

FIGURE 3-2   The relationship between annual increment in basal area (ΔB.A.) and radial thickness (Δr) of growth rings. The annular rings of two successive years are straightened out into rectangles to show the large increase in basal area increment that is necessary to maintain constant ring width.

It is erroneous to assume that successful thinning must cause dominant trees to show a shift from narrow rings to wide. Such pronounced increases in growth can be welcome but they actually indicate that the thinning might better have been done earlier. It is the trees of the subordinate crown classes that show the most spectacular response, provided they have not been suppressed so much that they fail to respond at all. Thinning is usually better directed at forestalling major decelerations of diameter growth than at remedying them after they have occurred. Ideally there should be an ambitious but realistic plan for the regime of diameter growth of final crop trees.

Variations in stand density, such as those induced by thinning, cause very large variations in diameter growth but remarkably little in height growth. Very light thinning may have no effect on diameter growth but, with drastic thinning, it is entirely possible to produce crop trees with twice the diameter that they would have attained in the same time without thinning. The height growth of the *dominant* trees, on the other hand, is changed scarcely at all; it is likely to be slightly decreased at the extremes of stand density. The height growth of the leading trees of a stand is, in fact, so independent of stand density and so closely controlled by the totality of growth-supporting factors of the site that it is used as an integrated expression of site quality. Site index, the total height of the leading trees at 50 years or some other standardized age, is thus, in this context, a manifestation of the fact that thinning affects height growth very little.

The height growth of dominants is not absolutely and always unaffected by stand density or thinning. Trees that have become exposed, isolated, or open-grown may become somewhat stunted. The reasons are not clear but may involve growth of branches and lower bole at the expense of height. Although thinning release usually accelerates height growth it sometimes has the opposite effect at least temporarily. Very high stand density often causes reduced growth of trees in all dimensions, including height, probably because of the high respiration presumably associated with a poor ratio of photosynthetic to aphotosynthetic tissue in each tree.

Values of average height calculated for *all* trees in stands are inversely related to stand density because the height growth of trees of the lower crown classes is stunted by competition. The greater the number of trees being submerged through competition within the crown canopy the lower is the average height.

The common opinion that trees grow taller in dense stands is, in general, incorrect. The slender trees merely look taller than the more tapering ones of a less dense stand; the effects of this optical illusion vanish if the heights are measured.

One rough but convenient index of the ability of the crown to nourish the remainder is the **live crown ratio,** which is the percentage of length of stem clothed with living branches. It approximates the ratio of photosynthetic to aphotosynthetic surface and is related to considerations of tree management such as the branch-free length and the taper of the stem. It is a parameter that is better measured than guessed at because the upper parts of a tree, being farther from the eye, always seem smaller than the lower parts. In closed stands, the bottom of the foliar canopy retreats upward with surprising rapidity. This is fine from the standpoint of natural pruning and reduction of stem taper but discouraging for the maintenance of good diameter growth. With most trees it is desirable to plan to let the base of the crown retreat to a chosen height and then to halt or slow down the retreat. This means that the live crown ratio decreases from 100 percent to a certain amount and then one attempts to make it increase again or at least to slow down any further decrease. This can be done only by thinning to keep the lower parts of the crown adequately illuminated. If the ratio is allowed to decrease to 30 percent or less the general reduction in vigor will cause substantial loss of diameter growth. If the ratio is very low, the recovery after thinning will, at best, be delayed or the tree may even succumb.

The loss of trees with small crowns after thinning may be caused by insect attack, sunscald, or even the cutting of other trees that had formerly nourished the unthrifty through root grafts. The most important cause may be merely the increase in respiration induced by the sudden increase in temperature caused by exposure. If the respiratory demand is great enough very little carbohydrate will be left for renewal of vital tissues. However, unthrifty trees are usually eliminated by bark beetles and other biotic agencies before they actually starve to death. Such difficulties are avoided by selecting crop trees from the dominant and codominant classes with acceptable live crown ratios.

Biological Basis for Quantifying Tree Development

It is easy to demonstrate that the accelerated crown expansion induced by thinning increases diameter growth but hard to develop reliable quantitative relationships necessary to predict how various thinning programs would affect the growth of individual crop trees. When such relationships have been established it becomes possible to simulate alternative thinning regimes in computers in minutes and thus to imitate field experiments that take decades. Ordinarily data derived from thinning trials that required decades or other observations of long periods of growth are necessary for such simulation. Most efforts to explain tree growth in quantitative terms seem to work best if tree diameter is bypassed in favor of direct relationships between amounts of foliage and wood volume growth. The models might work even better if dry weight were substituted for volume.

Even though new wood is laid down on old wood by the vascular cambium, it is not the wood but the foliage of the crown that "grows" more wood. It is fairly obvious that wood grows in layers laid on the surface of core that gradually expands. It is not so obvious that the foliage that ultimately produces the wood is also borne on layers wrapped around an expanding core because, unlike the wood, the old layers of leaves disappear when they cease to perform their function. The total cubic volume of the tree is a cumulative expression of all wood ever produced by the main stem. It is, at least theoretically, closely related to and controlled by the crown volume or, more precisely, by a kind of "historical" crown volume projected into the past as shown in Fig. 3-3; this is a cumulative or integrated approximation of all the foliage that the tree ever had.

According to the same line of reasoning, the growth of wood during any short period of years would be related to the expansion of the crown during that period. The growth of one year might have a close relationship to the amount of foliage or the living crown surface area of that same year. In trying to determine such quantitative relationships, it helps to distinguish between cumulative and current parameters and to draw relationships within these categories and not between them. For example, if things have to be simplified by the treacherous step of omitting the vertical dimension, the basal area of a tree, which is a cumulative parameter, can be related to the crown area, which is the area of ground directly under the crown and, in effect, a cumulative parameter. The changes in both basal area and crown area during a certain short period would be corresponding "current" parameters.

Much of the practice of thinning rests heavily on the importance of crown relationships. However, trees do not live by light and carbon dioxide alone. They also have roots so water and soil nutrients can be limiting factors. There are, in fact, sites so deficient in available (oxygenated) water or some nutrient element that thinning will *not* improve the growth of the trees unless coupled with fertilization, drainage, or irrigation. Furthermore, there are many sites on which improvements in growth from thinning do not come directly from crown expansion but from allowing the roots to expand and claim more of the limited supply of soil water and nutrients. In such cases, the concomitant expansion is a necessary, but secondary, consequence and not a primary cause. Light and carbon diox-

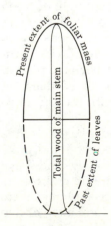

FIGURE 3-3 A two-dimensional representation of the relationship between the cumulative volumes of stemwood and space occupied by the foliage of a tree during all of its previous development. The present size of the crown is enclosed within solid lines and the space that has been successively filled with previous foliage is depicted by the dashed lines projected from the widest part of the crown down to the ground. This is intended to show how the total amount of foliage that the tree has produced can be related to the total amount of stemwood which that same foliage produced. The diameter of the tree stem is exaggerated.

ide become the main limiting factors only on sites that are "good" in terms of the supply of available water and nutrients in the soil.

Effect on Wood Quality

The structural requirements of tree stems are not met entirely by laying down wood of homogeneous density, structure, and strength. If wood were of homogeneous strength, trees would be much more tapering than they are. Instead the strength of the wood in a stem usually increases from the pith outward as well as from the top downward. Resistance to breakage at any point is determined largely by the product of the strength of the outermost fibers and the distance by which they are separated. The strength of the wood between is of little consequence other than in making the stem more rigid and reducing the actual amount of bending caused by a given wind load. The overall strength of the stem is still further enhanced by the fact that the outermost fibers are under tension and the inner ones under compression. These internal growth stresses contribute to the tendency of lumber to warp; however, they make the stem stronger just as prestressed concrete beams with stretched cables beneath the surface are stronger than reinforced concrete with inert rods in the same positions.

When a conifer is young and short, the loads placed upon it are small and it usually produces the weak **juvenile** or **core wood.** However, during its period of active increase in size and height, the specific gravity and strength of wood laid down tend to increase. When a tree of almost any species reaches maturity and ceases to increase much in height or size of crown, the annual rings added to the bole become very thin and of low strength, representing little more than enough xylem to renew the water-conducting system. It is at least tempting to conclude that this situation develops because the load on the stem ceases to increase much with time.

Growth Rate—Not a Causitive Factor

One erroneous view that long persisted was that the strength of wood is directly controlled by the rate of growth. According to this, rapid growth was reputed to produce strong wood in ring-porous hardwoods, weak wood in conifers, and no appreciable difference in diffuse-porous hardwoods. The main practical consequence of these ideas has been the persistent notion that thinning or any other measure that increased diameter growth of conifers weakened the wood; as is the case with human nutrition, the fattening things of life are perceived to be evil.

The idea that fast-grown conifers produce weak wood is based on little more than the observation that young trees have weak wood and grow rapidly whereas somewhat older trees grow more slowly and have stronger wood. It is now clear that the effects of growth rate and factors related to age were being confounded and that the age-related ones were more nearly the controls of wood strength and density. Correlation of two factors is not proof that one causes the other because the two may both be the result of some other more controlling factor. If a conifer is made to grow rapidly by thinning, it creates a greater volume of the same kind of wood that might have been laid down without release and does not shift to production of weak wood.

It is probably best to interpret differences in the strength, density, and structure of wood in terms of the main biological functions of xylem, which are mechanical support and water conduction. When trees are small and young their stems do not have to be particularly strong. It is noteworthy that there are fast-growing, weak bodied, pioneer species that never get out of this condition and topple over before they become very tall. So long as a tree continues to grow in height the load of its own weight and that of the wind increase and so does the strength of the stem. If the tree becomes very old, it shifts to the production of very thin annual layers of soft, weak wood of the kind prized for fine veneers, cabinet work, and finish grades of lumber. It may be suspected, but is not proven, that this kind of wood is produced by the necessary renewal of conductive xylem in trees that are no longer adding to the load on their stems because their height growth and crown expansion have nearly ceased; such trees have already developed the stem strength necessary to support the crowns.

A more refined indicator of the strength and density of wood is the relative proportion of early-wood, which functions mainly for conduction, and of late-wood, which is mainly for mechanical support. In conifers, the time of the annual change from formation of early-wood to that of late-wood appears to coincide with the end of the initial period of shoot and needle elongation (Larson, 1963). Any treatments, such as thinning, fertilization, or irrigation, or factors of soil and climate that prolong or increase diameter growth during the early summer usually increase the proportion of late-wood, which is produced mainly at that time.

Controlling Wood Properties by Thinning

The structure and anatomy of xylem vary tremendously between and within species as well as within tree stems; there are many important variations more subtle than those considered here (Haygreen and Bowyer, 1982). The choices that can be effected between good trees and poor ones in thinning improve the ultimate utility of the wood much more than they might be impaired by any vaguely suspected baneful effects of making the trees grow faster. In general, the larger the trees grow and the straighter their stems the greater is the usefulness of the wood, at least until the heart-rots of old age start to cause deterioration.

At least with intolerant species, such as the 2- and 3-needled or "hard" pines, it is often best to allow the trees to grow unchecked when they are young so that they will develop crowns large enough to continue good health in later stages. The best way to deal with the juvenile wood that this produces is to develop suitable uses for it consistent with its shortcomings.

This line of action will produce a greater amount of denser wood in later stages than if one attempts to restrict the amount of core wood and rebuild larger crowns later. Furthermore, it is desirable to avoid abrupt changes in the rate of diameter growth such as those that can occur when trees are released by long-delayed thinning. This is one effect of rapid growth that can cause problems. There may be shake (splitting along the plane of annual rings) or warping in boards cut from the transition zone. This is one of the reasons why it is best to thin to forestall declines in diameter growth than to correct them after they have taken place.

There are many shade-tolerant species that are adapted to start as advance growth and that grow very slowly in the early stages. They do not form any fast-grown juvenile wood, which can be an advantage. Once released they generally commence rapid growth of wood with good properties. However, because of their ability to endure long periods of suppression they are very subject to the problems associated with sudden accelerations in growth. Ideally they should be released when in the sapling stage and induced to grow steadily thereafter.

Some conifers, such as the spruces, true firs, and 5-needled or "soft" pines, which have no pronounced variation between early- and late-wood, produce very homogeneous wood. They adjust to structural demands more by increasing taper than by laying down stronger wood in the lower and outer parts of the stems.

Thinnings that are heavy enough to cause major changes in the form and taper of boles may be detrimental as a consequence. If the logs are converted into lumber there is increased waste in slabbing and the boards are more likely to be cross-gained. The trees may no longer meet the specifications for poles and piling if they taper too much or have excessively large branches high on the stems.

Thinning has a tendency to halt natural pruning and stimulate the development of large branches. If not remedied by artificial pruning, this effect will increase the size and number of knots in the wood. The only compensating effect is that the branches remain alive longer, thereby reducing the number of loose knots eventually produced. If a large amount of clear material is to be grown without artificial pruning, thinning should be delayed until natural pruning has proceeded to the extent ultimately desired. Regardless of how the pruning is accomplished, it is prudent to set some realistic goal as to the length of branch-free bole that will be developed; the remaining upper portion of the stem should ordinarily be kept clothed with living branches. The contemplated length of clear bole should be greater the better the site and the longer the rotation because it depends on the ultimate height and live crown ratio. If one plans to grow trees 100 feet tall with a live crown ratio of 40 percent, it is important to plan for the proper development of the 60-foot length below the live crown.

The criteria of quality for pulp depend on the pulping process and the characteristics desired in the product. If the paper needs to be strong, the procedures necessary to produce strong dense wood are appropriate. Fibers that are long, strong, and narrow interlace to produce strong paper. However, if the paper must be smooth, it is necessary to have some short fibers to nestle in the gaps between the stronger ones. These differing requirements are better met by mixing fibers of different species than by trying to make spruce fibers like those of maple and vice versa. In general, the farther the wood is from the pith, the greater is the length, density, strength, cellulose content, and anatomical structural quality of the fibers. If the wood is of high density, the volume that must be processed for a given yield of pulp will be low. The wood of knots has anatomy that is different from that of the main stems so it is usually an undesirable but unavoidable adulterant of pulpwood.

Not all wood is grown for pulp or structural material. Sometimes the most valuable is that which is sufficiently soft and uniform in texture to be shaped and finished easily for millwork, furniture, plywood, veneer, and similar products. Homogeneity of properties and freedom from defect are more crucial characteristics than strength and density. This kind of material now comes mainly from the very fine-ringed wood of the outer portions of large, ancient trees.

The problems of producing such wood in the future arise with species in which there is a sharp contrast between early- and late-wood. Such species include the oaks and other ring-porous hard-woods as well as the hard pines and Douglas-fir among the conifers. For these species, there is no good solution short of growing the trees to those advanced ages at which diameter growth becomes very slow. Ordinarily the rotations are thought of as being financially far too long. However, in the Spessart region of western Germany it has long been customary to grow oaks on rotations of 300 years to produce such wood and the prices of it are high enough to make such procedures worthy of sober analysis. Sometimes the proportion of hard late-wood in such species is low when, as is the case with some interior ponderosa pine, the trees are subject to summer deficiencies of soil moisture.

These difficulties are less serious with diffuse-porous hardwoods, soft pines, true firs, and spruces, which show little contrast between early- and late-wood. Consequently such species provide the best opportunities to grow soft or highly homogeneous wood on economic rotations.

PRUNING

Trees must have branches but the only good branches are the living ones. Part of the problem with branches is that they form knots, which are the most common defects of wood grown in managed forests.

Branches do not necessarily fall off when they cease to function; except to the extent that they act as infection courts for rotting fungi, their continued presence does not threaten tree survival. However, dead knots are more serious defects than ones from living branches and dead branches are unsightly on ornamentals.

Natural Pruning

Most of the branch pruning that takes place in forests is caused by physical and biotic agencies of the environment and is called natural or self-pruning. It proceeds from the ground upward and starts with the killing of branches by the shade of those above. If the object is to keep the branches small and soon shed, then it is desirable to maintain high stand density and even to refrain from thinning until some desired branch-free length has been developed. In fact, it often helps to distinguish between an initial period of "stem training" in which branch-free length is developed and a later one in which thinning is used to encourage crown expansion and to halt the dying of branches.

If branches die before they get large enough to contain heartwood, their bases are usually sealed by the formation of resins in conifers and gums in hardwoods. This tends to keep wood-deteriorating organisms out of the central stem. Fortunately the fungi that literally rot dead branches off trees are saprophytes of dead sapwood and are rarely capable of attacking heartwood or living sapwood. The rotten branches ultimately break off through the action of precipitation, wind, or the whipping action of subordinate trees. However, they tend to persist if they have become large or if either chemical substances or merely excessive dryness inhibit fungal action. It is especially awkward if the initial stand density is so low that many branches become large and then die when an unthinned stand becomes tightly closed.

When new shoots form they usually point vertically upward even if they are on the tips of old branches. If they sag from the vertical it is mainly because they do not build enough supportive tissue to remain erect. If the light reaching them is suddenly increased, they can form enough reaction wood to curve back toward the vertical. However, branches normally sag increasingly as they become older and more submerged in the crown. Their resistance to natural pruning is least when they are horizontal.

If a terminal shoot is killed and two or more branches turn upward to replace it, some deformities can result. Even if a single shoot quickly asserts dominance, it may remain curved enough to cause a crook in the main stem. So long as dominance is shared between two or more upturned branches, the main stem remains forked. Any upturned branch that competes for dominance for some years but is then subordinated may persist as a **ramicorn.** This is an abnormally large branch which projects at a small acute angle from the main stem; it tends to persist longer and is much more undesirable than an ordinary branch. Ramicorns can also form if the terminal bud dies or merely if some hormonal abnormality causes it to fail to assert apical dominance over the branch buds.

In most natural stands the branches of surviving trees tend to be larger in the middle and upper parts of the bole than in the lower. This is because inter-crown competition is usually severe before differentiation into crown classes takes place. Trees on poor sites usually have small branches but the vertical distance between them is short.

The final step in natural pruning is the occlusion or covering with new tissues of the short stub left by the dead branch. Whether the pruning is natural or artificial, this process can be likened to the submergence of a post set in a rising stream of water. The rate of submergence depends on that at which the stream rises; that is, the faster the radial growth the sooner the wound heals but the diameter of the branch makes remarkably little difference. As with the post, the upstream edge of the top, that is, the upper side of the stub, is engulfed somewhat sooner than the lower or "downstream" side.

The new wood that ultimately covers the old branch stub is the most dependable seal for excluding water, oxygen, and fungal spores from the interior of the stem. So long as dead branches protrude from the stem they can act as conduits for these materials. Even if rot has already started underneath a dead branch stub, it may cease to spread when water and oxygen are excluded.

Knots

The knots (Fig. 3-4) produced while branches are still alive are known as **live** or **intergrown** knots, and are usually **red** in color. Those formed after the branches die are called **dead,** or **encased,** knots; in conifers they often become **black** because of heavy resin deposits. The conductive elements laid down around live knots are termed intergrown because they bend outward and appear to be continuous with those of the branches. Actually this is really true only of the xylem and phloem of the lower half of the live branch. Those of the upper half actually sweep around the sides and top of the branch base without being continuous with it. The discontinuity can be observed by forcibly pulling a living branch out of the bole.

The elements laid down around dead branches bend inward and have no connection with those of the branches. In fact, the new xylem is separated from the old branch wood by an encasing layer of new bark. Dead knots are serious defects in lumber or veneer because they are apt to fall out when they dry, especially if they are long-dead parts of branches that had started to deteriorate when encased. In other words, live knots are tight knots, but dead knots have a strong tendency to be loose knots. Live knots remain tight because the attachment in the lower half of the former branch suffices to keep them in place.

Softwood lumber is graded on the basis of the size, soundness, number, and distribution of knots in the boards. While it is best to have clear material with no knots at all, they are permitted in and characterize the

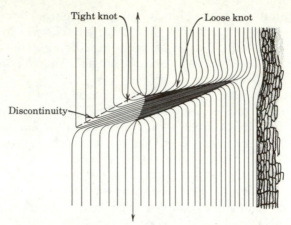

FIGURE 3-4   Radial section through a pine branch
that persisted for many years after its death. The
branch is thickest at the point where it joined the stem
at time of death. Note that the annual rings laid down
following the death of the branch turn inward without
joining the wood of the branch. Even when the branch
was living the fibers of the upper part of the branch
did not actually link with those of the main stem even
though they appear to do so if one examines a radial
section of this kind. The arrows at top and bottom
mark the last annual ring formed before the branch
died.

medium and lower grades. Since most hardwood lumber is used, fortunately in short lengths, for furniture and other highly discriminating purposes, it is usually graded on the basis of the number of completely clear or
defect-free pieces 2 feet long that can be cut from the boards being graded.
A small, tight "pin knot" is as serious a defect by this reckoning as a very
large rotten hole. The only reason why there is more concern about pruning with softwoods than with hardwoods is that softwood branches are
generally more persistent. Knots in hardwood boards are actually more
serious problems and it is merely good fortune that they are less numerous.

As far as plywood is concerned, knots are scarcely permitted at all in
the face veneer layers of high-grade material. Furthermore, one knot in
rotary-cut veneer can degrade a wide sheet at many repetitive points. On
the other hand, defective material can be either buried in the interior layers
or laminated together into construction grades in which the structural
weaknesses induced by knots are well compensated. As a result, in most
modern plywood manufacture it is desirable to have as much clear material
as possible but there are also good ways of utilizing poor material.

Dead branches of maples and certain other diffuse-porous hardwoods constitute a special kind of detriment to wood quality (Shigo *et al.*, 1979). Whenever any branch larger than about 3 cm in diameter dies, it becomes an entry point for bacteria that turn all of the existing stemwood below that point into undesirable, brown "pathological heartwood." Fortunately, gums almost immediately form inside the stem in such manner that the bacteria are sealed off inside the tree and the wood subsequently formed is desirable white sapwood at least until the next branch dies. This problem can be overcome by cutting the branches off before they die, provided one does not saw into the swollen collar around the branch. A more passive solution is to set some goal as to how tall and thick the pathological heartwood column is going to be and then thin hard enough to prevent branches from dying after the size of that column is determined.

Artificial Pruning

The training effects obtainable with manipulations of stand density are not always sufficient to eliminate branches early enough and it is sometimes desirable to resort to artificial pruning. This is an expensive and labor-intensive operation so it is necessary to be very discriminating about when and where it is done.

The most common tools (Fig. 3-5) are hand saws, usually with curved blades and handles of lengths appropriate to the height of pruning. It is also possible to knock dead branches off with clubs, which can be faster than and almost as satisfactory as saws. Sometimes special chisels with curved or notched cutting edges are effective; shears are usually too slow. Axe pruning is too dangerous but ordinary power saws can be used if the operators have firm footing.

It is easy enough to prune up to about 3 meters, especially with hand saws mounted on axe handles, but difficulties and costs increase rapidly above that level. Saws or clippers on long poles are effective to about 6.5 meters above ground but wobble too much to enable one to prune higher with them. It is also possible to use hand saws from ladders but the moving of ladders is cumbersome enough that it is hard to prune higher with them also. Another hand-powered device for high pruning is a saw-chain, which has teeth that cut in both directions and its end attached to long nylon ropes. One problem with this device is throwing the rope-end over the right branches high in the tree. Expert tree climbers, with safety belts, can get to almost any height using climbing irons or ropes; this is the one kind of pruning in which cost does not rapidly increase with height, but it requires agile people.

The use of personnel-carrying booms mounted on trucks is so costly that it must ordinarily be limited to roadside trees. The problems of delivering cutting power far above the ground are great enough that high pruning of trees in the forest has been very difficult to mechanize satisfactorily.

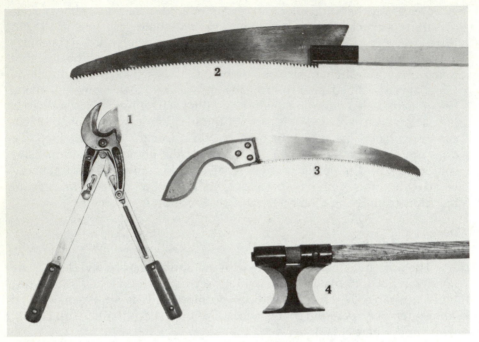

FIGURE 3-5   Typical instruments used in pruning forest trees: (1) A pair of shears that has both blades sharpened and is designed for close pruning. (2) The blade of pole saw. (3) A hand pruning saw. (4) The Rich pruning tool, typical of those that cut on impact. *(Photograph by Yale University School of Forestry & Environmental Studies.)*

Pruning for Timber Production

Artificial pruning is expensive enough that it must be limited to a small number of trees under rather special circumstances. There must be a substantial premium for knot-free wood. This usually exists with species that have wood suitable for cabinet work, interior finish, furniture, or high-grade surface veneer. Ordinarily the premium is insufficient in the case of species that are used primarily for framing timbers and building construction. Even though knots substantially impair the load-bearing strength of wood, the construction industry has usually found it cheaper to compensate for knots by using thicker members than to pay higher prices for wood with fewer knots. The same is also true of the construction grades of plywood that are hidden in use, but certainly not of those grades for which a smooth surface is required.

Among the conifers the species most worthy of pruning are those with low contrast between early- and late-wood that are soft and easily worked and finished. These include the 5-needled pines, the ponderosa pine of certain localities, and sometimes spruces. Most southern pines and Douglas-fir have wood that is strong but too coarse for finishing purposes so that pruning of these species would be chiefly for the surface veneer of plywood. Most hardwoods prune well naturally but there can be great advantage in the timely removal of a few ramicorns or other large, persistent branches from otherwise branch-free stems.

Provided that there is an adequate premium for clear material, pruning is most advantageous for species such as eastern white pine, which have rot-resistant branches. The removal of such branches has the effect of converting what would be loose-knotted lumber of the lowest common grades and minimal stumpage value to clear material that can have very high value.

It is essential that pruned trees be carefully selected and rigorously limited in number. It is easy to forget that most of the trees in a young stand are doomed to die of suppression before the survivors reach sawtimber size. On one hand it is undesirable to make the expensive choices involved until the trees are large enough to have expressed dominance; on the other, if the trees get too large there will not be enough time for them to produce shells of clear wood thick enough to repay the high cost of the pruning. It also helps to postpone the first pruning at least until the terminal shoots have grown to the height of the first log, thus avoiding the risk that forks or crooks may develop before that height is attained. In typical cases, pruning might be limited to 50–100 trees per acre in the D.B.H. class of 5–10 inches. If the object is to produce clear, sawn lumber, the clear shell must grow to at least 4 inches in radial thickness to be of any advantage. Pruning is cheap enough up to about 3 meters that one may sometimes justify pruning more trees but the number usually must be reduced when the time comes to carry the pruning higher.

Ordinarily pruning is done after the branches are dead and in two or sometimes more stages of height. Different tools, usually saws with handles of differing lengths, are used for each stage. The object is often to keep the knotty core of uniform thickness. Commonly the problem is not that the lower part of the knotty core becomes too thick but that the later stages of pruning are delayed so long that the upper part gets thicker than the lower part.

Pruning high above the ground is so costly that it is usually stopped at either the first standard 16-foot log length or the 6.5-meter limit set by the nature of pole saws. A very high proportion of the merchantable volume and value of a tree is often concentrated in this zone anyhow. There are many instances, however, in which it would be fine if the pruning could be carried higher; sometimes it is. The branches that develop somewhat higher on the stem are usually the largest; they form the poorest kinds of knots and the worst sources of heart-rot infections if they die. Artificial pruning at these heights probably awaits the invention of new machinery. The problem can be alleviated to some extent by thinning hard enough to halt or slow down the dying of these upper branches; large, live branches are better than dead ones but even large, tight knots can become so large that the only solution is to harvest the tree.

It is perhaps most important of all that artificial pruning be coupled with programs of thinning that are very positively aimed at maintaining diameter growth (Fig. 3-6). It usually takes rapid growth of substantial amounts of clear material to repay the high compounded cost of an operation that takes 10–15 minutes per tree with a wait of 20–40 years for the returns. The chosen trees should initially be as vigorous as possible, although not so vigorous as to have excessively large branches. It is generally best not to invest much in pruning in a stand until *after* the first thinning. This not only ensures that the chosen trees have been released but also reduces the risk of logging damage to them. Although it is possible to

FIGURE 3-6 Cross section through a node of an eastern white pine that was green-pruned 42 years ago when 3½ inches in diameter. The average diameter growth of the tree was about 2.6 inches per decade in the period after pruning. The small wounds healed swiftly and almost perfectly. (*Photograph by Forest Products Laboratory, U. S. Forest Service.*)

waste money on pruning, there are also circumstances in which the combination of pruning with thinning and good diameter growth can provide some of the highest long-term returns available in timber-production silviculture.

The ideal program of growing trees for timber would be one in which either natural or artificial pruning caused branches to be gone as soon as they died up to a certain deliberately chosen level. After the chosen branch-free length was achieved, it would be ideal if thinning were prosecuted so vigorously that no more branches died and diameter growth was maintained at the highest possible rate. It also helps to keep track of the rate at which branch death causes the base of the live crown to retreat upward. During the crucial stages in the development of closed stands this retreat is astonishingly and undesirably rapid. Any ideal thinning program should be aimed not only at controlling the diameter growth of the crop trees but also at regulating branch-free length, stem taper, and the sizes of branches.

It is generally desirable to refrain from pruning large branches (Fig. 3-7). They are costly to remove and may also contain heartwood that will not seal itself adequately against rot. Their cut stubs may also project so far out from the cambial zone that it takes a long time for them to be submerged by new wood. It is this phenomenon rather than diameter itself that makes large pruning wounds slow to heal.

It is usually best to delay pruning until a branch is dead or is about to die. The removal of living branches in what is called **green** or **live** pruning can cause wounding of the bole by tearing of bark from it. This risk can be reduced by making a cut on the lower side of the branch before completing the removal with a cut from above. It is also very desirable to try to confine pruning to dormant seasons when the bark is tight and fungal spores presumably less numerous. In some species, the bases of branches begin to be sealed by resin and gum deposits when they die so it is well to wait until this protection commences. The lowermost living branches can usually be cut without reducing the growth of the trees. Diameter growth may suffer considerably if the live crown ratio is reduced to 40 percent or less but height growth usually remains unaffected unless this ratio is less than 30 percent. Reductions in diameter growth slow down both wood production and healing of pruning wounds. Deceleration of height growth may cause a pruned tree to be suppressed by unpruned, faster growing neighbors.

While it is desirable to avoid ragged pruning cuts there is no need to go to the cost and effort of sawing off any raised collars that form about branch bases. The reduction achieved in the diameter of the knotty core is too small to justify the work and the practice may do more harm than good. In some diffuse-porous hardwoods, for example, cutting into the branch collar increases the amount of bacterially induced pathological heartwood (Shigo *et al.*, 1979).

There are cases in which it may be possible and desirable to substitute green pruning for the sort of natural pruning that is induced by crowding.

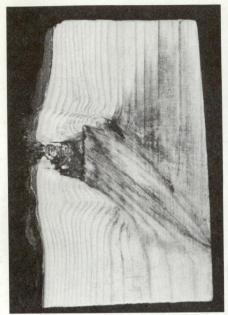 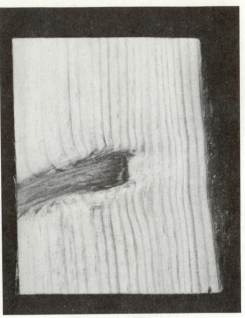

FIGURE 3-7 Typical radial sections through branches of red pines that were pruned 14 years ago. The branch at *left* was alive and 0.9 inch in diameter when pruned; the stub was 0.7 inch long, and the tree lapsed into a codominant position after pruning. The pitch pocket that formed as a result of the slow growth and rather large size of the stub would not have occluded for several years. The branch at *right* was 0.5 inch in diameter and dead when pruned; the stub was 0.5 inch long. The tree remained in a dominant position and grew rapidly so that healing was virtually complete 7 years after pruning. Note that both calluses grew more rapidly from above than from below. *(Photographs by Yale University School of Forestry & Environment Studies.)*

It is, for example, possible to plant new stands at a very wide spacing with a small number of trees and employ drastic green pruning not only to get rid of branches but also to prevent boles from becoming too tapering. It may be necessary to prune all of the trees in a stand so that there will be no faster-growing, unpruned ones to claim dominance.

Excessive green pruning may result in sunscalding of trees with smooth, thin bark. It can also lead to the development of epicormic branches on hardwoods as well as on those conifers, such as Douglas-fir, spruce, and true firs, that have numerous dormant buds.

Bud pruning is a method of very early green pruning in which the lateral buds are rubbed or clipped off annually. The purpose is to produce logs with no knots at all. Growth reductions and the serious results of injuries to the terminal shoots usually make the treatment disappointing. The same effect, called "foxtailing," can develop naturally in some tropical pines planted where the dormant season is not long enough. The branch whorls that develop at top or bottom of the long, straight, branch-free segment often produce large knots or other deformities.

Pruning of Ornamental Trees

It is often desirable to do various kinds of artificial pruning on open-grown shade trees. Dead or broken branches may have to be removed so that they will not fall on people or become infection courts for fungi. Live branches also have to be removed before they touch electric wires, structures, or passing people and vehicles. It is also possible to prune to improve the appearance of the crown or to make the branches stronger by thinning them much as one might thin a stand of trees. Pruning can also be used to lengthen the branch-free bole of small ornamental trees; sometimes it is much cheaper, more dependable, and almost as fast to develop a good shade tree in this way starting with a small one than to transplant a large, expensive, preformed tree that is apt to succumb in the process.

Shade-tree pruning often involves the removal of large, green branches that have formed heartwood. The branches have to be severed with initial undercuts or supported during the removal so that they do not tear the remaining stem tissues. The work requires special equipment and well-trained workers because it can be quite dangerous, especially near electric wires. Putting paint or other coatings on the wounds is of more aesthetic benefit than of any real protection against infections. Fungal spores are small enough to get through cracks in the coatings although it is possible that they might be killed by any fungicidal chemicals applied with the coatings.

It probably helps to make the cuts smooth and to shape them so that water flows off rather than onto the cut surfaces. If rotten cavities are too large to be sealed by growth of wood it is desirable to repair them in ways such that water will drain out of them through rust-proof pipes or by other means. Filling cavities with concrete or other substances may improve appearances but is as likely to speed decay as to slow it. Crotches that have started to split apart can sometimes be strengthened by screwing lag bolts into the inner side of each stem and connecting them with turnbuckles. Continuing diameter growth results in trouble with rods that go all the way through the stem or with modes of attachment that girdle the stems.

People get all kinds of erroneous ideas about the care of shade trees. Among these is the notion that it somehow helps to prune so that a projecting stub is deliberately left. This may stimulate the formation of wood gums or resins in a few species but helps only if the stub is soon removed. If it is not, it usually invites fungal or insect attack and delays healing for many years.

Shearing and Pruning of Christmas Trees

Conifers being grown for use as Christmas trees often require special techniques of pruning to correct the effect of excessively rapid or asymmetrical growth and produce dense crowns of the proper shape. The most common practice, referred to as **shearing**, is the clipping of both terminal and lateral shoots. The objective is to develop crowns that are narrow and globose at the bottom and conical in the middle and upper portions. Shearing is usually confined to the removal of part of the shoots of the current or most recent growing season. Buds formed just below the point of cutting must be depended upon for renewed elongation of the decapitated shoots. Species that have dormant buds throughout the length of each internode, such as firs and spruce, can be sheared at almost any season except early summer. Pines rarely have any dormant buds along the internodes and are best sheared when the shoots are actively elongating. This stimulates the prompt development of vigorous adventitious buds just below the cut ends (Bramble and Byrnes, 1953; Brown, 1960).

Excessively tall and rapidly grown trees can sometimes be converted into good Christmas trees in one operation simply by pruning off the lower three-quarters or more of their crown. The remaining tuft of crown at the top then grows much more slowly and may soon develop into a properly compact form without any shearing. This technique is successfully applied to fast-growing balsam firs of natural origin.

BIBLIOGRAPHY

The bibliography for Chapters 2 to 4 is at the end of Chapter 4.

C H A P T E R   4

Methods and Application of Thinning

METHODS OF THINNING

Four distinct patterns have evolved in the methods for determining which trees to favor and which to remove in thinnings. In the first three methods, the developmental status of the trees and their position within the crown canopy structure are the bases for choices. In the fourth, spacing or arrangement of trees in the stand is the first consideration. The four thinning methods are: (1) **low,** (2) **crown**, (3) **selection,** and (4) **geometric** or **mechanical**. It is also useful to recognize a fifth method, **free thinning,** which can be any combination of the other four simultaneously applied in a single operation, usually in stands of irregular development.

These methods will be considered in turn. Vertical profiles through stands are used to illustrate the basic patterns of choices made in the first three methods. It should be noted that such sketches lack the depth of three-dimensional representation with the result that they make the thinnings seem heavier than they actually are. The stands depicted are also single-canopied, that is, with all trees competing in a single crown stratum. Each profile shows a kind of stand for which the thinning method illustrated is especially appropriate.

Low Thinning

This method, the oldest, is sometimes called "thinning from below" as well as the "ordinary" or "German" method. In low thinning, trees are removed from the lower crown classes. This simulates, but also accelerates, the natural extermination of these classes in stand development.

There are four different grades of severity of low thinning designated by letters ranging from A for the very light to D for the heaviest. They are differentiated on the basis of the crown classes removed, as shown in Fig. 4-1. In A-grade low thinning, removals are confined to overtopped trees,

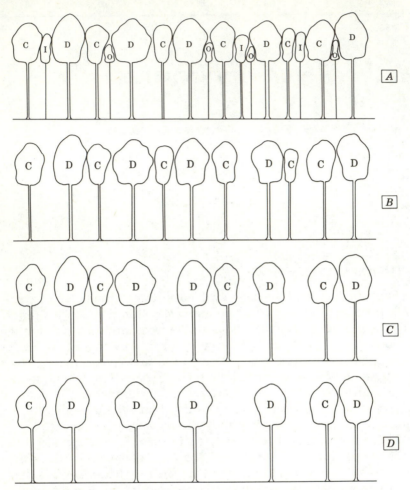

FIGURE 4-1 The B, C, and D grades of low thinning as they might look if each were *simultaneously* applied in the same middle-aged stand (*top*) of loblolly pine that had been rendered uniform by earlier thinning. The letters on the crowns denote crown classes and those at the side, the grades of low thinning. Note that previous treatments have promoted the crop trees to the dominant class. The nearly useless A grade is omitted.

or merely those dead or nearly dead of suppression. These are little more than salvage operations and the canopy remains unbroken as shown in Fig. 4-2. In the successively heavier grades, additional and higher crown classes are removed. In B-grade low thinnings, the intermediate crown class is also eliminated. The main virtue of A- and B-grade thinnings is that they are practically the only kind of thinning that can be done without any risk of reducing the gross production of wood by the stand. The trees of the overtopped and intermediate crown classes in pure, even-aged stands

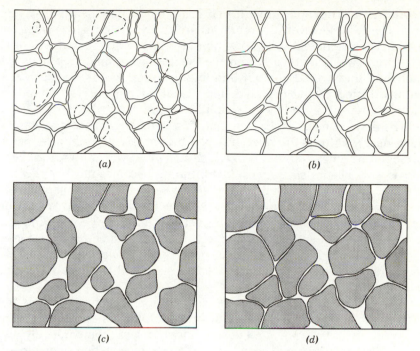

FIGURE 4-2 The outlines of the crowns of a previously unthinned stand as viewed from above. Overtopped portions of the crowns of trees of the lower crown classes are indicated by broken lines: (*a*) before thinning; (*b*) after an A-grade thinning in which overtopped trees were removed leaving the canopy unbroken; (*c*) immediately after a D-grade thinning in which overtopped, intermediate, and many codominant trees were removed leaving each remaining tree exposed on one or more sides; (*d*) the same stand 10 years after the D-grade thinning showing the canopy so nearly completely closed that another thinning is needed.

have done most of their growing already and either die soon or merely exist without much further growth.

Since the object of most thinnings is to stimulate the growth of some remaining trees and not merely salvage, the heavier C and D grades of low thinning are now much more commonly applied than the light A and B grades. Both of the heavier grades involve deliberate creation of temporary canopy gaps to accelerate crown expansion of the remaining trees (Fig. 4-2). In the C grade some codominants are cut along with the lower crown classes; in the D grade, many but not all codominants are cut. Any E-grade thinning in which only dominants were left would ordinarily leave so much growing space for new regeneration that it would be better thought of as shelterwood regeneration cutting.

Low thinning has a simple, close, and logical relationship to the natural course of stand development. It is easy to pick the trees to remove; those which remain are, at least in pure stands, sure to include those likely to continue the superior growth that has already put them in the upper crown classes. However, the removals are concentrated in the small trees least likely to be marketable. Low thinnings have to be very heavy or else be frequent and started early to make the important final-crop trees grow faster than they would without thinning. In other words, so much attention is devoted to the negative step of eliminating losers that positive actions to make the leaders grow still better are somewhat indirect.

The general result is that low thinning is most applicable to stands in which nearly all trees are merchantable. It is no accident that the method developed in times and places in which fuelwood was in demand. Where there is little use for small trees, low thinning becomes most applicable in the later stages of the rotation after all trees have become merchantable. The trees of the lower crown classes have by then lost so much foliar surface that their growth dwindles irreversibly and they become financially mature. Intolerant species are so likely to suffer reductions of live crown ratio that there is less latitude for applying methods other than low thinning to them than is the case with tolerant species that can carry foliage at lower light intensity.

Theoretically low thinning should be more appropriate than other methods for sites where moisture or other soil factors are seriously limiting. Similarly, but on any kind of site, it would seem comparatively favorable to the development of accessory understory vegetation, which may be desirable or undesirable from the management standpoint. However, the amount of foliage temporarily eliminated may effect these phenomena of competition for soil growing space more than the thinning method.

Crown Thinning

To overcome the limitations encountered in applying low thinnings, a method was developed in which trees are removed from the middle and upper portion of the range of crown and diameter classes rather than from the lower end. This technique is best referred to as **crown thinning.** It is also known as the "French" method of thinning because of its origin and early use in France. The terms "thinning from above," "high thinning." and "thinning in the dominant" have also been used for crown thinning, but these are too easily confused with selection thinning.

In crown thinning, trees are removed from the upper crown classes in order to open up the canopy and favor the development of the most promising trees of these same classes (Fig. 4-3). Most of the trees that are cut come from the codominant class, but any intermediate or dominant trees interfering with the development of potential crop trees are also removed. The trees to be favored are chosen, if possible, from the dominants and, where necessary, from the codominants. Crown thinning favors nearly the same kinds of trees as low thinning but more by removing a few strong competitors than by wholesale elimination of the weak. Low thinning and crown thinning differ radically from selection thinning in which dominant trees are cut to favor the lesser crown classes.

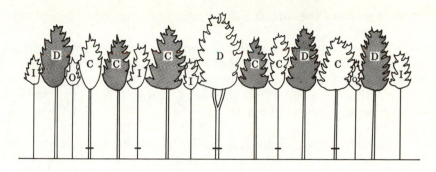

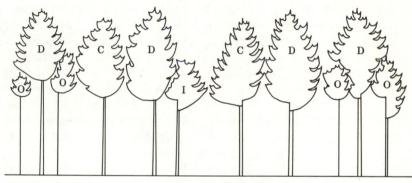

FIGURE 4-3 The upper sketch shows a coniferous stand immediately before a single crown thinning. The trees to be cut are denoted by horizontal lines on the lower boles and those with shaded crowns are the crop trees. The lower sketch shows the same stand about 20 years after the crown thinning and reclosed to the point where a low thinning would be desirable.

Crown thinning is basically different from low thinning in two respects. First, no matter how lightly applied, the principal cutting is made in the upper crown classes. Second, the bulk of the intermediate class and the healthier portion of the overtopped class remain after each thinning.

The question of whether individual dominant or codominant trees are favored is settled according to the relative potentialities of adjacent trees. If the choice lies between a promising codominant and mediocre dominant, the codominant is favored. A situation of this kind occurs most often where the codominant has a straighter, smoother bole and smaller branches than the dominant. Where all trees are of good health, form, and species, codominants interfering with the growth of dominants are removed, on the premise that position in the crown canopy is the best index of past and future vigor.

Theoretically, overtopped trees and intermediates that do not interfere with crop trees are not cut in crown thinning. In practice, there is little reason to leave such trees if they can be harvested profitably and their continued presence will add value neither to themselves nor to the stand as a whole. If many of the smaller trees are removed, the operation becomes an incomplete low thinning.

The immediate cash return from crown thinnings is greater than that from low thinnings of equal severity, because the material removed is larger and of greater utility. The smaller trees of the subordinate classes, which would be removed in low thinnings, can be left to grow to larger size. Any losses of material too small to be utilized profitably are of no consequence.

One result of crown thinning is the division of the residual stand into two categories of trees. The first consists of the favored dominants and codominants, which are destined for removal either in reproduction cuttings or in later thinnings. The second category is made up of the subordinate trees, which are favored only by indirect action and are gradually removed as they grow up and interfere with the trees of the first group. After a series of crown thinnings it is possible that a stand of two distinct stories, but only one age class, may be created (Fig. 4-4). This development is not likely to be persistent unless the subordinate trees are of tolerant species. Actually it is common to have the lower story of trees vanish as a result of the combined attrition of natural suppression and a series of crown thinnings. Under such conditions, a program of crown thinnings pursued to its logical conclusion will be replaced almost automatically by low thinning. With some very intolerant species the first crown thinning may prove to be the last because of the failure of the subordinate trees to survive after release. A result of this kind represents poor judgment only if the trees that die could have been utilized profitably at the time of thinning. It is at least theoretically possible that the trees of the subordinate stratum may restrict epicormic branching on the crop trees and the development of understory vegetation.

The effect of the subordinate trees left after crown thinning could conceivably be undesirable if root competition were the most critical factor limiting growth. They are occasionally a hindrance in the work of cutting and transporting products within the stand.

One important advantage of crown thinning lies in the opportunity to stimulate the growth of selected crop trees without sacrificing quantity production. The favored trees grow to a given size in a shorter time, or to larger size in a given time, than they would with a regime of low thinnings, owing to the freedom for expansion of the crowns. Meanwhile any growing space not taken up by the selected trees is given over to the subordinates. Most evidence indicates that the total yield, in terms of cubic volume, is no greater from crown thinnings than from a comparable series of low thinnings. With crown thinnings, however, this volume is harvested in fewer trees of greater average diameter than with low thinnings.

Crown thinning is a more flexible method than low thinning, and it demands greater skill on the part of the forester. Since it is not feasible to distinguish different grades of crown thinning, the severity of cutting must be regulated in terms of basal area or some other index of stand density.

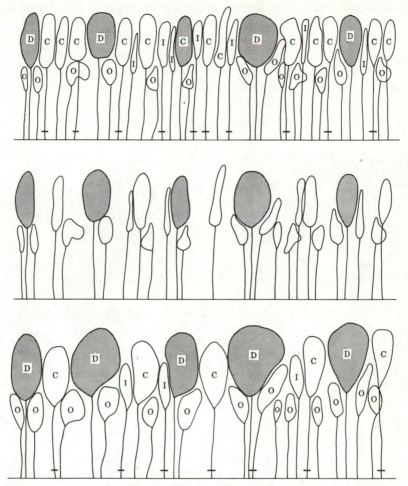

FIGURE 4-4 Part of a series of crown thinnings in a hardwood stand. The upper sketch shows the previously untreated stand immediately before the first crown thinning. The trees to be cut are denoted by horizontal lines on the lower boles and those with shaded crowns are the final crop trees. The middle sketch shows the stand immediately after the first thinning. The lower sketch shows it advanced to readiness for the second crown thinning. Note that a lower story is being developed by these thinnings. Presumably a third crown thinning could be conducted in this stand.

Applications of Crown Thinning

This thinning method can be applied uniformly throughout a stand or concentrated very specifically on the release of a limited number of chosen crop trees. The uniform application is most feasible if neither markets nor labor considerations are restrictive of if it is desirable to remove a large volume of wood in a single operation. The concentrated approach is useful where the object is to stimulate the crop trees as much as possible with the least amount of cutting.

In any crown thinning, it is most logical to identify the potential final-crop trees and then be sure to eliminate the most vigorous competitor of each (Fig. 4-5). While other trees may also be removed, the leading competitors are the most important. When a stand is young and crown expansion rapid, it may be possible and desirable to free the crop-tree crowns on all sides. In older stands canopy gaps do not close so quickly and it becomes

FIGURE 4-5   A 32-year-old plantation of red pine in Connecticut just after completion of a third crown thinning. While the crop trees have been thoroughly released the vigor of the subordinate trees is so poor that they will be removed in a subsequent low thinning. (*Photograph by Yale University School of Forestry & Environmental Studies.*)

impossible to release the crop trees completely without seriously reducing the volume growth of the stand.

Crown thinning often represents one of the most expeditious means of restricting the investment represented by value of growing stock without reducing the growth in value that represents the income on this investment. The codominants that are typically removed in crown thinning subtract more from the investment in growing stock than the smaller trees that would be removed in low thinning. At the same time the potential for increase in value remains essentially undiminished because the dominant trees are favored. Selection thinning, in which numerous dominants are cut, may reduce the investment more but it usually reduces the growth in value as well. If all of the trees in a stand are of good quality and value, differing mainly as to rate of growth, low thinning is likely to represent the logical way of manipulating the investment represented by the stand, because the dominants will increase in value most rapidly.

The principles of crown thinning can be followed in almost any situation where economic conditions are compatible with the application of thinning. This method provides such direct and consistent means of fostering and regulating the development of the chosen trees of the final crop that is has become very common. It is also sufficiently versatile in application that the details of procedure are subject to a wide variety of modifications; many of these modifications are recognizable as crown thinning only if the method is interpreted rather broadly.

Crown thinning works best, or at least can be repeated most often, in mixed stands or in pure stands of tolerant species; both kinds of stands are likely to have trees capable of forming a subordinate stratum. It can be applied at appropriate stages in handling pure stands of intolerant species, but must soon be succeeded by low thinnings if any use is to be made of the trees of the subordinate crown classes. One of the shortcomings of crown thinning is that it makes no provision for the ultimate disposition of the subordinates. This drawback is readily correctable by subsequent use of low thinning or other techniques; the subordinate stratum of trees is detrimental only where root competition is crucial.

Selection Thinning or Thinning of Dominants

This method of thinning differs radically in principle from the two methods already discussed. In selection thinning, dominant trees are removed in order to stimulate the growth of the trees of the lower crown classes (Figs. 4-6 and 4-8). The same kind of vigorous trees that are favored in crown and low thinning are the very ones that are likely to be cut in selection thinning. It is obvious then that selection thinning is suitable only for rather limited purposes and, if not carefully used, readily degenerates into high-grading.

The term "selection thinning" comes from the very superficial resemblance of this kind of cutting to the selection method of regeneration, which will be discussed in Chapter 15. Both are characterized by the removal of the largest trees in a stand. However, selection thinning involves the cutting of scattered trees from even-aged stands or aggregations of trees without the intent or result of regenerating new age classes. The selection method of regeneration cutting, on the other hand, is applied to uneven-aged stands to make vacancies large enough to allow new age classes to develop; this is actually likely to require removal of groups of trees rather than individuals. The similarity of the terms causes semantic confusion and, worse yet, confusion about the purpose of some kinds of partial cutting. "Thinning of dominants" is a less ambiguous term that might replace "selection thinning."

There are really a number of different kinds of selection thinning and the method can be understood only in the light of these various approaches and the situations in which they are applicable. These approaches differ with respect to the objectives sought and the number of times this kind of thinning is repeated.

Different Forms of Selection Thinning

In the first and most common approach, poorly formed dominants are eliminated in favor of satisfactory crop trees chosen from the highest possible level in the lower crown classes (Fig. 4-6). *There is no point in procrastinating about the removal of the kind of coarse dominants that become poorer and cause more harm to better trees the longer they are allowed to remain.* The removal of such trees takes advantage of the fact that codominant and intermediate trees as well as the smaller dominants often have smoother, straighter boles and smaller branches than the most vigorous dominants. This advantage can be pursued to absurdity because it is impossible to grow trees without branches regardless of how high the quality of the stems might be.

The trees selected for retention to the end of the rotation should have live crowns sufficiently deep to enable them to respond to release rapidly

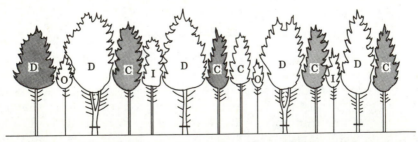

FIGURE 4-6   A coniferous stand marked for a selection thinning aimed primarily at the elimination of rough, poorly formed dominants and the release of less vigorous trees of better form. The next thinning would not come until the large holes in the canopy had nearly closed and it could not logically be another selection thinning because the crop trees (*shaded crowns*) would then be the dominants.

and then develop into thrifty, fast-growing individuals. Trees with live crown ratios less than 30 percent are rarely suitable. In general, the more tolerant the species the greater is the possibility that appropriate trees can be found in the lower crown classes.

Selection thinnings designed to improve the quality of the crop trees are best carried out as early as possible in the life of the stand and replaced by other thinning methods as soon as the crop trees approach the dominant position. Occasion arises for such treatment only where the irregularity and sparsity of initial stocking has caused rough dominants to develop (Fig. 4-7). An outstanding illustration is to be found in stands of eastern white pine that have been attacked by the white pine weevil. Since the dominant trees are the ones most likely to be deformed by this insect, it is advantageous to remove them as soon as the straight, potential crop trees from the lower crown classes have attained a height equal to the desired log length.

This kind of treatment is best regarded as a necessary evil dictated by imperfect initial stand structure, especially with intolerant species. The importance of this point should not be minimized because so many valuable species are relatively intolerant. With such species, it is rare that any additional advantage can be secured from selection thinnings after rough dominant trees have been eliminated in favor of better codominants.

In the second approach (Fig. 4-8), selection thinnings are continued until the point is reached when further removals from the main canopy would open holes too large to be filled by the expansion of the crowns of the remaining trees. At that point attention is turned either to low thinning or, more often, to whatever measures are necessary for regeneration. The objective of this kind of treatment is not to develop large trees but to grow as many trees as possible to medium size for pulpwood, small saw logs, posts, poles, or piling. The procedure differs from diameter-limit cutting only to whatever extent discretion is excercised in choosing the trees to cut.

The stands with the greatest capacity to endure repeated selection thinning are of species, such as spruces and true firs, that are both shade-tolerant and negatively geotropic. The subordinate crown classes of such species are likely to retain high live crown ratios, to respond promptly to release, and to have remained straight. Stands of intolerant species cannot withstand more than one or two selection thinnings before most of the trees capable of useful response are gone. Positively phototropic species, such as many hardwoods, which have stems that bend toward the light as they grow, generally become deformed if they have grown underneath more vigorous trees.

Repeated selection thinning probably does not significantly reduce production of total cubic volume or tonnage of wood, but it greatly diminishes that measured in units as merchantable cubic volume and especially board-foot volume.

A program of repeated selection thinnings can have the virtue of holding the investment represented by merchantable growing stock at a low

FIGURE 4-7 An old-field stand of loblolly pine in which a selection thinning has just been completed. Most of the trees removed were large, rough dominants that produced the kind of material typified by the butt log appearing in the foreground. The men are standing in front of a gap in the canopy created by the removal of trees of this kind. (*A. F. I. Photograph by South Carolina State Commission of Forestry.*)

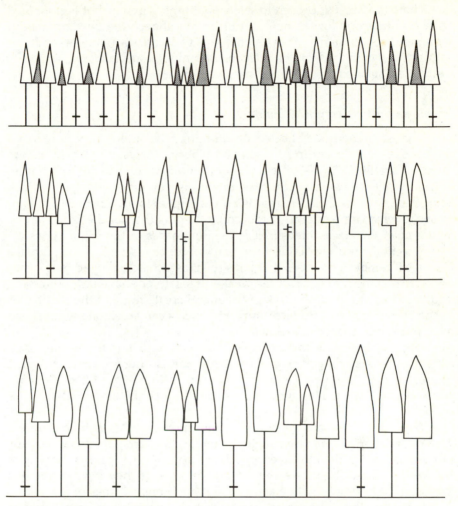

FIGURE 4-8 A series of three selection thinnings in a stand of balsam fir and red spruce being grown for pulpwood on a good site. Each sketch shows the stand immediately before a thinning. The trees with the shaded crowns in the first sketch are those that persist until the end of the rotation. The choice of trees to be removed in the second and third thinnings is somewhat modified by the necessity of avoiding enlargement of gaps caused by the earlier removal of dominants. The advanced reproduction that would undoubtedly become prominent by the time of the third thinning is not shown.

value. The value of the growth obtained then tends to return a high rate of interest on this part of the investment in the forestry enterprise. This can, of course, be an empty virtue if the annual cash return per acre is so low as to constitute a low rate of interest on the capital represented by the whole enterprise. The rotation will be very significantly prolonged, but if the overall returns can be kept high throughout the rotation this can actually be advantageous.

There is a third approach to selection thinning in which it is combined with simultaneous low thinning in order to enhance the development of remaining codominants. This technique is of some advantage in counteracting the baneful effects of selection thinning, especially in previously unthinned stands that have become overdense.

Problems with Selection Thinning

Almost all methods of selection thinning have a tendency to increase the possibility of losses to physical and biotic agencies. The trees that are released are likely to be of less than optimum vigor and to have rather weak and slender stems. Their reservation is, therefore, a calculated risk until they develop thriftier crowns and stronger stems. The risk of major losses to wind can be slightly reduced by keeping the stand edge as strong as possible; this necessitates *leaving* the dominants in the stand edges.

Selection thinning tends to aggravate problems with wind damage because it removes the trees that have the most well-developed butt swells and are thus the strongest. If the weaker trees do not have strong trees on which to lean, they are apt to blow over and create more of the small gaps in the crown canopy in which funneling effects can accelerate wind speed and cause even more blow-down.

If preoccupation with logging costs leads to addiction to selection thinning that causes wind damage to residual stands, there is a psychological tendency to get desperate and cut even more of the larger trees and leave even weaker stands. This syndrome is one in which the first mistake seems to justify making worse ones.

The better solutions would be to thin more lightly, or by a different method, or simply not to thin at all. Sometimes this problem arises because of the mistaken belief that trees with large crowns are the most threatened by wind because they have the largest sail area. If they have developed pronounced butt swell, they are actually the strongest trees of a stand.

In theory, the choices made in favor of slower-growing trees would be undesirable from the genetic standpoint while the discrimination against poorly formed ones would be desirable. The degree of selectivity and of heritability is probably not great enough to cause any major change in the average characteristics of any ultimate progeny, but the rare, best genetic combinations are very likely to be eliminated.

One insidious advantage of selection thinnings is that they are more likely to return an immediate profit than any other kind of thinning. The trees that are cut are ordinarily the most salable to be found in the stand at the time of cutting. This advantage is gained at the expense of some others normally associated with thinning; for example, the time required to grow crop trees to a particular diameter is increased.

The proportion of the cubic volume removed in any one selection thinning should generally be as low as possible. There is a tendency to make low thinnings and crown thinnings too light; the reverse is generally true of selection thinnings. The intervals between successive selection thinnings are longer than for thinnings of other kinds because the holes created by cutting dominants close so slowly.

The stratum of trees of the lower crown classes left in selection thinning has much the same effect as that left in crown thinning.

There is a desperate sort of cutting, sometimes euphemistically described as one of the kinds of "selective cutting," that is best regarded, at least in its early stages, as selection thinning. Such cuttings involve attempts to extract the fullest potentialities for growth from the subordinate crown classes by successive cutting of trees that have been brought to the dominant position. This often results from despair over the problem of replacing existing stands with desirable reproduction. If the problem is evaded by very light periodic cuttings of the largest trees, the residual stand grows poorly and an undesirable understory may burgeon. One way to break this vicious cycle is to destroy the understory and open the overstory enough to allow reproduction of the desirable intolerant species.

This kind of cutting can also result from single-minded concern about avoiding the high costs of harvesting small trees. Even more short-sighted is the policy of regarding the stand merely as a magic warehouse into which one ventures sporadically attempting to find trees that will meet the specifications for current orders. Both procedures ultimately leave stands of poor trees that cannot be harvested economically by the most ingenious logging or the most astute salesmanship.

A common American term for harvesting the best and leaving the poor is **"high-grading,"** that is, harvesting only material of high grade.

Geometric Thinning

In the fourth general method of thinning, the trees to be cut or retained are chosen on the basis of some predetermined spacing or other geometric pattern with little or no regard for their position in the crown canopy. The older, ambiguous designation of such thinning as "mechanical" refers to the mechanistic mode of choices and not to any use of machinery.

Geometric thinning can be advantageous in treating young stands that are densely crowded and previously unthinned. Rather arbitrary choices of trees can be justified if the number of present dominants greatly exceed the ultimate number or if the stands are so uniform that there has not yet been much differentiation into crown classes. Geometric thinning is also becoming increasingly common as a means of enabling the use of large or cumbersome machines to fell and extract trees in thinning.

Ordinarily geometric thinning is employed only in the first thinning of a stand. If properly done, it usually sets the stage for exercising a much higher degree of selectivity in any subsequent thinnings and also reduces congestion enough to allow large machines to be maneuvered effectively.

Geometric thinning patterns are often, but not always, applied in **pre-**

commercial thinnings; these are thinnings made purely as investments in the future growth of stands so young that none of the felled trees are extracted and utilized. **Commercial thinnings** are most simply defined as those in which all or part of the felled trees are extracted for useful products, regardless of whether their value is great enough to defray the cost of operation.

There are two general patterns that may be followed in mechanical thinning. In the first, referred to as **spacing thinning**, trees at fixed intervals of distance are chosen for retention and all others are cut. In the second, which is usually called **row thinning**, the trees are cut out in lines or narrow strips at fixed intervals throughout the stand. The rigidity of these specifications may be modified as occasion demands. In general, the justification for adherence to an arbitrary pattern decreases as the variation in size and quality of the trees increases.

Spacing thinnings are most commonly applied in overcrowded stands that have developed from dense natural regeneration or artificial broadcast sowing of seed. The simplest mode of application involves adherence to a rigid spacing. This approach can be modified by leaving the best tree in each square defined by the desired spacing. The amount of work can often be reduced by removing only the most active competitors of trees designated for release. The removal of the adjacent tree with the largest crown causes the greatest improvement in growth of the released tree and would thus be the best one to remove. However, the thinning effect of such limited kinds of removals does not long endure.

Thinning to a predetermined spacing is ordinarily employed only in young stands. The idea of thinning to a certain *average* spacing is often followed in older stands but as a numerical guide to the severity of the thinning and not as a way of selecting trees to be reserved.

Row and Strip Thinning

The removal of trees in rows, lines, or strips is another kind of geometric thinning done for a variety of reasons. The fact that this kind of thinning usually represents a compromise between doing a perfect job or doing nothing is not a reason to disdain it.

This kind of thinning plays a very important role in the precommercial thinning of excessively dense stands. A variety of motor-driven devices can be used to destroy trees in narrow strips so that the ones remaining along the edges of intact strips can grow faster (Fig. 4-9). Among these are rolling brush cutters, bulldozers, large mowing devices with horizontal rotating blades, and some of the other kinds of machines also used in mechanical site preparation (see Chapter 8). If devices are used that cut or kill trees one-by-one, it is usually best to employ methods of selecting the trees that are more sophisticated than eliminating lines or strips, but there is no reason why this must be the case.

FIGURE 4-9 Strip thinning in a dense, stagnating stand of ponderosa pine, nearly 50 years old, in eastern Oregon. The trees in the cleared lanes (*right*) were crushed with a bulldozer, leaving the narrow strips of trees visible in the aerial photograph (*left*). Such treatment is crude but effective; the only serious danger is that *Ips* beetles will breed in the slash and attack the standing trees. (*Photographs by American Forest Institute.*)

Row or strip thinning has been successfully used, with or without modification, to facilitate the harvesting or products from dense stands that have not been thinned previously. The first thinning in a stand is the most difficult. Congestion in the crown canopy and the light weight of the trees make felling difficult. The close spacing of the trees, both cut and left, impedes log transportation. The relatively small size of the trees results in high cost and low value per unit of product removed. These problems often make it virtually impossible to initiate thinnings in the very stands that need them most.

With row or strip thinning, however, it is possible to proceed such that trees are successively felled into the vacancies already created by the cutting of adjacent trees. The products can then be easily removed along the lanes that have been cut through the stand; the logging debris can either be stacked along the edges out of the way or else used as a kind of roadbed to protect the soil from damage by the log-moving machinery. This pattern of initial thinning overcomes many of the objections, real or imaginary, that are raised against thinning because of logging problems. A second cycle of this or any other kind of geometric thinning is rarely necessary or advisable because of the improved accessibility resulting from the first thinning.

Row thinnings are most readily applied in plantations that are already laid out in straight rows. In its purest form, this involves removal of every third row (Fig. 4-10). Every residual tree is freed on one side and the removal of one-third of the stand is a good approximation of the normal severity of most thinnings in young stands. The removal of every other row would ordinarily be too severe. It is sometimes found that spacing of

FIGURE 4-10 A row thinning being started in a dense, 25-year-old plantation of red pine in New York. Every third row is removed arbitrarily. (*Photograph by U. S. Forest Service.*)

the trees can be adjusted more satisfactorily in subsequent thinnings if the residual strips alternate in width between those that are two and those that are three rows wide.

The strips can also be shifted to go around especially desirable trees. In closely spaced stands the lanes might be made two rows wide in order to allow the passage of equipment used in transportation. The severity of the thinning can be reduced by spacing the cut strips at wider intervals. This course would be desirable where the objective was to stimulate the growth of trees destined to be removed in the final crop without heavy expenditure of labor or the production of an excessive amount of small material for which there was a limited demand.

Thinning in strips and rows can also be used effectively in combination with other methods of thinning simply to create avenues of transportation for logging equipment. This approach is becoming increasingly common because of the development of large, powerful, and rather cumbersome machines for felling, bunching, or transporting trees and logs. Even when most timber harvesting was done with hand tools and animals, it was difficult to thin on any highly selective basis without giving some consideration to the problems of felling each tree and moving the products out of the woods. As motor-driven logging equipment has become less maneuverable, it has become even more necessary to adjust the patterns of tree removal to the logging equipment. Unfortunately efforts to adjust logging machinery to various silvicultural objectives seem to get less emphasis than those designed solely to expedite logging. It is often necessary to use some sort of row or strip thinning in the first entry into a stand or else not to thin at all.

The lanes that are cut across a stand can be straight and at some arbitrary spacing or they can be gently curved and placed at intervals that will fit the stands and sites better. If the object is thinning and not regeneration, the lanes should be as narrow as possible. The spaces between them are best kept as wide as they can be consistent with extracting trees from them to the transportation lanes. It often helps to fell the trees in a herringbone pattern with their butts (or sometimes the tops) pointing in the direction along which they will be pulled out into the lanes and in the general direction along which they will be pulled to some loading point.

This general approach helps overcome the logging problems encountered in the first thinning of almost any kind of stand. If avenues are cut through the stand at appropriate intervals, it is then often simultaneously possible to cut and leave trees in much more appropriate patterns between the avenues. It is sometimes possible to thin whole stands with a high degree of selectivity simply by proceeding in straight or gently curving lines from one tree that ought to be cut to another, marking them as one goes along. Sharp curves cause difficulties with timber extraction, especially if whole tree lengths are removed. If such special measures are taken in the first thinning, the avenues remain useful for subsequent thinnings that can be much more selective.

Potential Problems

Row and strip thinning can lead to the development of lopsided crowns that can, if they remain that way for long periods, lead to the development of stems that are elliptical at the base. Except to the extent that unequal crown loading may predispose trees to later damage by wind or frozen precipitation, this effect is seldom of much consequence. However, it can help to try to develop more symmetrical crowns in later thinnings.

The main disadvantages of arbitrary selection of trees in various kinds of geometric thinning patterns lie in the cases where there is substantial variation between individuals. The removal of too many of the dominant trees can have the same undesirable results as selection thinning. Among other things, it usually causes a more definite and longer-lasting depression in the production of a stand than do the more uniformly distributed vacancies in growing space caused by low or crown thinning. Therefore, geometric thinning is best carried out when stands are young and plastic, before the deep-crowned trees become too few and the crowns of the trees so broad that wide gaps result. As with all methods of cutting, geometric thinning must be conducted with discretion and will give poor results if applied indiscriminately.

Free Thinning

Cuttings designed to release crop trees without regard for their position in the crown canopy are **free thinnings,** in the sense of being unrestricted by adherence to any one of the other methods of thinning. However, in any free thinning the mode of treatment accorded any one of the trees cut or released can be identified as representing one of the other four methods. The greatest need for combining several methods in a single thinning is encountered in stands that are somewhat irregular in age, density, or composition. Plantations are usually sufficiently uniform that free thinnings are unnecessary in them.

Technical terminology should convey as much meaning as possible. Therefore, vague terms like "free thinning" ought to be avoided if there is any reasonable possibility that one of the other terms might apply to a particular operation. It is more informative to use terms like "modified crown thinning" or "crown and selection thinning" than to throw all variants into a miscellaneous category. It should be noted that all of these terms refer to single operations and not sequences thereof. A program of treatment is not referred to as free thinning simply because crown thinnings are, for example, followed by low thinnings.

Conditions requiring free thinnings are most likely to exist at the time of the first thinning in previously untreated natural stands. This approach is distinctly advantageous for bringing a stand into shape for efficient production. If the objectives of the thinning program are clearly defined and consistently pursued the irregularities begin to fade and the likelihood increases that a logical thinning operation will conform to a single method. From the standpoint of efficiency of logging and administration, it is desirable to achieve regularity of treatment as soon as uniformity can be imposed without undue sacrifice of the growth potential of the stand.

A typical free thinning operation in an unevenly stocked but even-aged stand might simultaneously include: (1) selection thinning to eliminate scattered undesirable dominants, (2) crown thinning to release crop trees drawn mainly from good dominants and secondarily from thrifty codominants in the more sparsely stocked portions, and (3) low thinning to salvage all merchantable overtopped trees throughout the stand and to thin the well-stocked portions to the severity of the D grade.

Quantitative Definition of Thinning Methods

The ratio d/D, where d is the average D.B.H. of trees removed in thinning and D is, the average before thinning, is sometimes used to characterize methods of thinning quantitatively (Bailey and Ware, 1983). The ratio is more than 1.00 for selection thinning, 1.00 or less for other methods, and least for light low thinning. It is exactly 1.00 for perfect row thinning in plantations.

APPLICATION OF THINNING

A schedule of thinning should be a systematic plan for a whole rotation based on deliberate decisions about the kind of vegetation, benefits, and products desired. One should reason from these chosen goals backwards to the schedule of treatments designed to achieve them. The ultimate objective is some concept of the stand that will exist near the end of the rotation. The details of a thinning program should remain flexible enough to be settled in the light of conditions prevailing at the time of each treatment.

Although thinning can certainly be used for other objectives of managing forest vegetation, timber production is the chief purpose and also the one that requires the most precise adjustments. Prospective markets for wood products set most of the long-term objectives. Markets are, in a sense, also the tools that are used, often quite opportunistically, to achieve ultimate goals. Fortunately, the most rapid product innovations tend to apply to the small, poor trees of the kind that are best removed in thinnings, even though improvements in the technology of their harvesting tend to lag.

Changes in utilization sometimes seem to cast doubt on the need to grow the kind of good, moderately large trees usually envisioned as the final goal of thinning. However, the prices paid for the products of such trees have shown more tendency to soar than to remain stable. The most important choices usually lie in determining how much of the total production to channel into trees large enough for sawn or sliced products and how much into bulk commodities such as pulpwood, chips for reconstituted products, and fuelwood. Small trees are almost as good as large ones for the latter uses.

Choosing Methods of Thinning

Three interacting considerations enter into the formulation of thinning programs, which may, more than incidentally, include deliberate programs of not thinning. One set of choices has to do with *timing*, that is, rotation length, time of first thinning, and intervals between subsequent thinnings, if any. Another is over the *method or methods* of thinning employed at each stage. It is only under the least complicated circumstances that a single method can be followed through an entire thinning program. The most difficult set of choices is about the *regulation of stand density* or the amount of growing stock to be left after each thinning.

If a stand is thinned only once and is not very uniform, it is likely that the necessity of doing so much with such limited opportunity will dictate simultaneous use of two or more methods in a free thinning operation.

The likelihood that repeated application of a single method might be found best for a long series of thinnings is rather limited. The stand would have to be very uniform so that the crop trees would be alike and treatment would not have to be varied in different parts of the stand. Then, if markets were excellent and if the crop trees were of good quality and always in a clearly dominant position, one might conceivably employ a simple series of low thinnings like that shown later in Fig. 4-12. A series of selection thinnings might be applicable to growing stands of very tolerant conifers for pulpwood, although there might be some problems of securing reproduction and disposing of unmarketable residual trees at the end of the rotation. With species of sufficient tolerance, it might be possible to apply a series of what could be called crown thinnings if the ultimate disposal of any lower layer of trees were not regarded as a deviation from the method.

Ordinarily the crown classes that should be removed to favor the crop trees and those that can be harvested profitably will change as a stand grows older. This calls for changes in the method of thinning even in the absence of differences in condition of various parts of the stand.

The high labor costs and low product values associated with small trees often make it necessary to change sequentially from one thinning method to another to evade or postpone the need for cutting small trees. Figure 4-11 illustrates the point that the four well-defined methods involve harvesting trees from different segments of the distribution of diameter classes in a pure, even-aged stand. From this it can be seen that if selection, crown, and low thinnings followed in succession as a stand grew older there would be some possibility of avoiding cutting trees when they were small. If removals from the dominant crown class are undesirable, the same can be accomplished by crown thinning that is followed with low thinning. Such measures are not perfect solutions to the problem because it is so difficult to make the trees of the lower crown classes increase rapidly in diameter. Geometric methods such as row thinning reduce logging costs mainly by facilitating extraction and not by postponing the handling of small trees; the trees removed are an essentially random sample of the

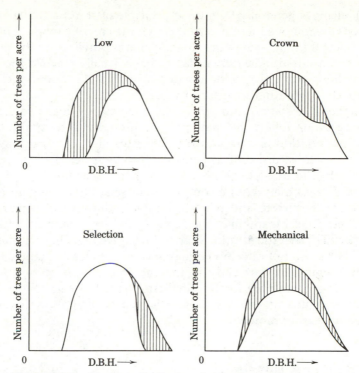

FIGURE 4-11 Diameter distributions for the same pure, even-aged stand showing, by cross-hatching, the parts that would be removed in four different methods of thinning. In each case about one-third of the basal area is represented as having been removed. It is assumed that no overtopped trees are salvaged in the crown and selection thinnings, that the stands have not been treated previously, and that D.B.H. is closely correlated with crown class.

existing diameter classes. Nevertheless, the increasing role of heavy machinery for timber harvesting increases the desirability of using some sort of row or strip thinning initially. The dilemma faced in the need to thin out small trees to capture the subsequent advantage of large trees (and the lower cost of harvesting them) can sometimes be overcome by using light, low-cost machinery in the early thinnings and switching to larger machinery in the later ones.

The changes in thinning methods appropriate to developing good crop trees often follow the same sequence by which the costly logging of small trees is evaded. It is logical that any coarse dominants be removed in early selection thinnings. The enlargement of the crowns of crop trees is, in the early stages, most readily expedited by crown thinnings. Once the crop trees have been successfully promoted to clearly dominant positions, their competitors can be removed in low thinnings. One might even start such a sequence with a geometric thinning.

Such changes generally follow the principle of shifting the level from which trees are removed downward through the crown canopy. The compulsion to low thinning late in the rotation comes purely from the fact that the trees of the subordinate crown classes have usually reached the end of their usefulness and only the dominants or vigorous codominants retain much capacity for further increase in value.

Even though a stand is uniform it is not necessarily desirable that each thinning operation follow the same pattern throughout the stand. For example, if the rotation is long and the thinnings start early, it may at 30 years be useful to accord different treatment to trees destined for removal at 60 years than to those to be held for 100 years (Macdonald, 1961). The "short-term" trees may require complete release so that they will lose no more live crown surface and will grow rapidly during the remainder of their allotted time. Meanwhile, the "long-term" trees may be left somewhat crowded so that the attrition of their lower live branches will continue and their boles will not develop large branches or excessive taper. In general, when thinnings start early in the rotation, it is necessary to plan proper treatment not only for the final-crop trees but also for the rather numerous ones that can develop substantial value even if they are destined for somewhat earlier removal.

Timing of the First Thinning

In theory the first thinning can be made as soon as the crowns or the root systems of individual trees close together and start to interfere with one another. Competition of this sort commences early in the sapling stage in any but the most poorly stocked stands. The best single criterion for determining when to apply the first thinning is the live crown ratio of the potential crop trees. As long as a satisfactory ratio is maintained, a stand need not be thinned until the thinning will pay an immediate profit. When the ratio approaches an amount at which undesirable losses of diameter growth are in prospect, it is time to consider the first thinning. Some planned dwindling of the live crown ratio is usually desirable to restrict branch size and the degree of stem taper as well as for natural pruning. The practice of thinning and regulation of stand density usually involves compromises of this sort between diameter growth and various "stem-training" effects.

The various factors that affect timing of first thinnings are considered mostly in terms of their economic consequences. If there are ample funds for long-term silvicultural investments, the first thinning can perhaps best be done when the value of ultimate benefits, discounted to the present at compound interest, equals the cost of the operation. If such investments cannot be made, the first thinning is delayed until it will give an immediate profit. In some cases the appropriate time never arrives.

Precommercial Thinning

If thinning is done as an investment, it is desirable to proceed in such a manner as to achieve the greatest possible increase in diameter growth with a small amount of work. This line of attack differs from commercial thinning in which it is usually advantageous to harvest the highest possible proportion of those trees that will not contribute to the future increase in value of the stand.

One useful approach to precommercial thinning is that of "crop-tree thinning" in which only the trees likely to form the final crop are released. Usually most of the potential benefits are gained merely by eliminating the one or two most serious competitors of each crop tree. Reukema (1961) found, for example, that the cutting of additional competitors caused little further increase in diameter growth of Douglas-fir and that dominant trees responded better than those of lower crown classes. However, the period of enhanced diameter growth is longer the more drastic the reduction of competition. If an investment is required to carry out the first thinning, it is desirable to make it heavy enough to ensure that no further treatment will be required before the stand reaches the stage where a profitable cutting can be made.

The cost of precommercial thinning can be substantially reduced by use of power-driven equipment or chemicals.

The policy of postponing the first thinning until it will return an immediate profit can be defended on the ground that many fine stands have developed in nature without treatment of any kind. In most stands, failure to thin should be looked upon as a lost opportunity rather than as an invitation to disaster. However, in many of these same stands the net return from the whole crop could be increased if an outright investment were made in thinning at the appropriate time. In fact, economic analyses have repeatedly shown that precommercial thinning nearly always has the highest present net worth of any long-term investments that can be made in silvicultural treatment. This is because the investments are not made at the beginning of the rotation and it is not long before the treatment pays off through increased production of merchantable volume.

Precommercial thinning can also reduce the cost of subsequent harvesting by getting rid of small, unmerchantable trees that impede logging. If stands are made more open there is less reason for skidding equipment to damage the remaining trees because there is more room to go between them.

A strong case can be made for investment in early thinning with stands that would deteriorate without treatment. The most common cases of this kind involve very dense, pure stands, usually from regeneration by natural or artificial seeding, and especially those of species with low genetic variability in factors that affect height growth. The hard pines, especially red, jack, and lodgepole, are notorious for these tendencies but they can exist in some degree with any species. Some species, on the other hand, such as eastern white pine, appear to have sufficient genetic variability to express dominance even in dense stands.

The likelihood that trees will stagnate or stall in diameter growth or, worse yet, growth in height is greatest on poor sites. Sometimes fertilization will favor strong trees and hasten suppression of the weak and thus produce the same effect as thinning. There are even cases of extreme deficiency of water or nutrients in which thinning alone will not produce any effect without fertilization, irrigation, or drainage.

Effect on Stand Structure

The trees of planted, pure stands are exceptionally uniform in age and spacing so they tend to be evenly matched in size and vigor. Therefore, it is necessary to anticipate that their crowns or root systems will close together rather suddenly and that the time of this event may be the ideal time for the first thinning. Ordinarily it is wise to make the initial spacing of the planted trees wide enough so that all of the trees will have grown to minimum merchantable size at this time, provided that need for competitive training effects does not make this impossible. This general kind of problem can also be mitigated by deliberate avoidance of extreme genetic uniformity in the stand. The genetic uniformity that is sought in agricultural annuals is usually undesirable in woody perennials. The dwindling of numbers of trees is a process that must be anticipated; it is also a desirable one that is put to planned use in thinning.

Stands can also be so irregular that some trees race ahead to become branchy, malformed super-dominants. If their removal in a first thinning is too long delayed they are likely to suppress their better-formed subordinate neighbors.

Even if an untreated stand does not stagnate or develop into a collection of malformed dominants, the stage is ultimately reached at which it may be dangerous or fruitless to initiate thinning or any other form of partial cutting. Even dominant trees can eventually become weak, slow-growing, and slow to respond to release. If thinning is delayed too long, the residual trees will be slow to reclaim the vacant growing space and the production rate of the stand will suffer a major, but not necessarily permanent, reduction.

In other words, stands can become so crowded that it may be questionable whether it is wise to thin them even when such treatment would be immediately profitable. Investment in early thinning may provide the only way in which the important financial benefits of later thinnings can be captured.

Timing of Subsequent Thinnings

The effects of a single thinning are not maintained indefinitely. After a few years, the gaps in the canopy close together and before long the same crowded condition that existed before thinning redevelops. As the crowns expand the number of trees that can occupy an acre to best advantage decrease and the surplus volume available for removal in the next thinning accumulates.

The rate of growth of the crop trees is the best single criterion for determining when thinnings should be repeated. It is logical to set some realistic rate of diameter growth as a goal and thin when the growth of the crop trees falls below it.

It is also likely to be necessary to wait until the volume that can be removed has accumulated enough to support a profitable operation. If the stand closes too tightly and diameter growth slows down too much before sufficient volume accumulates, it is a sign that the previous thinning was either too light or was not done in such a manner as to give the crop trees enough room to grow. In this respect close attention must be paid to manipulation of the diameter distribution of trees destined to be removed in thinnings. If too many of the larger ones are cut at once, the next thinning may have to be postponed until long after the small trees have choked the crown canopy in the process of growing to merchantable size.

The proper balance between severity, method, and timing of thinning is not an easy one to strike. This kind of adjustment is still highly intuitive because there are so many considerations to take into account. The diameter growth of the crop trees is by far the best and simplest guide; unfortunately it indicates the status of only one component of the stand.

The heavier the thinnings the longer is the interval between them. Heavy, infrequent thinnings tend to reduce the total yield of a stand because of the long periods during which parts of the growing space remain unoccupied.

It is seldom possible to thin lightly and frequently enough to avoid losses in production of gross, total cubic volume or tonnage of wood. Ordinarily it is logical to thin in ways that will increase diameter growth enough to increase the actual yield of merchantable volume or value of wood even if this does involve deliberate sacrifice of the gross production of wood. Almost any kind of commercial thinning augments actual yield through harvest of material that would otherwise be lost to natural suppression. The interval between otherwise judicious thinnings would have to be very long indeed to preclude some sort of increase in the value of wood harvested from a stand because there are so many ways of achieving such increases.

The practice of repeating thinnings at equal intervals simplifies administration but has no other virtues. Actually stands should be thinned more frequently when young than when old because they close more rapidly.

The height of dominant trees of a stand is an integrated expression of age and site quality. Therefore, it is an excellent parameter to use in developing thinning schedules. Provided that there are no large differences in site quality, a stand of a given species will go through the same processes while it is growing from 12 to 15 m of height regardless of whether this takes 3 years or 10. Dominant height becomes the index of the stage of stand development while increase in height can be used to measure the rate of development. Since height growth and crown expansion both depend on shoot elongation, the relationship between the two is generally close (Mitchell, 1975).

If one planned to thin at equal intervals of attained dominant height, the time intervals between thinning would automatically and logically increase with stand age and be longer on poor sites than on good. Knowledge developed on one site would be applicable to others, but one should anticipate that trees on good sites might culminate in rate of height growth at greater height than those on poor. Height can be translated into age and height growth into time by use of site-index curves of height over age, but this kind of arithmetic can be postponed until some late stage in the development of thinning schedules.

Rotation Length

The planning of thinning schedules and setting of rotation lengths are strongly interactive kinds of analysis.

There is one kind of biological limit on rotation length. When the dominant trees cease to grow much more in height, shoot elongation and crown expansion also become very slow. At this time the trees cease to respond much to thinning so there is not much point in continuing such treatments. This development is one of the causes of the declines in production rates that set economic limits on rotation lengths. The only reason to have longer rotations is some kind of desire to allow trees to grow very slowly to large diameters. The culmination of height growth comes very early in some species, especially those that grow fastest in height in youth; slash pine and aspen are good examples. At the other extreme are species such as eastern white pine, many western conifers, and yellow-poplar, which continue height growth to substantial heights and ages.

While the difference between them is often blurred in practice, there are two different economic theories about rotation length. One, which may be termed the **physical rotation,** is aimed at maximizing physical yield under long-term sustained yield under the forest rent principle in terms of the mensurational units of yield. The other, the **financial rotation,** has the goal of optimizing money return on capital under the soil rent principle; calculations of it are characterized by use of compound interest. The idea of the physical rotation fits best when it is necessary to guarantee a perpetual supply of timber to some voracious local or regional forestry industry from a fixed and limiting amount of forest land. That of the financial rotation applies best when the amount of capital available for long-term silvicultural investment is the limiting factor.

The physical rotation length is that of the maximum mean annual increment in terms of the mensurational unit *chosen* for measurement of yield. The financial rotation is that of highest present net worth of all costs and returns discounted back to the beginning of the rotation at some *chosen* rate of compound interest. Both kinds of choices are guesses about distant futures and rotation lengths can vary widely depending on which mensurational units or interest rates are selected. The full range of possibilites can be great enough to make the difference between the two basic principles seem trivial. Nevertheless, the usual situation is that financial rotations based on soil rent are shorter than physical rotations based on forest rent.

Under the soil rent principle, the higher the interest rate the shorter are the rotations and the greater the advantage from making commercial thinnings early in the rotation. In the case of physical rotations, the analogous variable effect comes mainly from the way in which tree diameter affects the proportion of total production that is counted as yield. The higher the minimum diameter that is set on what is reckoned as yield, the longer is the rotation and the greater it will be prolonged by kinds of thinning that increase diameter growth.

Under any given set of assumptions, the effect of not thinning is usually to shorten rotations because there is no salvage of prospective mortality and no acceleration of diameter growth. However, this generalization would be false if there were the simple goal of growing trees to some large, fixed diameter.

Bookkeeping Problems

Some effect of thinning on rotation length are for practical purposes the same regardless of which economic theory applies. Any thinning that converts prospective mortality to yield lengthens logical rotations. This effect is magnified by a "bookkeeping" artifact; the mortality losses ascribed to a given later part of a rotation count the full volume of the dead trees and these are subtracted from the increment of the living trees during that same period.

The choice of mensurational units is important in either kind of calculation. If wood is to be chipped up for bulk products such as pulp or particle board or if it is used for fuel, the appropriate units would be merchantable volume or dry tons ordinarily with small diameter limits. With this kind of utilization, rotations would be short and any advantages of thinning small but not necessarily absent. If the trees are to be made into sawn or sliced wood products, the units are more likely to be thought of as American board feet or merchantable cubic volume, each with comparatively large diameter limits set on how much wood is counted. With this approach, the rotations are comparatively long and the effects of thinning on diameter growth become crucial and controlling.

However, none of these approaches really solve the problem of taking the economic effect of tree diameter fully into account because all lead into the trap of the simple assumption that all units of a given kind of product are equally valuable. The only way out is the difficult one of using studies of logging costs and product values to determine the true net value of units of product from trees of differing size (and, for that matter, quality). If this line of logic is followed to its end, it is seen that the most logical physical rotation would be that which maximized mean annual increment of money, which is a better measure of social utility than cubic volume. Financial rotations would be determined by the same calculations except that their length would be shortened by taking the time value of money into account.

Regardless of which economic theory or mixture thereof is followed, there is the general effect that the simpler the assumptions about mensurational units and net value the shorter will be the rotations compared with those that would prevail in a state of full knowledge. The only simple assumption that logically shortens rotations is the selection of a higher rate of compound interest under soil rent and the logic of this depends on whether the chosen rate is appropriate to the circumstances. One pragmatic way of dealing with this general situation might be to calculate a rotation from simplistic assumptions and then lengthen it on the basis of intuitive judgments about those less knowable factors that operate mostly to lengthen rotations anyhow.

Use of Financial Maturity Analysis

Another imperfect way of approximating rotation length involves use of financial maturity analysis (see "Optimum Use of Capital and Growing Stock" in Chapter 12). The realizable value of a stand at any point in its development is treated as an investment and the rotation is ended when the increase in value of the stand ceases to exceed some desired rate of compound interest. The return on this particular kind of investment is zero until the first day one tree in the stand is suddenly worth one penny of net value. On that day, the compound interest earned with that one penny, on an investment of zero of the day before, is infinitely high. From that day onward, except for important effects of trees improving in value because of quality improvement or becoming suitable for products of higher value, the rate of compound interest earned on this peculiar kind of investment represented by the stand sinks inexorably downward. However, this rate of decline can be slowed or temporarily reversed by commercial thinning.

Such thinning definitely reduces the size of the investment in trees standing in the woods usually without markedly reducing the amount of money that the trees earn through growth in a year or decade. This somewhat passive effect by itself either increases the rate of compound interest earned by the stand or at least slows the rate at which the rate declines. Beyond this, if the diameter growth of the trees is increased or kept at a high rate, there can be a positive effect that increases the rate of interest return or slows its decline. In the application of this approach, repeated thinnings are applied until the rate of return seems unlikely to bounce back to an acceptable level; when this time comes the stand is regenerated.

This approach is really another kind of application of the soil rent principle. It has the shortcoming that it does not take into account any money actually spent. It is probably best viewed as one of several ways in which biological, financial, and even political factors are analyzed in the complicated problem of determining rotation lengths. The important point to note here is that thinning schedules are actually more likely to control rotation lengths, usually by extending them, than to be controlled by them.

Regulation of Stand Density and Thinning Intensity

Thinning programs should be governed by reasonably definite schedules indicating the density of stands at all stages of development (Fig. 4-12). Such schedules help ensure final crops that consist of trees of the qualities and sizes desired. They are also necessary for purposes of forest regulation to predict yields from intermediate and final cuttings.

Without thinning schedules there are psychological tendencies to thin too lightly, at least if the trees to cut are the ones that are marked. A thinning usually looks much more severe after the trees are marked than it does after they are cut; within a few years the rotting remains of the cut trees are the only readily apparent evidence of any cutting. Conscientious foresters are chronically fearful of running out of trees and often fail to anticipate how rapidly the stands will close up again. There are tendencies to make thinnings too severe if the trees to be left are the ones that are marked.

Thinning schedules based on appropriate studies of growth and development after thinning are the best source of assurance. It is desirable for those who mark stands for thinning to check themselves occasionally with the use of prisms or other angle-gauges under the Bitterlich point-sampling scheme.

No one schedule is optimum for all stands, even within a given forest type, because economic as well as natural factors must be considered. Consideration of the details of any thinning schedule requires recognition of the fact that, at the time of each thinning, the trees of the stand are segregated into three categories. The first and most important group consists of the trees destined to form the final crop. The second comprises those trees that will be removed in subsequent thinnings but will be needed in the meantime to utilize the growing space that will eventually be occupied by the final crop. These two categories, which do not always have to be clearly distinguished, make up the growing stock left after each thinning. The final category consists of the surplus trees to be eliminated in the thinning.

In the kind of analysis that leads to development of schedules it is generally desirable to approach the matter from two directions. The first, and often the most important, is consideration of the development of the final-crop trees, that is, the amount of growing space that they will need at

FIGURE 4-12   A plantation of eastern white pine at Biltmore, North Carolina, at various stages during a series of five light, low thinnings. The plantation was established at very close spacing to check serious erosion and was first thinned at the twentieth year (Wahlenberg, 1955). Subsequent thinnings were made at intervals of about 6 years, reducing the basal area to approximately 100 square feet per acre each time. (*a*) 20 years–numbered trees being left in first thinning. (*b*) 26 years–freshly numbered trees being left in second thinning. (*c*) 32 years–just before third thinning. (*d*) 45 years–just after fifth thinning. (*Photographs by U. S. Forest Service.*)

each stage to meet whatever goal is set for their growth. The second approach guards against failing to see the forest (or stand) for the trees. It involves determining how much growing stock should be left after each thinning to give an optimum yield, over the whole rotation, in terms of amount of product per hectare or acre.

One approach cannot be followed to the exclusion of the other. It would be easy to get so concerned about the crop trees that one might end the rotation with a few fat trees and sacrifice too much total yield. Conversely, preoccupation with yield of cubic volume per unit area almost always leads to sub-optimal diameter growth. If both approaches are used, desirable compromises often become obvious.

Many of the basic principles underlying these approaches were discussed in Chapters 2 and 3. What follows is a discussion of the ways that are, or have been, used to create thinning schedules based on these principles.

Parameters of Stand Density or Stocking

Stand density is a measure of the amount of tree vegetation on a unit of land area. It can be the number of trees or the amount of basal area, wood volume, leaf cover, or any of a variety of less common parameters (West, 1983). **Stocking** is the proportion that any such measure of stand density bears to any of a wide variety of norms expressed in the same units and chosen for differing purposes. Density tells what is and stocking how this density relates to someone's notion of what ought to be, usually in terms of percentages. Full stocking might, for example, be the basal area per hectare of trees larger than 25 cm D.B.H. that was deemed necessary to maximize production of sawn boards. A thinning schedule is a guide to what stocking, expressed as some kind of stand density, should be at each stage of development.

No parameter of stand density has yet been devised that would, if held constant after each thinning of a whole rotation, define a series of thinnings that met any plausible set of objectives. Many of these parameters can be used to quantify stand density or stocking in thinning schedules, but neither they nor any other form of simple mathematical magic seem able to define a schedule by themselves. It may help to consider the utility of each of these parameters as measures of density or stocking.

The simplest parameter of all is **number of trees** per unit area. This takes no account of either the sizes of the trees or the space they occupy. However, there is no fundamental reason why a thinning schedule could not be presented in these terms, for a given species and site quality, if it had already been developed in terms of more meaningful units. Such a schedule would have to be adhered to rigidly because the trees would develop sizes different from those contemplated if there were significant departures from the program.

Basal area per unit is by far the most commonly used parameter, although there is probably more tradition than biological reason for this. Its main virtue is that it is a kind of integrated expression of numbers of trees and their sizes. It was originally devised not so much as a parameter of stand density but as a crude indicator of cubic volume of stemwood. Since it measures the cross-sectional area of physiologically dead wood it has no direct biological significance. However, the basal area of a tree is fairly well correlated with the cross-sectional area of the crown. This means that if one removed 75 percent of the basal area of a fully closed stand one would also be making approximately 25 percent of the crown growing space vacant.

It is quite easy to measure basal area per unit area by various point-sampling techniques. However, there is seldom any virtue in continually thinning back to some constant basal area even though it often seems convenient to think so.

Theoretically the cubic volume of stemwood per unit area should be better than basal area as an indicator of the amount of foliage per acre, but it is more difficult to determine and there is no indication that the refinement would help. Because it is weighted by tree diameter, board-foot volume is good as a means of assessing the results of thinning programs but, for the same reason, it would not be good as a parameter for expressing them.

There is strong evidence that good correlations may exist between the cross-sectional area of sapwood and the amount of foliage borne by a tree. This might well develop into a good surrogate parameter for foliage in devising thinning schedules. However, it is so difficult to determine that it would not be a practical parameter for carrying these schedules into effect.

The amount of stem or **bole surface** of trees on an acre or hectare has the virtue of approximating the amount of growing and respiring meristematic or aphotosynthetic surface of the stand. If the amount of foliage that a stand of given species and site can produce is fixed, it would seem that one could predict the average diameter growth of the trees by knowing how much bole surface area was linked to that foliage.

The handy index of bole surface per unit area is the **sum of the diameters** of trees. This is because bole surface is the sum of the products of circumferences and heights of tree stems modified by some function of stem taper. If one makes the sweeping assumption that the trees of an even-aged stand do not differ in height or taper, and notes that *pi* is a constant too, then the sum of diameters becomes the chief variable function of bole surface. In this same way one may deduce that one could change basal area to the sum of the squares of the diameters and get just as good a surrogate for stem volume per unit area. As far as the present purpose is concerned, people have done much analysis in terms of basal area but little in terms of bole area.

There is one difficulty that lurks behind the use of breast-high diameter in assessments of stand density or growth. The problem is that trees are taller than people and breast height usually falls in the zone of the butt swell where the effects of fast or slow growth of whole trees are greatly exaggerated. This is the point discussed in Chapter 3 and depicted in Fig. 3-1. What this means is that one unit of basal area stands for more crown area with a smallish tree in a stand than does that unit in a dominant.

This problem could be solved, at least in research work, by measuring diameters higher up on the trees. In American observations, the best fixed point would be 17.5 feet, the level at which inside-bark diameter is measured to determine Girard Form Class. This mode of quantifying stem taper has been used in constructing many volume tables.

Figure 4-13 is a nomogram depicting the fixed relationship between (1) numbers of trees per acre, (2) basal area, and (3) average D.B.H., which is, by convention, the quadratic mean diameter or that of the tree of average basal area. While this nomogram does not define any thinning schedule, it is very useful for describing them and focusing attention on the important relationship between stand density and tree diameters (Gingrich, 1971). Although this nomogram does not show any time dimension, one could inscribe lines on it that describe the development and thinning of a stand over time. Each point on a line, proceeding from right to left, could be

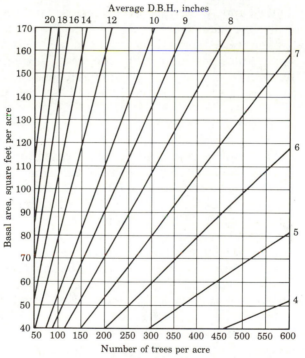

FIGURE 4-13 A nomogram for relating the average D.B.H. of stands to basal area and numbers of trees per acre in the guidance of thinning schedules. Note that the average D.B.H. of a stand is, by convention, the quadratic mean or the D.B.H. of the tree of average basal area and not the arithmetic mean. The largest trees in the stand would usually be considerably larger than the average tree. Almost any thinning schedule could be described on this diagram by a saw-toothed line along which the stand "moved" over time from right to left. However, there is nothing inherent in this diagram that tells what the schedule should be or how rapidly stands could be made to "move" across it.

labeled with an age determined from appropriate observations of stands under different regimes of thinning. The harder one thinned, the lower would be the line on the diagram and the faster should the point of current density move from right to left.

Tree Spacing Guide

The stand-density parameters just discussed can be used to describe almost any thinning schedule that has been decided upon but they do not necessarily define any. The same is true of various ways of using the spacing between trees. Spacing itself is really the same as numbers of trees per acre. However, this can be modified to produce some "rule of thumb" regarding the proper spacing of the trees left after thinning.

According to the handy $D + x$ rule, the average square spacing in feet should equal the average stand diameter D in inches plus a constant x. The values of x that are suggested for various situations range from 1 to 8, the commonest being 6. Rules of this form provide steady increases in basal area. Rules of the form Dx provide for constant basal area; the value of the factor x for basal areas of 90 and 120 square feet of basal area per acre are 1.61 and 1.4, respectively.

Another way of regulating the spacing of trees left after thinning is to set the average spacing equal to a constant fraction of the height of the dominant trees. Wilson (1955) suggested that the spacing for tolerant species such as spruce and fir be one-sixth of the height and that for intolerants such as red and jack pine be one-fourth of the height. Dominant height is an excellent indicator of the stage of development of a stand and, unlike D.B.H., is independent of the effects of thinning.

In the foregoing and in most cases it is assumed, purely for convenience of calculation, that trees stand at the corners of squares. This means that if the crowns closed perfectly they too would be square in horizontal cross section. Since they are closer to being round it may be recognized that, were the trees uniform and evenly spaced, they would stand at the corners of equilateral triangles and the crowns would nest together like hexagons (Fig. 4-14). This means that, with crowns of uniform width and circular cross section, the more efficient equilateral spacing would give 15.47% more trees per unit of ground area than square spacing.

This phenomenon should be taken into account if one has determined the crown widths of trees of some desired D.B.H. and wants to know how many might stand on an ideally stocked acre or hectare. Table 4-1 gives numbers of trees per unit area for the range of tree spacings that might exist in stands being thinned and in final-crop stands. For closed stands of equilateral arrangement, spacing and crown width are synonymous. However, a good, closed stand will, in practice, have more trees per unit area than that with square spacing but less than that with equilateral spacing (Assmann, 1970). There are always gaps; the crowns will not be perfectly round or of exactly the same size.

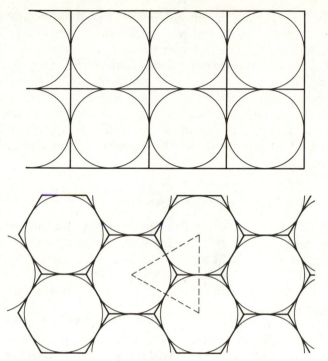

FIGURE 4-14 Schematic diagram contrasting the arrangement of circular tree crowns with trees standing at the corners of squares (*above*) and equilateral triangles (*below*), with the same distance between trees and the same crown diameters in each case. Note that the arrangement of nested hexagons with the triangular spacing allows the crowns to cover more of the ground area such that there are also more trees of the same size per unit of area.

Table 4-1 Numbers of trees per unit of area with square and equilateral spacing in English and S.I. systems

Trees per acre		Spacing	Trees per hectare	
Square	Equilateral	(ft, m)	Square	Equilateral
4840	5589	3	1111	1283
1742	2012	5	400	461
889	1027	7	204	236
681	786	8	156	180
538	621	9	123	143
436	503	10	100	115
302	349	12	69	80
151	174	17	35	40
99	114	21	23	26
48	56	30	11	13

Note The stand densities for the English half of the table are *not* equivalent to those on the same lines in the S.I. or metric half.

One useful parameter for the guidance of thinning practice, developed by Abetz (1982) for Norway spruce plantations in Germany, is the ratio of total height to D.B.H., with both measured in centimeters. While this does not define stand density, it does indicate when to thin and also quantifies two very important goals. First, it focuses attention on the rate of diameter growth, the crucially important economic parameter. Second, it very simply defines the degree of butt swell that governs the resistance of trees against the mechanical loads of wind and frozen precipitation. The goal in the case of spruces in southwestern Germany is to keep the *h/d* ratio from dropping below 70.

Thinning Guides Based on Crown Expansion

The basic reason for using tree spacing to guide thinning is that it defines the area, along a horizontal plane, that can be occupied by the crown. The size to which the foliar sugar factory can grow is thus defined. There are high correlations between crown diameter and stem diameter and crown volume and stem volume, as well as between crown expansion and diameter growth.

The key to using these relationships to predict the growth of trees is determination of the rates of crown expansion (Mitchell, 1975). Given the large data storage capacity of electronic computers it is possible to "grow" tree crowns by computer simulation and present the results in the form of crown maps that change with time. It is even possible to "grow" the crowns in three dimensions and for sufficiently large portions of simulated stands. The dimensions of the stems can be calculated from the crown dimensions and their growth from the changes. This approach has the virtue of predicting the growth of wood on the basis of the amount of foliage, which is the real source of the wood. Any predictive method that is based on correlating the periodic growth of a tree with its initial D.B.H. or basal area is founded on the fallacy that wood grows wood; the results amount only to extrapolation of past growth.

With this general approach, as well as others, it becomes possible to do thinning "experiments" in minutes that would take decades in the woods. One can test different thinning regimes and consider the results not only in terms of the volume yield but also with respect to important considerations such as the prospective assortment of tree diameters. The best data to put into the computers come, however, from long-term observations of the growth of trees and their crowns in thinning experiments.

Relationships between Numbers and Sizes of Trees

Studies of very fully closed, pure, even-aged stands have shown tendencies for certain simple straight-line relationships between the average sizes of trees and their total numbers. These are the kinds of stands described in so-called "normal" yield tables. The stand density index developed by Reinecke (1933) is based on inverse straight-line relationships between the logarithms of (1) average D.B.H. and (2) numbers of trees per acre. The index value is the number of trees for stands that have average D.B.H. of 10 inches. This value is higher the more shade tolerant the species.

An essentially similar relationship for such stands holds that the numbers of trees per unit area vary inversely as the volume or mass of the average tree raised to the 3/2 power (Drew and Flewelling, 1977). These relationships may define the self-thinning process that takes place in crowded stands as the trees grow larger and older. It should be noted that just such mathematical relationships have long been used in "smoothing" the data to construct yield tables. It can be dangerous to deduce "laws" from artificially constructed models that are at least once removed from the original data. However, in this case it is probably safe to infer that the relationships do exist in crowded stands and must have some biological basis. What is not clear is whether thinning schedules should logically be guided by parallel lines that were drawn on the graphs to specify lower stand densities but faster diameter growth.

There is a similar approach that has been made to the problem from the opposite part of the possible range of stand density, the open-grown tree (Gingrich, 1967). In this it is postulated that the very lowest density to which a stand might logically be thinned is that in which each tree was left occupying the space that its crown could expand to cover if it were open-grown. The statistical relationships between the diameters and the numbers per acre of such trees are determined for hypothetical stands that consisted of open-grown trees of uniform stem sizes that were just barely closed. The upper limit, beyond which stand density should not be allowed to increase, is defined by similar relationships between numbers and sizes of trees in fully stocked stands. The results are usually plotted on nomograms like that of Fig. 4-13.

These relationships represent attempts to develop mathematical relationships between tree size and stand density. It is not clear whether parallel relationships would exist with trees grown by thinning regimes that kept stand density away from the open-grown condition or that of overcrowding.

Multiple Regression Analysis

The purpose of all of this is to predict the growth of trees or stands on the basis of measurable factors altered by thinning. Many of the efforts that have just been described are based on some preconceived hypotheses.

The statistical techniques of multiple regression analysis have been used (Fries *et al.*, 1978; Clutter and Jones, 1980; Brown and Clarke, 1981) to attack the problem in more open-minded fashion. In this procedure, as many factors as possible that might govern growth, the single dependent variable, are tested as independent variables in multiple-term equations.

Whatever equation explains the greatest amount of variation in growth (of tree or stand) is taken as the best quantification of the phenomenon. Among the independent stand variables commonly used are age, basal area, numbers of trees, site index, average diameter, wood volumes, and various indexes of stand density based on the aforementioned relationships between average D.B.H. and numbers of trees.

Where the goal is to explain the growth of individual trees the variables often include the initial sizes of the trees themselves, site index, crown dimensions, distances to other trees, live crown ratios, or parameters designed to quantify competitive effects. Each analysis is typically confined to data for some species in a particular locality and is designed more to predict growth or yield than to explain the results.

This kind of analysis can be very useful in providing the basis for thinning schedules if it is done with due regard for both biological principles and those of statistical analysis. The results should not be used without critical appraisal and consideration of why the equations come out the way they do. The fact that the multiterm equations have a superficial resemblance to those of physics does not mean that they explain everything.

If such analyses are set up so that the goal appears to be maximizing the production of cubic volume per unit area, the equations will tend to indicate that thinning should be very light or not done at all. If the initial diameters of trees in even-aged, pure stands are used as variables to explain the subsequent growth of the trees, this will obscure the importance of other more fundamental variables. The reason for this is that such analysis will always show that the trees that have grown the best in the past are likely to continue to grow the best in the near future. The factors that caused the tree to grow well are obscured. The resulting equation is mostly an extrapolation of the past and it will apply well only with low thinning. The purpose of these warnings is not to deprecate the mode of analysis but only to argue against uncritical acceptance of its results.

Formulation of Thinning Schedules

The general result of these considerations is that thinning schedules are based partly on the best biological, economic, and mathematical analysis available and partly on intuitive art designed to integrate these considerations. They are not etched in stone, and it is fortunate that stands that are still vigorous enough to thin are flexible enough to respond to changes in plans.

An approximation of a program of intensive, frequent thinnings is depicted in Fig. 4-15. It shows how the interval between thinnings would lengthen with advancing age. The most important line on this graph is that defining the residual basal area left after each thinning; in this case, it is shown as gradually increasing. If the thinnings were less frequent they would be heavier and stand basal area would fluctuate within a wider range.

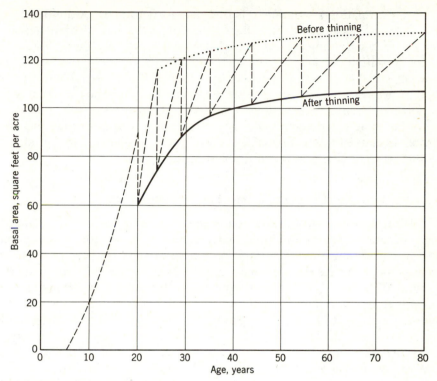

FIGURE 4-15 A hypothetical example showing how basal area should vary during a series of thinnings. The solid line (—————) indicates the lower guiding limit of basal area reserved after thinning. The dashed line (————) shows the fluctuations in actual basal area. The dotted line (. . . . . .) represents an approximation of the level to which basal area is allowed to increase before thinnings are repeated. The interval of time elapsing between thinnings lengthens with age because the rate of growth in basal area declines as the trees grow larger. The arbitrary assumption is made that, in this case, the basal area left after each thinning should increase gradually from the middle of the rotation onward.

The amount of standing tree vegetation, basically the amount of foliage, that should be reserved in thinning varies widely, depending on species, site, and management objectives. The rates of diameter growth that are sought will greatly influence the choices. Ordinarily this consideration will dictate that during the middle and later stages of a rotation stands will be thinned to residual basal areas somewhere between 40 and 175 ft.^{2} per acre or 9 and 40 m.^{2} per hectare. Appropriate levels probably depend heavily on the total amount of foliage that can be supported by a given species on a given site. This calls for higher residual stand density with evergreens than with deciduous species. The more shade tolerant the species, the greater is the logical amount. Good sites can carry more foliage than poor, so the amount left on them should be higher. On sites that are so poor that the root systems close but the crowns cannot, the stand density may have to be extraordinarily low.

In formulating a schedule it is first necessary to plan for the degree of initial crowding that will be required to control the size of branches and achieve other training effects in the early stages. Ideally there should be some definition of the stage at which further dying of branches should be slowed or halted. Ordinarily it is at this stage that it is desirable to thin as heavily as possible to maintain good diameter growth of the trees that were previously "trained" by deliberate crowding. When all of the trees have been brought to merchantable size, thinnings would usually become lighter because the creation of excessively large vacancies in the stands would be more likely to waste usable production.

Relation to Forest Uses Other than Timber Production

Thinning plays a role in the achievement of most purposes of management because it represents the primary means by which forest stands can be controlled and altered during the course of their development. It increases the water yield of forested watersheds by temporarily reducing the amount of foliage that transpires and intercepts water. The transpiration effect is the more important and universal one both in magnitude and also because it is most significant at the times when transpirational demands are greatest and stream-flow the least. Losses of water from interception by tree crowns result from direct evaporation of precipitation, especially snow and ice, such that some never reaches the soil. The importance once ascribed to interception losses was greatly diminished when it was learned that much temporarily intercepted water ultimately reaches the soil by flowing down the stems. It is seldom possible to increase water yield even temporarily by more than about 10 percent by thinning. Regeneration cuttings cause larger and less temporary increases, but they cannot be repeated as often.

Thinnings are also useful in enhancing the development and controlling the composition of understory vegetation that may provide forage, browse, and seeds for herbivorous animals, both wild and domestic. Tightly closed crown canopies usually keep this food supply above the reach of all animals but birds. The vegetation that is valuable from the standpoint of wildlife management and the grazing of domestic animals is sometimes the same as that which causes problems in the regeneration of species for timber production. Therefore, the understory vegetation must be manipulated with due regard for all the effects that is has on the economic use and overall ecological balance of the forest community. For example, attention must be given to the possibility that an understory layer ultimately harmful to reproduction and temporarily beneficial to wildlife may also lure browsing animals away from other stands actually being regenerated. The lesser vegetation is an integral part of the stand and should be regarded as something that can be usefully manipulated for either increase of benefits or reduction of harmful effects. No progress is made in this direction if it is looked upon only as an annoying impediment composed of plants unworthy of identification.

Thinning, like all forms of cutting, poses problems for intensive recreational use because slash is a far greater hazard and eyesore to the layman than it is to the forester. However, it is better to dispose of the slash than to refrain from thinning in areas of heavy recreational use. Trees must be vigorous if they are to endure the trampling of the soil by their admirers or the more active attacks of vandals. Reduction of the numbers of trees is inevitable and is better used as a means of stimulating the residual trees than as a last resort for protecting the public from injury by falling dead trees and branches.

Conclusion

Thinning represents the primary means by which the yield of stands can be increased beyond the best that might be achieved under purely natural conditions. As a consequence, it is the one technique that, more than any other, distinguishes intensive silvicultural practice from extensive.

It is possible that the essential generalities about thinning have been buried in the flood of qualifications that must inevitably enter into any discussion. The most important fact to be noted is that the total cubic volume of wood that can be produced by a given stand in a given length of time may be reduced but rarely increased by thinning. However, by skillful control of the growing stock the forester can marshal this production so as to capture the highest net return possible under the economic circumstances. The trees that are cut represent a salvageable surplus not needed to ensure full utilization of the growing space; they are, therefore, available for harvest. In the act of removing them it becomes possible to build more crown surface on selected individuals so that they will attain larger diameter, become more valuable, and be less vulnerable to damage.

The positive action of choosing the kind and number of trees to remain for future growth should be emphasized much more than the selection of trees to cut for immediate use.

BIBLIOGRAPHY

Abetz, P. 1982. Zuwachsreaktion von Z-Bäumen and Durchforstungsansätze bei Auslesedurchforstung in Fichtenbeständen. (Growth of crop trees and application of thinning treatments using the selective thinning method in Norway spruce stands.) *Allgemeine Forstzeitchrift* **47**:1444–1446, 1448–1450. (Engl. sum.)

Assmann, E. 1970. *The principles of forest yield study.* Translated by S. H. Gardiner and P. W. Davis. Pergamon Press, Oxford. 506 pp.

Bailey, R. L., and K. D. Ware. 1983. Compatible basal-area growth and yield model for thinned and unthinned stands. *Can. J. For. Res.* **13**:563–571.

Bennett, F. A. 1980. Growth and yield in natural stands of slash pine and suggested management alternatives. *USFS Res. Paper* SE-211. 8 pp.

Berg, A. B. 1976. *Managing young forests in the Douglas-fir region, 1973 symposium proceedings.* Vol. 5. Oreg. State Univ., School of Forestry, Corvallis. 147 pp.

Bramble, W. C., and W. R. Byrnes. 1953. Effect of time of shearing upon adventitious bud formation and shoot growth of red pine grown for Christmas trees. *Penn. AES Progress Report* 91. 6 pp.

Brown, J. H. 1960. Fall and winter pruning of pines in West Virginia. *W. Va. AES Current Report* 26. 7 pp.

Brown, K. M., and F. R. Clarke (Eds.) 1981. *Forecasting stand dynamics.* Lakehead Univ., Thunder Bay, Ont. 261 pp.

Burger, H. 1919–1953. Holz, Blattmenge und Zuwachs. I–XIII. *Mitt d. Schweizer. Anstalt f. d. forstliche Versuchswesen* **15**:243–292, **19**:21–72, **20**:101–114, **21**:307–348, **24**:7–103, **25**:211–279, 435–493, **26**:419–468, **27**:247–286, **28**:109–156, **29**:38–130.

Cannell, M. G. R. (Ed.) 1982. *World forest biomass and primary production data.* Academic, New York. 400 pp.

Cannell, M. G. R., and F. T. Last. (Eds.) 1976. *Tree physiology and yield improvement.* Academic, London. 567 pp.

Chapman, A. G., and R. D. Wray. 1979. *Christmas trees for pleasure and profit.* Rutgers Univ. Press, New Brunswick, N. J. 212 pp.

Clutter, J. L., and E. P. Jones, Jr. 1980. Prediction of growth after thinning in old-field slash pine plantations. *USFS Res. Paper* SE-217. 14 pp.

Drew, T. J., and J. W. Flewelling. 1977. Some Japanese theories of yield–density relationships and their application to Monterey pine plantations. *For. Sci.* **23**:517–534.

Duvingneaud, P. 1971. Productivity of world ecosystems. *UNESCO Ecology and Conservation Series* No. 4. 707 pp.

Fries, J., H. E. Burkhardt, and T. A. Max (Eds.) 1978. *Growth models for long-term forecasting of timber yields, IUFRO Proc.* School of Forestry and Wildlife Resources, Va. Polytechnic Inst. and State Univ., Blacksburg. 249 pp.

Fujimori, T. 1971. Primary production of a young *Tsuga heterophylla* and some speculation about biomass of forest communities on the Oregon Coast. *USFS Res. Paper* PNW-123. 11 pp.

Gingrich, S. F. 1967. Measuring and evaluating stocking and stand density in upland hardwood forests in the Central States. *For. Sci.* **13**:38–53.

Gingrich, S. F. 1971. Management of young and intermediate stands of upland hardwoods. *USFS Res. Paper* NE-195. 26 pp.

Grey, G. W., and F. J. Deneke. 1978. *Urban forestry.* Wiley, New York. 279 pp.

Hallé, F., R. A. A. Oldemann, and P. B. Tomlinson. 1978. *Tropical trees and forests, an architectural analysis.* Springer, New York. 411 pp.

Haygreen, J. G., and J. L. Bowyer. 1982. *Forest products and wood science: an introduction.* Iowa State Univ. Press, Ames. 495 pp.

Holsoe, T. 1950. Profitable tree forms of yellowpoplar. *W. Va. AES Bul.* 341. 23 pp.

Jacobs, M. R. 1939. A study of the effect of sway on trees. *Australia, Commonwealth For. Bur. Bul.* 26. 17 pp.

Larson, P. J. 1963. Stem form development of forest trees. *For. Sci. Monogr.* 5. 42 pp.

Long, J. N., and F. W. Smith. 1984. Relation between size and density in developing stands: a description and possible mechanisms. *For. Ecol. and Mgmt.* **7**:191–206.

Macdonald, J. A. B. 1961. The simple rules of the "Scottish electric" thinning method. *Oxford Univ. For. Soc. J.* **5**(9):6–11.

Mar:Møller, C. M. 1947. The effect of thinning, age, and site on foliage, increment, and loss of dry matter. *J. For.* **45**:393–404.

Mar:Møller, C. M., *et al.* 1954. Thinning problems and practices in Denmark. *State Univ. N. Y., Coll. For., Tech. Publ.* 76.

Mar:Møller, C., D. Müller, and J. Nielson. 1954. Graphic presentation of dry matter production in European beech. *Forstl. Forsøgsvaesen i Danmark* **21**:327–335.

Mitchell, K. J. 1975. Dynamics and simulated yield of Douglas-fir. *For. Sci. Monogr.* 17. 39 pp.

Nelson, T. C., *et al.* 1961. Merchantable cubic-foot volume growth in natural loblolly pine stands. *Southeastern For. Exp. Sta., Sta. Paper* 127. 12 pp.

Oliver, C. D., and M. D. Murray. 1983. Stand structure, thinning prescriptions, and density indexes in a Douglas-fir thinning study, Western Washington, U.S.A. *Can. J. For. Res.* **13**:126–136.

Ovington, J. D. 1956. The form, weights and productivity of tree species grown in close stands. *New Phytologist* **55**:289–388.

Ovington, J. D. 1957. Dry-matter production by *Pinus sylvestris* L. *Annals of Botany* **21** (n.s.):287–314.

Ovington, J. D., and W. H. Pearsall. 1956. Production ecology. II. Estimates of average production by trees. *Oikos* **7**:202–205.

Reinecke, L. H. 1933. Perfecting a stand-density index for even-aged forests. *J. Agr. Res.* **46**:627–638.

Reukema, D. L. 1961. Response of individual Douglas-firs to release. *PNWFRES Res. Note* 208. 5 pp.

Reukema, D. L., and D. Bruce. 1977. Effects of thinning on yield of Douglas-fir: concepts and some estimates obtained by simulation. *USFS Gen. Tech. Rept.* PNW-58. 36 pp.

Sassaman, R. E., J. W. Barrett, and A. D. Twombly. 1977. Financial precommercial thinning guides for northwest ponderosa pine. *USFS Res. Paper.* PNW-226. 27 pp.

Satoo, T., and H. A. I. Madgwick. 1982. *Forest biomass.* Nijhoff/Junk, The Hague. 152 pp.

Schlesinger, R. C., and D. T. Funk. 1977. Manager's handbook for black walnut. *USFS Gen. Tech. Rept.* NC-38. 22 pp.

Shigo, A. L., *et al.* 1979. Internal defects associated with pruned and non-pruned branch stubs in black walnut. *USFS Res. Paper* NE-440. 27 pp.

Tomlinson, P. B., and M. H. Zimmermann. (Eds.) 1978. *Tropical trees as living systems.* Cambridge Univ. Press, London. 675 pp.

Wahlenberg, W. G. 1955. Six thinnings in a 56-year-old pure white pine plantation at Biltmore. *J. For.* **54**:331–339.

Waring, R. H., K. Newman, and J. Bell. 1981. Efficiency of tree crowns and stem-wood production at different canopy leaf densities. *Forestry* **54**:129–137.

West, P. W. 1983. Comparison of stand density measures in even-aged regrowth eucalyptus forest of southern Tasmania. *Can. J. For. Res.* **13**:22–31.

Whittaker, R. H., and G. M. Woodwell. 1969. Structure, production, and diversity of the pine–oak forest at Brookhaven, New York. *J. Ecol.* **55**:155–174.

Wiley, K. N. 1968. Thinning of western hemlock: a literature review. *Weyerhaeuser Forestry Paper* 12. Centralia, Washington. 12 pp.

Wilson, B. F. 1970. *The growing tree.* Univ. Mass. Press, Amherst. 152 pp.

Wilson, F. G. 1955. Evaluation of three thinnings at Star Lake. *For. Sci.* **1**:227–231.

Woodwell, G. M., *et al.* 1978. The biota and the world ecosystem. *Science (Washington)* **199**:141–146.

Zimmermann, M. H., and C. L. Brown. 1971. *Trees, structure and function.* Springer, New York. 505 pp.

CHAPTER 5

Release Operations and Herbicides

Release operations free young stands of desirable trees, not past the sapling stage, from the competition of undesirable trees that threaten to suppress them. Distinction may be made between two kinds of release operations, **cleaning** or **liberation**, which differ mainly as to the ages and sizes of trees eliminated.

In the release of most young stands, the operation is, in effect, an uncovering of them through elimination of overtopping trees. The basic objective is to give the trees that are released enough light and growing space to grow adequately and develop into trees of the main canopy. Sometimes whole layers of overtopping trees are removed; sometimes they are merely interrupted over especially desirable small trees. The degree of release sought depends on the method and cost of release, the minimum amount of vacant growing space that must be created, and other considerations.

Release cuttings are most readily visualized in terms of freeing the crowns of existing desirable trees, but there are several additional considerations. The growing space in the soil may be fully as important as that in the crown, particularly on dry sites. Even when the crown of a tree is fully released its growth may still be hampered by competition for moisture and nutrients with the root systems of adjacent trees. Some techniques used for releasing stands are very effective in killing the tops of competing trees but do not prevent sprouting if the roots remain alive. Entirely new vegetation of aggressive competitors may claim new vacancies faster than the released trees can expand to fill them.

As with any silvicultural treatment, release cuttings should be conducted with a clear understanding of the way in which all the vegetation of the site will develop after treatment. This is especially true of the release of young stands. There the desirable trees command such a small fraction of the growing space that they must expand considerably to preclude recapture of the site by other species. If the work is conducted with little fore-sight and much wishful thinking, discouraging developments are likely.

Before the consideration of techniques of release cutting, it is first logical to examine the ways in which plants can be killed and some charac-teristics of the chemicals that provide one of the ways.

Killing Vegetation in Place

The harvest of useful products is the cheapest way of killing trees to guide the development of forest vegetation so it is chosen whenever possi-ble. Some silvicultural treatments, especially those employed to mold stands when they are young and plastic, require killing trees and other plants that are not worth harvesting. The rest of this chapter is devoted to techniques of dealing with such vegetation, except that prescribed burning and mechanical methods employed in conjunction with preparing sites for regeneration are discussed in Chapter 8.

The techniques of killing unwanted vegetation depend on (1) whether the undesirable plants can reproduce by vegetative sprouting, (2) the de-gree and pattern of selectivity desired, and (3) the size and other character-istics of the stems of the unwanted plants. Virtually all of the ways of killing woody plants depend on killing the roots directly or indirectly. This was first learned by fur-clad scientists who girdled trees with stone axes to make way for primitive agriculture. Sprouting plants cannot be killed with-out killing the roots. Many angiosperms sprout and most conifers do not.

The surest and most direct way to kill a plant is to uproot it. This is still a dependable way of dealing with perennial grasses and other small plants that sprout profusely from the roots, although there are chemical tech-niques of killing such species. Sizable trees can be ripped out with some kinds of site preparation machinery (see Chapter 8).

Ordinary cutting merely decapitates the plant. It is effective against those that do not sprout but sometimes worse than useless against those that do. However, sprouting is somewhat reduced if cutting is done early in the growing season just after rapid formation of new tissues has de-pleted carbohydrate root reserves.

Girdling is the removal or killing of a ring of bark around the tree stem so that the flow of carbohydrates from crown to roots is blocked. The roots die and the whole tree is killed. Unfortunately girdling is almost as effec-tive as cutting in stimulating sprouting.

Fire kills plants by girdling and also causes sprouting. In some kinds of forests the vulnerability of different species and sizes of plants varies enough that prescribed burning can be used as a selective technique (see Chapter 8).

In the ways in which they are used to control woody perennials, herbicides often operate by killing the stem cambium and phloem thus producing a girdling effect somewhere between foliage and roots. As will be described later, herbicides kill woody plants mostly by killing the roots directly or by starving them.

Tactics

Two or more methods of tree killing, sequentially applied, are frequently more effective than one alone. For example, one highly effective sequence involves burning or some sort of cutting soon followed by herbicide spraying; the first treatment stimulates sprouting and the second kills them, preferably while they are still succulent and have yet to form new dormant buds. Another effective sequence starts with some treatment that covers an area rather completely and is followed by a different one concentrated on those localized or resistant plants that escaped the first treatment.

Emphasis on preventing sprouting does not mean that the goal is always complete killing of an undesirable plant. It may be if there is some reason to eliminate root competition quickly. Sometimes it suffices merely to cripple the undesirable vegetation so that the desirable can gain ascendency in competition. The objective is to *control* the living system of the forest; any killing is a means and not an end. It should also be noted that it is not practically feasible to achieve total eradication of any undesirable species from a stand or site. Eradication would require effort pursued beyond the point of diminishing returns, to say nothing of possible harm to the environment.

Cutting Small Trees and Shrubs

Using hand tools to cut small woody plants is so laborious that it is done mainly where labor is cheap or the amount of work is so small that it is not expedient to bring in chemicals or motorized equipment. The best tools for cutting very small stems are those that can be held in one hand, are well balanced, and have straight or concave cutting edges. The axe, with its convex edge, is a better tool for stems thicker than 5 cm.

Where muscle power is used, it ordinarily takes less time, effort, and risk of injury to inject or spray a chemical than to sever a stem even "with one quick stroke of an axe or power saw." This objection does not necessarily apply to brush-cutting saws powered by motors of the kind that the operator can carry.

It should also be noted that one can use tractor-mounted devices with heavy, horizontally mounted, rotating blades, similar to rotary lawnmowers, to thin young stands by cutting narrow swaths through them. The same can be done by crushing trees with rolling brush-cutting drums or by shearing them off with the kinds of bladed equipment described in Chapter 8.

The main objection to cutting is that it causes much resprouting of the very sizes and species of woody plants that one is likely to be trying to kill. The most important exceptions exist where nonsprouters, such as most conifers, are to be eliminated. The selectivity of herbicides has not been improved to the point where broadcast foliage spraying can be used to favor one conifer over another. Therefore, either cutting or some sort of chemical treatment of single stems is necessary for release operations in young conifer stands.

HERBICIDES

Toxicity of Herbicides and Other Pesticides

The pesticides used for killing the woody weeds of the forest are called **herbicides** because of their wider use against the herbaceous weeds of agriculture. Since they are pesticides, about which there is always concern, the forester must know as much as possible about the nature, use, and toxicity of the compounds. There are many more comprehensive accounts of these matters such as those of Newton and Knight (1981), Weed Science Society of America (1984), Klingman Ashton, and Noordhoff (1982), U. S. Forest Service (1984), and Walstad and Dost (1984).

Most problems with pesticides arise from two categories of ignorance about their chemistry and biological effects. One form is hysteria on the part of the public and the other is heedlessness of stupidly complacent users. Education is the best solution to both problems. The public is not required to educate itself but users are. Any forester or farmer who is going to use pesticides, other than those safe enough for anyone to purchase, must pass an examination to demonstrate enough knowledge to get an applicator's license. The manuals that licensing agencies prepare to help users qualify are often excellent sources of information about the necessary precautions.

Registration of Herbicides

In most countries, no pesticide can be marketed until a governmental regulatory agency has "registered" or approved it for some specified kind of use. To get a compound registered the developer must show that many different kinds of tests have been reliably performed to demonstrate its efficacy and patterns of toxicity as well as whether it is safe enough to use. The label on the container is a document with the force of law that must be read and followed by the user. Among other things it tells the chemical name and concentration of the compound, necessary precautions, allowable uses, and detailed instructions about matters such as dilution rates and modes of application. Sample labels are commonly distributed by manufacturers and are convenient and reliable sources of information.

The greatest hazards with legal pesticides are to the people who handle and apply them. If they protect themselves properly and also confine the pesticides to the places where they are supposed to be applied, the risks of other damage are reduced.

There is also an unplanned safeguard that reduces the general hazard from pesticides used in forestry. The tests of new compounds now required are very costly. The agricultural market is generally regarded as the only one large enough to allow recovery of the large investments involved. Therefore, the chemical industry tends to avoid compounds so toxic that they have no prospect of use on food plants. While this greatly reduces the potential for problems with forestry pesticides, it has the ironic effect of paralyzing the development of pesticides that are narrowly and ideally specific for any particular forest pest.

Quantitative Toxicology

Regardless of whether one is concerned about target or nontarget organisms, it is necessary to know some elementary toxicology. The popular fear of "chemicals" ignores the simple fact that all substances are chemicals. Furthermore, all chemicals, natural or synthetic, can be poisonous to any organism if administered in sufficient amount by some route that reaches a vulnerable tissue. For example, air-breathing mammals are killed by high liquid dosages of the most common oxide of hydrogen if it is administered to the lungs, that is, they drown. No substance is absolutely safe or nontoxic and none is absolutely poisonous, except possibly for certain radioactive and carcinogenic substances about which there is both legal and scientific debate.

The best way to interpret matters involving the toxicity of registered pesticides is in terms of the kind of curve (Fig. 5-1) that depicts the basic relationship between *dosage* of a compound and biological *response* over the whole range of possible dosages. In the case shown, dosage is that of some

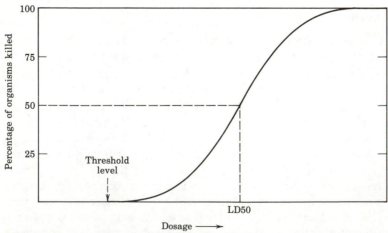

FIGURE 5-1 The typical form of the sigmoid curve by which the dosage of a biologically active substance is related to the response of a population of organisms.

pesticide and response is the percentage of mortality in a series of populations of one species exposed to the various dosages. Below a certain "threshold" or "no-response" level, there is no mortality at all; in fact, it is even possible that a compound might have a beneficial effect at some dosage much below the threshold level.

The most important thing to note is that the dosage–response curve is asymptotic at both ends. This is mainly because there is variation within the tested populations such that some individuals are much more resistant than others, although other factors contribute to this kind of statistical variance.

The asymptotic nature of both ends of the curve warns of at least two things. First, it is impractical, expensive, and environmentally unwise to seek 100% or even 99.8% mortality of target organisms by increasing the dosage. It is much better to be content with something like a 96% kill and either tolerate the survivors or attack them later by some other means. Second, it can be presumed that a curve of the same shape applies to humans and similar organisms so its asymptotic nature near the no-response level becomes important. Biological variance, risks of accidents, and the possibility of hazards still unknown are factors so important that the maximum allowable dosage for these organisms should be much less than the no-response level. For example, in the extreme case of human food, the legal "tolerance" or maximum allowable amount of a given pesticide is 1% of the highest dosage at which no harm is observed in laboratory test animals.

It is also necessary to distinguish between **acute** toxicity, which involves single exposures and short-term effects, and **chronic** toxicity, which is associated with repeated exposures and long-term effects. Pesticide registration requires determinations of both kinds. Recognition of chronic toxicity serves not only to guard against danger but also as a possible way of using pesticides more efficiently. Two or more applications at low dosages may kill more target organisms than a single high-dosage application. This is mostly because not all organisms are simultaneously at the same stage of development and vulnerability. However, a nontarget organism that is not damaged by one exposure may suffer from several. If the dosage–response curve of Fig. 5-1 is for acute toxicity, one for chronic toxicity with repeated exposures would lie far to the left and perhaps be steeper.

The sigmoid curve form of Fig. 5-1 is useful for revealing the nature of the dosage–response relationship but could not be plotted without a very large amount of data for low and high dosages. Such curves can usually be converted to straight lines if the dosages are expressed as logarithms and the responses as probits. Probits are statistics that convert cumulative normal-distribution curves to straight lines. Since straight lines can be extrapolated dependably, this kind of statistical transformation enables the dosages of the threshold and 99.9% response levels to be predicted with reliability from moderate amounts of data covering the range of responses.

The abbreviation LD50 denotes the dosage that kills 50% of the test organisms; it is the most common expression of the relative toxicity of different chemicals. It must, of course, refer to the organism used in the tests; most tests rest on such inferences as the assumption that minnows react like trout, beans like alders, and people like rats. The expression LC50 means "lethal concentration, 50%" and might be used to denote the concentration of some substance in water that killed 50% of the minnows in it.

Fate of Herbicides

Another important toxicological consideration is the degree of *persistence* of a pesticide in the environment. The ideal pesticide would be one that decomposed into such substances as water, carbon dioxide, and chloride ions after it had done its work. Most modern pesticides break down from the actions of sunlight, water, and soil microorganisms. The best of them are, in other words, biodegradable. There are, for example, bacteria that can decompose 2,4-D by using it as an energy source for respiration.

Since life processes require water, it is desirable that pesticides have enough affinity for water that living organisms can react with them chemically. DDT and other chlorinated hydrocarbon insecticides were banned basically because their insolubility in water caused them to be very persistent. Their solubility in oils allowed them to accumulate in potentially active form in the fats of animals and to be concentrated as they were passed along the food chain to eagles and other carnivorous birds. The threatened extinction of these organisms through the thinning of egg shells served as an early-warning symptom of harm that might result from long-term accumulation in people and other animals.

Most evidence indicates that fire is more likely to break herbicides down into simple oxidation products than to convert them into dangerous compounds.

While risks to nontarget plants must be considered, herbicides (and fungicides) are generally less dangerous to animals, including humans, than insecticides and rodenticides. This is because herbicides have been selected for toxicity to green plants, which have physiological processes quite different from those of insects and other animals. Potential dangers are likely to lie as much with additives, diluents, and contaminants (such as dioxin in 2,4,5-T) as in the herbicides themselves. It is prudent to regard any substance, whether it be a pesticide, tree sap, or printing ink, as being possibly more hazardous than it is known to be.

Occasionally it is possible to use tactics that improve the effectiveness of low dosages of herbicides. Two low-dosage treatments done in succession can be more effective than applying the same amount all at once. The application of two different herbicides may be more effective than a large total dose of one alone; such effects are **synergistic** in that the effect of one reinforces that of the other. The combination of different measures, such as herbicide treatment and fire, can be synergistic. However, it should not be presumed that combinations are always more effective than single treatments.

Modes of Entry and
Movement of Herbicides in Woody Plants

Some herbicides can be effective if sprayed on intact leaves and bark but some must be injected through the bark. Others can be applied to the soil and absorbed by the roots. Many of these used to control herbaceous weeds in agriculture kill only those plants with which they come in contact or render the soil surface lethal to the roots of germinating seedlings. These agricultural herbicides, of which there are very many, are used in forest tree nurseries and sometimes to prepare spots for tree planting but are otherwise not useful in silviculture. The killing of woody plants depends on herbicides that can move within them.

In practical application, the main concern is with the physical properties that determine how herbicides enter plants and move after entry (Ashton and Crafts, 1981). Most herbicides will kill the appropriate vital tissues once they get to them; the problem is causing them to get there. Also crucial are those properties that may govern the degree of differential killing or selectivity among species that can be achieved with carefully developed techniques of broadcast spraying.

Penetration

The surfaces of land plants are almost always covered with waxes or other lipoids that serve to reduce transpiration. The most important of these materials are the cutin of leaf surfaces and the suberin that impregnates bark. The structure and thickness of the cutinous coverings of leaves vary tremendously between species and in different parts of the leaf. Electron microscopy reveals that cutin consists of woolly strands of wax and is not a solid coating. Molecules that are small enough can pass through the gaps between the extrusions. The stomata or guard-cell coverings are also potential entry points, but it is not easy to get herbicides to move into leaves. The variability of their entry is indeed a possible way of getting them to kill some species and not others.

After it has suberized the bark of woody plants is practically impermeable to water or herbicides dissolved in it. However, the absorbing tissues of the roots offer an open pathway.

Water-soluble herbicides can be introduced into plants by injection through cuts in the bark or by root uptake from the soil, but only some of them can enter through the leaves and then under special circumstances. The waxy coverings of plants not only resist passage of water but also tend to prevent aqueous solutions from sticking to them; the droplets may instead ball up into little spheres that merely rest on the surfaces or roll off. If herbicidal materials are to stick to such surfaces and penetrate them they must be lipophilic or oily, which usually means that they are insoluble in water and do not ionize.

Translocation

After herbicidal materials have penetrated into plants they cannot move to vulnerable tissues unless they are water-soluble. Those that are not work only because they have been formulated such that they hydrolyze to become water-soluble. Some move readily throughout plants; these tend to be so lethal that it is difficult to use them for selective effects with broadcast foliar spraying. Such effects often involve a very complicated pathway typified by that which 2,4-D follows when sprayed on the leaves of woody plants.

This compound must move in the living phloem from the leaves to the roots and arrive there at times when the roots can be killed. The movement of the active substance is with the so-called mass flow of carbohydrates manufactured in the leaves and to whatever "sink" such movement is directed at a particular time. If the leaves themselves are still expanding, they are the "sinks" and the herbicide will merely remain in the leaves and probably kill them. If the shoots are elongating, the herbicide will move to them but not down to the roots. Therefore, application must be delayed until leaves and shoots have ceased their growth. Then the flow may be directed toward the roots, which are likely to be killed if there is enough soil moisture to cause them to be growing actively.

The compounds involved are most lethal to actively dividing meristematic tissues. This is why they can move in the living phloem but not block their own movement even though they may be killing the immediately adjacent cambial tissues as they move along. If they arrive in the roots when the meristems are not active, they seem to be diverted into the parenchymatous storage cells. This may prevent the desired root killing, although sometimes it is merely delayed until the roots become active again. The selective effects of foliar spraying sometimes result from the fact that this complicated pathway gets broken in some species but not others.

Formulation of Herbicides

The physical chemistry of herbicides varies substantially in ways that affect methods of application, movement within plants, and selectivity among species.

Highly water-soluble herbicides can be made to enter plants through the roots, gaps in the waxy armor of leaves, or cuts made in the bark. They are quite mobile within the plants. However, it is not easy to achieve selective effects with them unless they are applied to plants individually. Selective effects with foliar application are easier to achieve with herbicides that are less mobile and less soluble in water.

Oil-soluble materials can stick to the waxy surfaces of bark and leaves and penetrate such surfaces better than water-soluble ones. However, they must change to water-soluble form after penetration if they are to move to vulnerable tissues, especially if they are to move from the leaves to the roots. This difficulty of movement is, paradoxically, part of the basis of the selectivity among species that can sometimes be achieved by applying these compounds in broadcast foliage spraying.

Herbicides are usually made oil-soluble by converting them to esters. This is done by combining them with alcohols of sufficiently high molecular weight that their volatility is low. This reduces the risk that phytotoxic vapors will drift out of the target areas. (Water-soluble herbicides may drift as droplets but not as gases.)

Sometimes such materials are simply dissolved in oil, but the esters can also be modified to provide enough affinity for water to make them emulsifiable. In ordinary emulsions, droplets of oil (in which the herbicide is dissolved) are dispersed within a matrix of water. Emulsions are used mainly because water is much cheaper than oil and often available close to the site. If the active oil-soluble ingredient can be spread over the target surface to an adequate extent, one might as well use the cheapest carrier.

Use is sometimes made of "invert" emulsions; these are dispersions of water droplets in matrices of oil, similar to mayonnaise. The creamy color of the sprayed surfaces makes it easy for aerial applicators to guide themselves; the oil with the herbicide also comes in closer contact with the leaves than with oil-in-water emulsions. However, invert emulsions easily clog spray equipment.

There are herbicides of very low solubility that have some specialized uses in silviculture. Most of them are applied to soil surfaces, as in nurseries, to kill herbaceous weeds as they germinate. Some of them kill the foliage of plants on which they are applied. Such "contact-killers" can be useful in cases, as in killing annual weeds, where it is sufficient to kill the foliage and there is no need that the herbicide be translocated to the roots. Herbicides of low solubility are formulated as "wettable powders" or "flowable liquids," both of which are devised to be sprayed not as solutions but as suspensions in water. These compounds can also be applied as dusts, but the risk of drift is so great that any such use is best confined to nurseries.

The efficacy of these materials can also be modified with various additives normally made by the manufacturers. These may include wetting agents (often actually household detergents) that cause the sprays to spread over surfaces and stick to them better. Other additives may confer emulsifiability, improve solubility, or stabilize the spray mixtures. Sometimes highly water-soluble herbicides may be formulated as pellets or as coatings on sand grains so that they can be scattered in dry form and go into solution when it rains.

Water-soluble herbicides may have to be formulated in ways that neutralize acid reactions enough to reduce tendencies to corrode metal spraying devices, although it may help to use plastic devices for corrosive materials. The physical chemistry of the various additives is complex enough that applicators should follow the instructions on the container labels and not presume that they know more about the materials than do the manufacturer's chemists. Impromtu mixtures are not only illegal but can also leave intractable precipitates in spray tanks.

Methods of Applying Herbicides

Stem Injection

If stems are more than about 5 cm in diameter it is convenient to kill them individually by injecting *water-soluble* herbicides into their xylem. If the herbicides are highly mobile and phytotoxic, they act as systemic poisons and kill many parts of the plant. However, the problem with this is that most of their movement is upward with the sap stream in the xylem and not downward to the roots. If the herbicide could be depended upon to move throughout the plant, it would be necessary to inject it at only one point.

Because this effect is either not attainable or is undependable, the herbicide is usually injected at several points around each stem or even in a complete frill-girdle. The purpose of this is to achieve a kind of chemical girdling in which the cambium and phloem are killed in a ring around the stem; this killing may be from a combination of the effects of both cutting and poisoning of tissues. If the injections are made in bark incisions that have spaces between them, the sideways movement of herbicides through pits in the sides of the conductive elements of the xylem may cause complete girdling at some place higher on the stem than the point of incision. However, with any sort of girdling effect, mechanical or chemical, it must be noted that it will not be effective unless it takes place *below* all of the foliage.

If spaced incisions are used, they are usually made at intervals of 7–15 cm around the circumference of the trees being farther apart with the larger trees. However, it is sometimes stated that the spaces between the cut ends of incisions should not be more than 5 cm wide. Injection into continuous incisions around the stem may be necessary during the dormant season or in diffuse-porous angiosperms.

The simplest way of doing the work is to make the cuts with a hatchet held in one hand and to put about a milliliter of solution into each cut with a squirt oilcan held in the other hand. There are faster devices that make the cut and apply the solution simultaneously; these can be "hypohatchets" or devices such as that shown in Fig. 5-2. They are more expensive and difficult to maintain but can still be the most economical equipment to use.

There are circumstances under which downward movement can occur in the xylem after stem injection. If the herbicide can be applied almost

FIGURE 5-2   A tree-injector being used to liberate a young stand of pitch pine in New Jersey. The valve of this injector is opened by pressing it against the stem of the tree above the incision; the chemical then squirts onto the blade and into the incision. (*Photograph by Cranco Co.*)

simultaneously with the cutting, the breakage of the sap columns, which are normally under tension, can suddenly pull some of the herbicide downward. Solutions can also diffuse very slowly downward in the xylem at night or at other times when transpiration is not actively pulling water upward. In fact, problems can arise if trees are connected by intra-specific root grafts and one tree pulls the herbicide up from the roots of another. This phenomenon, called "flash-back," is most common when the herbicide has been injected into some small, overtopped tree and pulled into another by its large, actively transpiring crown. The best way of avoiding this is to make the injections higher on the stems and at times when transpiration is active. It is also a reason not to poison small, feeble trees in precommercial thinning; this is usually a time-consuming waste of effort anyhow.

It is sometimes observed that basal sprouting is less common when herbicides are applied in spaced incisions than in complete frill-girdles, but it is not clear why.

This technique has the highest degree of selectivity of any. It can be used at any time of year when it is warm enough to allow the solutions to flow, although it is most effective during the growing season. It cannot be used at times when watery sap flows out of the cuts because that flushes the herbicide away. Although it is a common means of precommercial thinning and release cutting, it is too costly for killing undesirable plants that are small and numerous. One must also be able to walk under a stand to use this method of killing trees.

Stump-Surface Treatment

A closely related method involves application of water-soluble herbicides to the surfaces of freshly cut stumps. In such cases the work must be done within a few minutes after the cuts are made because downward movement will be blocked by air bubbles or the swift formation of wound gums and resins. Motorized circular brush-cutting saws are available with attachments that apply herbicides almost simultaneously with the cutting.

The herbicides are best applied to rings around the cut surface that are as close to the cambium as possible; there are no tissues to be killed in the inner parts of the stump. All effects depend on slow downward diffusion of the herbicide. The only purpose of stump treatment is the prevention of sprouting; unless the bark is too thick, this can also be done by basal spraying as described below.

Basal Bark Treatment

Woody plants can be girdled by applying *oil* solutions of certain herbicides in continuous rings around the basal bark. The purpose is to kill not only the plant but also any basal buds from which sprouts might arise. Except for this, the solution might just as well be applied at some higher and more convenient height where the bark is thinner. Because of the difficulty of getting satisfactory penetration of thick bark, basal treatment is usually limited to control of shrubs and trees too small for stem injection.

Directions usually specify that the oil solutions be applied so that the entire circumference of the lower 15–30 cm of stem surface is wetted to the point of runoff. This is to ensure that enough runs down over the buried root-collar surfaces to kill the buds beneath them; it is not actually necessary to kill such a wide ring of cambium. There is no fundamental reason to *spray* the solution; cheap devices, often with split nozzles, that cause it to flow under gravity in fine streams that encircle stems can suffice. If there is no concern about sprouting, the encircling solutions can be applied at a convenient height with paintbrushes.

Sometimes effects similar to basal spraying result from deliberate spraying of oil-soluble herbicides on the upper stem surfaces of woody plants. This is associated with and will be considered with foliage spraying.

This treatment can also be used to prevent stumps from sprouting, if the bark is thin enough. In this case the oil solutions are applied to the sides of the stumps; there is no reason to put them on the cut surfaces as there would be with water-soluble herbicides.

The main advantages of basal bark treatments are that they can be done with high selectivity at any time of year and are not likely to cause leaves to die and turn to an unsightly brown color. This is because the girdling effect starves the roots slowly and scarcely any herbicide moves to the leaves. The main drawback is the high cost of materials and labor. As will be described in the next section, the localized application of newer herbicides to the soil at the bases of sprouting shrubs and trees is often a superior technique, but it does not prevent "brown-outs" of the foliage.

Soil Application

Some herbicides are sufficiently water-soluble, phytotoxic, and mobile within plants that they can be applied to the soil and taken up by the roots of the unwanted vegetation. They are distinctly different from some relatively immobile, "preemergent" herbicides of agriculture; those kill the emerging rootlets of germinating herbaceous weeds by sterilizing the soil surface.

If soil herbicides are to reach deeper roots and be absorbed by them in lethal quantities, the powerful effect of the soil in diluting and immobilizing them must be overcome. The first 6 inches of an acre of forest soil weighs about 1.6 million pounds so 1 pound per acre of herbicide mixed into it would be diluted to 0.625 ppm. Furthermore, a major portion of that might be adsorbed on the tremendous surface area of mineral and organic colloids in the soil. It is for this reason that only highly mobile, very phytotoxic compounds can be used. Suitable herbicides must be very low in toxicity to animals and not excessively persistent or mobile in soils.

Soil herbicides are applied either as water solutions or as pellets in which they are mixed with inert ingredients. Either form can be applied directly and selectively at the bases of the plants to be killed. Convenient squirt guns are available to spray metered amounts of solution on the bases of small trees or clumps of woody stems from distances of several meters. Pellets can also be deposited just above the roots. In each case the herbicide moves to the roots in water solution.

The pelletized forms can also be broadcast over whole areas for certain special purposes. If application is from the air, there is virtually no risk that wind will blow the material out of the target areas. No spraying equipment or liquid is necessary since rain is depended upon to dissolve the herbicide. This technique is suitable for achieving the nearly total kill of vegetation sometimes sought in preparing sites for regeneration, but aerial spraying is usually much cheaper.

Under the right circumstances selective effects can sometimes be achieved by broadcasting pelletized soil herbicides. One situation of this kind is where there is a good stocking of small seedlings that must be released from the cover of much larger undesirable trees with extensive root systems. If the desirable seedlings have narrow root systems and the pellets are scattered sparsely, the chance is great that the larger trees will be killed but adequate numbers of seedlings will escape.

Foliage Application

Herbicides can also be applied to the leaves in ways that enable them to be translocated to and kill vital tissues in the roots and stems of woody plants. The work is usually done with helicopters because that is faster, cheaper, and requires far less herbicide than spraying with mist-blowers from the ground. It is very important to do the spraying quickly because it must be closely synchronized with the development of the vegetation and the proper periods are short.

If spraying is done from above the foliage is covered much more uniformly than when it is done from beside or beneath. Aerial application generally requires lower volumes, higher concentrations, and smaller amounts of total active ingredients than ground application. However, there is less risk of drift beyond the target area with ground application. Because of this problem, fixed-wing aircraft are seldom used.

Foliage spraying is relatively uncomplicated if the purpose is to prepare sites for regeneration by killing most of the existing vegetation without need for discrimination. Several modern water-soluble herbicides are suitable but so phytotoxic and mobile within plants that species selectivity is not easy to achieve with broadcast application of them. Another uncomplicated kind are the herbicides that are not translocated and kill only the foliage to which they are applied. These "contact" herbicides are used mainly to kill grass and other herbaceous plants on spots where trees are to be planted.

The difficulties arise when the purpose is to kill a sufficient amount of undesirable vegetation without unacceptable harm to the desirable. There are various ways of obtaining such selectivity; none work in all cases and it takes much experimentation to determine which one, if any, will produce the desired results. Selective foliage spraying often, but not always, requires use of older herbicides such as 2,4-D, which must be in oil-soluble form if they are to stick to and penetrate leaves. The highly useful selectivity of these compounds seems to depend on their poor and variable mobility in the phloem of trees.

Differential wetting is the most common source of selectivity in foliage spraying. Most important is the simple fact that spray droplets are much more likely to fall off narrow leaves of conifers and grass than off the broad leaves of angiosperms. This is why foliage spraying is so commonly used to release conifers from hardwood overstories. Any physical properties of the spray materials or of the surfaces of leaves that alter the retention and penetration of herbicides may also alter the pattern of selectivity among species. The wooliness or waxiness of leaves can cause important differences; not enough advantage is taken of possibilities of modifying the physical chemistry of spray materials to change their ability to stick to and penetrate leaf surfaces.

Morphological selectivity results from interspecific differences in the location of vital meristems. Part of the resistance of grasses results from the fact that their intercalary meristems are thoroughly ensheathed. Woody plants do not differ much with respect to this kind of selectivity.

Biochemical selectivity begins to operate after a herbicide gets inside a plant. Some species have enzymatic systems that can inactivate a herbicide but others may convert an inactive form to an active one. Such phenomena are most common and useful with herbicides that are sufficiently similar to naturally occurring growth regulators that they operate as counterfeits and fatally derange normal processes. Members of the genera *Fraxinus* and *Acer* seem to inactivate several common herbicides and are thus resistant to them. On the other hand, some species can activate a herbicide and kill themselves while those that do not remain undamaged. Detection of useful patterns of biochemical selectivity depends mainly on trial and error so developments along this line are slow but also promising.

Phenological selectivity is possible if there are times when desirable species are significantly less susceptible to damage than the undesirable. The ordinary cases are ones in which undesirable plants are susceptible because they have vital tissues that are developing actively while those of the desirable species are not. For example, if soil moisture conditions are still good, some angiosperms remain vulnerable late in the growing season after conifers have become resistant because their shoots have "hardened" and formed buds. It takes close observation to disclose useful patterns. In applying the knowledge, one must be guided more by the phenological stages of plant development than by the calendar. The appropriate stage and time for treatment are not the same with all herbicides, species, or sites, so it is necessary to apply all information and experience available.

Finally, there is some possibility of **selective placement** of herbicides even with foliage spraying. If spraying can be done from the ground it can be aimed to a limited extent. Large, light containers also make easily movable temporary protective covers for seedlings. If spraying is done from the air there is the possibility of using an undesirable overstory to shield the desirable understory being released. Decisions should be made about whether helicopters should or should not be operated so as to drive the sprays down through crown canopies.

Techniques of Foliage Application

Aerial spraying can be done only by highly skilled and licensed aviators operating under detailed contracts. The work is usually done under nearly calm conditions early in the morning. The size of spray droplets is critical. If they are too fine they are apt to drift; if too coarse, the foliage may not be covered with adequate uniformity.

The boundaries of areas to be treated must be clearly visible from the air; maps alone are insufficient. Prominent features such as roads, ridges, and edges of clearcut areas should be used as much as possible but it is usually necessary to have other markers as well. Flags or other markers placed on trees must project above them to be visible. Balloons can be used but smoke drifts too much. It may also help to mark some edge trees in advance by injecting them with some highly mobile herbicide that will kill their foliage. During the operation it is useful to have workers with flags moving along to designate the ends of treated swaths.

It is so costly to move in aircraft for aerial spraying projects that operations are feasible only if there are at least 200 hectares to be treated at once. One helicopter does about 25 hectares an hour. The individual treatment areas must be at least 2 hectares.

If smaller areas must be treated it is best to do the work from the ground. There are mist-blowers that can be carried on one's back; they can cover foliage to heights of only about 4 m. More powerful equipment can be mounted on tractors. Spraying from the ground is especially effective as a means of controlling understory vegetation underneath stands of trees that might be damaged by foliage spraying. This kind of vegetation control is done to free growing space either for regeneration or, where soil moisture is limiting, to increase the growth of the overstory trees.

Small spraying devices similar to those used in gardens can be used for small-scale operations such as applications of contact herbicides to grass in spots where trees are about to be planted. With this or any other kind of equipment it is necessary to guard against contamination that may cause problems when it is used for other pesticides. Mixtures of herbicides and insecticides may kill crop plants as well as their insect pests. Physically incompatible mixtures may also clog spray equipment. Sometimes the equipment is so hard to clean completely that it can be used only for one kind of pesticide.

Herbicides can also be applied to low vegetation with moplike devices that are used to wipe the materials over the leaves. This kind of equipment, mounted on tractors, can be used to kill strips of grass or similar vegetation prior to planting or between rows of planted trees. Placement of the herbicide can be very accurate and there is no risk of drift, but the method cannot be used on tall plants.

Stem or Stem-Foliage Spraying

In foliage spraying it is almost inevitable that the surfaces of small branches as well as those of leaves will be coated. If the herbicide is oil-soluble this can add significantly to the phytotoxic effect. In fact, it is sometimes advantageous to treat the stems of small deciduous trees when they are leafless. This kind of "dormant" spraying can, for example, be used in spring and fall to release Douglas-fir, hemlock, and spruce from red alder and other angiosperms in the Pacific Northwest (Newton and Knight, 1981). The effectiveness of the treatment appears to depend on what amounts to a large-scale girdling of virtually all conductive tissues at the top of the plant as well as wholesale killing of dormant buds.

The great advantage of foliage spraying in general and aerial spraying in particular is that they enable quick, large-scale release operations. One hour of aerial spraying can accomplish as much as months of hand brush-cutting and prevent resprouting as well. As with all techniques it can be done unwisely and unsafely. Among other precautions, it must be borne in mind that the purpose of foliage spraying is to kill the roots; killing leaves is incidental and the work may fail if they fall off before the herbicide moves out of them.

Herbicide Compounds

As has been the case with all pesticides, improvements in herbicides have led in the direction of ones that are not only more effective but also safer. Developments continue and it should not be presumed that understanding of the compounds or their use is easily obtained. This chapter merely introduces the subject. Better understanding can be obtained from Newton and Knight (1981), Ashton and Crafts (1981), Buchel (1983), Garner and Harvey, (1984), and other sources such as the *Herbicide Handbook* issued by the Weed Science Society of America (1984).

Herbicide compounds have such complicated chemical names and structures that handier *generic* names are devised for the convenience of users. For example, *N*-(phosphonomethyl)glycine is called **glyphosate,** a word that is handy but should not be construed as telling anything about chemical structure or physiological activity. The trademark name Roundup has been applied to this herbicide and registered by the Monsanto Agricultural Products Co. Trademark names must always be capitalized and are owned by the registrants. The generic names are best for exchange of information so there will be no other use of trademark names in this chapter.

The first herbicides successfully used in silviculture had comparatively simple structure and were water-soluble. **Sodium arsenite**, which contained the exceedingly toxic trivalent form of arsenic, was so poisonous to all organisms and dangerous to handle that it has gone out of use. **Ammonium sulfamate** (generic name, **AMS**), developed in the 1930s, is remarkably safe to use because it decomposes to ammonium sulfate, a nitrogen fertilizer. However, it has to be applied in large dosages and is so corrosive of metal spray equipment that its use is costly and difficult. Sometimes it is still used as a foliage spray to control vegetation close to sources of drinking water.

Chlorophenoxy Herbicides

The first modern herbicides were developed in the late 1940s when the pseudo-auxin **2,4-D** (for 2,4-dichlorophenoxyacetic acid) came into agricultural use (Bovey and Young, 1980). The oil-soluble esters of the chlorophenoxy herbicides proved to be very useful for selective killing of broad-leaved woody plants in releasing conifers by broadcast foliage spraying. This type of release depends on herbicides entering the leaves without killing them, converting to water-soluble form, and then moving to the roots with the sugars at times when the roots are vulnerable. This works only if the spraying is done when the soil is moist enough that the roots are growing but the time of year has come when the shoots are not expanding and drawing sugars from the leaves. It is also often important to delay treatment until some time late in the growing season when the desirable conifers have formed dormant buds but when root growth still continues.

A slightly more complicated compound, **2,4,5-T,** was soon found to be more effective on woody plants because it was slightly less toxic and less likely to kill leaves or phloem and thus block its own movement to the roots. The esters of 2,4,5-T were widely used to release conifers from hardwood brush until it was decided that a highly toxic contaminant dioxin compound, which can develop during the manufacturing process, made the compound unsafe to use.

The esters of 2,4-D and some more modern herbicides can be used as broadcast, selective foliage sprays but their higher phytotoxicity makes the selectivity harder to achieve than was the case with 2,4,5-T. They can also be used in basal spraying and treatment of dormant stems and buds. Certain amine salts of chlorophenoxy acids are soluble in water and are often applied by stem injection, although some newer water-soluble herbicides seem superior.

Chlorophenoxy herbicides act by deranging respiration, growth of tissues, and many other biochemical processes in plants. The animal toxicity of the compounds themselves is low and they are readily broken down by soil microorganisms. 2,4-D is comparatively low in price and widely available.

Other Auxin Herbicides

Triclopyr has many of the same properties as 2,4-D except that it is so much more phytotoxic and mobile within plants that selectivity is not easily achieved with foliage applications. It has a water-soluble amine form that, unlike that of 2,4-D, can be sprayed on the foliage; the timing of foliage spraying is about the same as that of the chlorophenoxy compounds. The amine can also be used for stem injection. An oil-soluble ester is available for basal spraying. Grasses are resistant and so are conifers in limited degree if dosages are low. Animal toxicity and persistence are low. Triclopyr is not as good as broad-spectrum plant killers for site preparation.

The most outstanding of those is **picloram**, which has amine salts that can be taken up by leaves or roots and esters that can be used for dormant sprays on stems. Both forms can be mixed with 2,4-D esters for applications on leaves or stems. The water-soluble amines, often mixed with amines of 2,4-D, are very effective in stem injection. Picloram is easily translocated within plants, which are so readily killed that no selective effects are sought from broadcast applications of sprays or pellets; these treatments are used chiefly in preparing sites for planting. Picloram is very low in animal toxicity but can persist for about a year, which is longer than the other herbicides now in silvicultural use.

Organic Arsenicals

Water-soluble arsenic compounds, **cacodylic acid** and **MSMA,** are used for stem injection, especially for precommercial thinning of conifers. These are pentavalent forms of arsenic that are toxic to plants but so low in animal toxicity that some of them are injected intravenously in humans to combat bacterial infections if antibiotics cannot be used. They move readily in the sap stream of plants and can form gases poisonous to bark beetles; this can help prevent trees killed in thinnings from becoming sources of beetle infestations.

Triazine Herbicides

Triazine herbicides disrupt photosynthesis and nitrogen metabolism. They vary greatly in their water solubility and their uses. They have the useful property of being toxic to grasses but not highly toxic to conifers.

Simazine is very low in solubility while **atrazine** is only slightly more soluble; they are applied as wettable powders suspended in water. They are rather immobile and kill only those plant tissues with which they come in contact. For this reason they are used as preemergence herbicides in nurseries; the herbicides stay at the surface and kill the emerging roots of germinating weed seeds. They can also be used as contact killers of grass and herbaceous weeds in spots being prepared for planting. Since they do not move readily into the soil, trees can be planted on the spots soon or immediately after the treatment.

A much more soluble triazine herbicide is **hexazinone.** It can be sprayed on the foliage at low dosages to release conifers from broad-leaved species although the main purpose of such application is in site preparation. Hexazinone is taken up by roots effectively enough that solutions of it can be sprayed at the bases of trees and shrubs to kill them; squirt guns are used for this kind of sharpshooting. It can also be applied in pellet form as was described under "Soil Application."

Phosphonalkyl Compounds

Phosphonalkyl compounds include **glyphosate** and **fosamine,** which are water-soluble and highly phytotoxic. Both can be taken up as water solutions by foliage and even by buds and twigs but they are inactivated in the soil so that they do not reach the roots by that pathway. However, they are so readily translocated to all parts of most plants that they can move from leaves to roots and thus kill them. Both are effective in the unselective kinds of foliage spraying often used to prepare sites for planting; selective effects can be achieved with directed foliage spraying. For complex reasons that have to do with their breakdown in actively photosynthesizing leaves, they are usually applied in late summer or autumn. Carefully prescribed broadcast foliage spraying with glyphosate at that time can sometimes produce selective killing of broad-leaved species to release conifers. Glyphosate can also be used to kill some deciduous plants if they are sprayed shortly before the buds burst. These compounds are very low in animal toxicity and swiftly decompose.

Special-Purpose Herbicides

Amitrole and **asulam** are water-soluble and used as foliage sprays. Asulam kills bracken fern and some herbaceous weeds such as those that are a problem in Christmas tree plantations. Amitrole kills a wider variety of woody and herbaceous plants and is sometimes used to release conifers, which are somewhat resistant to it.

Dalapon and **pronamide** are used to control grasses before or after establishment of conifer plantations. **Dinoseb** is used as a desiccant to kill foliage and dry it in preparation for prescribed burning; it kills the leaves so fast that it does not enter and kill other parts of the plant.

The development of new and better herbicides should be anticipated. *Furthermore, the information presented here about pesticides should not be construed as providing directions for the use of any of them.* Additional sources should be consulted and it is illegal not to read the container labels. Foresters who direct the application of many of the compounds must be specially licensed.

CLEANING

A **cleaning** is a cutting made in a stand, not past the sapling stage, to free the best trees from undesirable individuals of the same age that overtop them or are likely to do so. The need for cleaning arises in situations where the methods of cutting and site preparation employed during the regeneration period have created conditions favorable to undesirable as well as desirable plants.

The principal purpose is to regulate the composition of mixed stands for the advantage of the better species. Oftentimes the composition of the

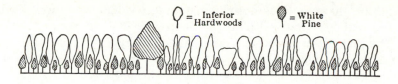

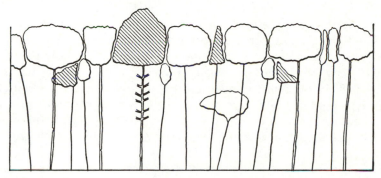

FIGURE 5-3 A stand of eastern white pine before (*top*) and after (*middle*) the removal of overtopping hardwoods in a cleaning. All of the overtopping stratum must be removed in this instance because the young hardwoods grow much faster than the pines. This treatment would best be done by complete foliage spraying, although the large pine would have to be cut or killed by other means. If left it would develop into a limby wolf tree. The lower sketch shows the stand as it would look 40 years later if there had been no treatment.

stand is radically changed as a result of these cuttings (Fig. 5-3). Cleanings are rarely essential in pure stands of reproduction, although they are sometimes carried out to remove trees of poorer form than those that are overtopped.

The term **weeding** is conventionally reserved for the more thorough removal of all plants competing with the crop species, regardless of whether their crowns are above, beside, or below those of the desirable trees.

Types of Vegetation Removed

The only plants that really need to be eliminated are those that are going to suppress, endanger, or hamper the growth of trees needed to constitute the new stand. The reduction of undesirable vegetation should not be so great that it causes even less desirable plants to invade growing space. If those trees that are truly wanted have been released in sufficient numbers, it is possible that remaining subordinate plants may have a beneficial effect in excluding very undesirable ones. It is a guiding principle of silviculture that one does not eliminate any vegetation without having a good idea about what will fill the vacated growing space.

The plants eliminated in cleanings may include not only overtopping trees of undesirable species but also poorly formed specimens of desirable species that are likely to compete with potential crop trees. It can be important to control climbing vines, overtopping shrubs, and sometimes rank herbaceous growth.

Sometimes young stands have scattered but isolated trees that are good but taller than those that are to form the main stand. If these are going to race so far ahead of the main crop that they will develop large branches and wide-spreading crowns the temptation to leave them should be rejected. If they represent a potential source of seed of some species that will be undesirable in the next rotation, it may be especially beneficial to get rid of them when they are small. However, if they are of good species and the trees of the main stand will catch up with them and restrict their crown development such trees can be left to form a useful component of the new stand. This situation is most likely to arise where the scattered trees are older than those of the main crop and of species that grow more slowly.

Cleanings are sometimes aimed at eliminating the alternate hosts or the favored food plants of dangerous pests of the main crop, such as species of *Ribes* adjacent to five-needled pines. Diseased or insect-infested trees can be removed if such action will really reduce the risk that desirable trees will be attacked.

Timing of Cleanings

If cleanings are needed they are best done even earlier than precommercial thinning, although there is no fundamental reason why the two operations cannot be combined.

In humid tropical climates it is often necessary to do the cleanings during the first year, perhaps even several times in the first year as well as afterwards. In climates with less luxuriant growth, cleanings can be delayed for as much as a decade or two, especially if the desirable species are shade-tolerant.

It may be desirable to delay release operations while certain developments take place. Sometimes the taller vegetation that may later hamper the development of desirable trees may act initially as a **nurse crop** by providing crucial protection against sun, frost, insects, or other damaging agencies. Occasionally nurse crops may choke out even more undesirable plants or, if of leguminous species or *Alnus*, improve the supply of soil nitrogen.

It is also important to recognize that some kinds of pioneer vegetation may seem rampant and overwhelming in the early stages but then simply die off without really causing significant harm to the desirable seedlings beneath them. This is especially true of herbaceous annuals, which may flourish during the first year after some soil-baring disturbance and then vanish; such plants actually retire to the status of dormant seeds that can remain stored in the soil until another disturbance causes them to germinate.

Proper timing of release operations should be based on past observations of how the initial vegetation develops and how the constituent species behave. In fact, it may help to allow it to develop until the competitive status of different individuals and species begins to be asserted. This is especially true if the cleaning is to involve treatment of individual plants. If such treatment is involved, it can also be crucial to schedule the treatments before or after the new stand may be difficult to penetrate on foot. The best time may come when the undesirable vegetation is slightly more than waist high. Very small stems can be awkward to treat, but the difficulty and expense of treatment increase rapidly after they attain 5 cm D.B.H. The cost may decrease again when it is possible to walk under the stand, but the desirable plants may have been lost by then.

Occasionally it is advantageous to delay cleaning until the desirable trees have become tall enough to remain free of any regrowth of competing vegetation after treatment. It may also be necessary to delay until the trees being released have developed roots and conductive systems capable of supporting rapid height growth.

Cleaning by aerial foliage spraying can be done at almost any stage of stand development. If any protective effects of nurse crops are at an end, there is seldom any reason to delay treatment because access to the stand is not a problem.

Provided that their roots extend more than several centimeters below the surface and their tops require no more shading, seedlings are likely to accelerate in growth after release. It is true that shaded trees may have shade-leaves which suffer some injury and turn yellow when suddenly exposed to the sun. However, such injury is usually only temporary and when buds next produce new leaves they will be sun-leaves adapted to the new conditions. In other words, an alarming yellow color does not necessarily denote permanent injury. However, large saplings that have developed poor live crown ratios may succumb after release, presumably because they are unable to supply enough sugars for the accelerated respiration that results from increased solar radiation and temperature.

Any benefits from protective cover by overtopping vegetation are soon outweighed by reductions in growth. Continued suppression may cause the death of the better trees. These effects are the result of either direct competition or the mechanical effect of the overtopping trees in whipping buds from the twigs of the desirable trees.

Ordinarily it is most expedient to try to do all the cleaning necessary in one operation by releasing the desirable trees enough to allow them to grow satisfactorily until the time of the first thinning or even to the end of the rotation. There is no subtle advantage in gradual release if all protective effects have ceased. Repeated cleanings can have a high cost of administration and of moving workers and equipment to the stand.

Repeated cleaning is likely to be needed if the undesirable vegetation resprouts and resumes excessive competition with the desirable crop. One of the advantages of herbicides over cutting is the greater opportunity to prevent resprouting. As has been the case in agriculture, herbicides have obviated much of the menial labor that formerly had to be expended on weed control. Society is now seldom willing to pay the price that cleaning with hand cutting tools requires.

Cleaning can be used to favor shrubs and other lesser vegetation over trees. Shrub cover may be much more desirable than trees in areas such as roadsides, wild gardens, near electric transmission lines, and patches managed for wildlife food or cover. In fact, it may be desirable to develop dense cover of shrubs in such places in order to prevent the invasion of tree species. In some cases, shrubs may be better ground cover than grass because they exclude conifers as well as hardwoods.

Extent of Release in Cleaning

With broadcast foliage spraying the extent to which the desirable vegetation is released is controlled only by the selectivity of the chemical applied. Some control is possible with hand-directed foliage spraying but it is usually neither possible nor necessary that it be very precise.

Where it is necessary or desirable to resort to comparatively expensive treatment of individual stems there is rarely any justification for bringing through more trees than can reach merchantable size. Ordinarily it is sufficient to release no more than 150 trees per acre. This number will usually be ample to provide trees for both the final crop and one thinning, as well as a margin for losses. The release of single trees at an average interval of 17 feet will give 150 trees per acre. The less desirable trees left in the process serve to fill out the stand and prevent the favored trees from developing into wolf trees.

In determining how much to remove around each favored tree, one must know the relative rates at which both desirable and undesirable trees grow in width of crown and height. The most serious competition will come from the undesirable trees adjacent to the individual to be released. Trees that are farther away do not offer an immediate threat to the dominance of the favored tree but may eventually overtop it by the lateral expansion of their crowns. Therefore, it is usually best to think in terms of removing the trees that project upward into an inverted cone around the top of the tree to be favored (Fig. 5-4).

The magnitude of the angle represented by the cone should vary directly with the difference in rate of height growth between the released tree and its competitors as well as the extent to which the tree has been suppressed. If the desirable tree will grow faster than the weed trees, the

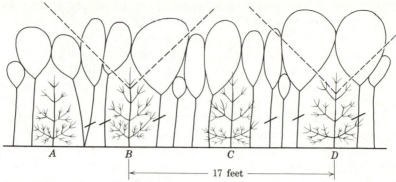

FIGURE 5-4 An example of how the intensity of cleaning might be regulated in freeing a stand of conifers from overtopping hardwoods. The objective chosen here is to remove all hardwoods that project upward into an inverted cone with an apex which is 2 feet below the tops of the crop trees and has an angle of 90°. The trees to be removed are marked by dashes. Trees *B* and *D*, which are spaced to give a stand of about 150 conifers per acre, are the only trees deliberately released. Trees A and C are ignored. In a heavier cleaning the angle would be increased or the apex of the imaginary cone lowered. The extent of release would depend on the relative growth rates of the different species present and the amount of time that would be allowed to elapse before another cleaning.

imaginary cone can be narrow, such that only the trees currently overtopping it are removed. More often the disparity in growth that caused the unfavorable condition to develop is likely to continue; in such cases the angle must be fairly broad.

Trees that have been overtopped for a long time require more drastic release than those that are still vigorous. No useful purpose is served by releasing small-crowned trees that have been suppressed for a long time, especially if they are of distinctly intolerant species.

The kind of partial reduction of competing vegetation that has been described does not necessarily cause the released trees to grow at rates anywhere near as great as they might if all competition were eliminated. This is especially true if the site is dry or infertile and root competition remains serious. The main object of such treatment is merely to assist the released trees in surmounting their competitors by subsequent growth. The only justifications for not releasing more thoroughly are the higher cost, any beneficial effect of moderate competition on the ultimate form of the crop trees, and the possibility that the otherwise unwanted vegetation will resist the invasion of even more undesirable plants.

LIBERATION OPERATIONS

Liberation, in the silvicultural sense, refers to the release of young stands, not past the sapling stage, from the competition of distinctly older, overtopping trees. The trees that are cut or killed in place are those left standing when the previous stand was harvested or, for other reasons, present long before the natural or artificial establishment of the young trees (Fig. 5-5). Liberation cuttings are somewhat like the removal cutting of the seed-tree and shelterwood methods of regeneration in which overstory trees are removed after the new crop is established. The trees removed in liberation were not, however, ones that had been left intentionally to provide seed, shelter, or additional growth.

Isolated trees have so much growing space that they often become stocky wolf trees with little clear length and wide, spreading crowns. The shade that they cast soon ceases to be beneficial and may cause the younger trees beneath them to die or become deformed (Fig. 5-6). Liberation operations can be conducted with most benefit and least damage to the new crop when the seedlings are not more than knee high.

Methods of Liberation

The undesirable trees can be eliminated by cutting, girdling, or chemical treatment. Cutting is, of course, the most advantageous method if the trees can be utilized at a profit or at a loss no greater than the cost of killing them. After an operation of this kind it may be necessary to cut away any parts of tops that interfere with the young growth. In considering treatments that do not involve utilization of the trees, it is important to decide

FIGURE 5-5   Unreleased (*left*) and released (*right*) parts of a 15-year-old plantation of red pine that had been established under slow-growing oaks on a deep sandy soil in Michigan, five years after herbicidal treatment of the plot on the right. (*Photographs by U. S. Forest Service.*)

FIGURE 5-6 The old post oak (*left of center*) was present when this 50-year-old stand of loblolly pine became established on an old field in North Carolina. The pine tree in the foreground and the trees in the right background have grown without interference from the wolf tree. An improvement cutting would be highly desirable in this stand, although it should have been conducted as a liberation cutting at least 40 years ago. (*Photograph by U. S. Forest Service.*)

whether markets are likely to improve sufficiently to make them merchantable in the near future. Destruction of sound trees of sawtimber size and quality, merely because they are of currently unwanted species, has often proven wasteful.

Cutting

The felling of large, undesirable trees without utilizing them is the most expensive form of liberation cutting. It may also result in far more damage to the young crop than would occur if the trees were killed and allowed to disintegrate in place. Therefore, such trees are felled, limbed, and left on the ground only where absolutely essential. This course of action is desirable where standing dead trees constitute a serious fire hazard. Cutting is also necessary along roads and in other situations where falling limbs and snags are likely to be a dangerous nuisance.

Girdling

The traditional method of killing trees without felling them is girdling. This involves severing the bark, cambium, and, sometimes, the sapwood in a ring extending entirely around the trunk of the tree. The materials already stored in the roots can still be carried up to the crown in the xylem, unless the sapwood has been interrupted. The roots die when the carbohydrates in them have been exhausted, a process that may take several years. If the sapwood is severed, the death of the crown is rendered more certain by the reduced flow of water, stored substances, and inorganic nutrients to it; the vigor of sprouting is, however, increased because the roots are not exhausted.

If any vertical strips of cambium are left intact, bridges of callus tissues are almost certain to form across the girdled ring. These bridges render the entire treatment ineffectual unless they are subsequently invaded by fungi. Even if the cambium is completely severed with a series of single incisions, callus tissue will often bridge over the gap. Therefore, it is desirable to cut out chips of bark and sapwood so that the cambium is actually removed in a visible ring rather than merely severed in a single cut. Trees that have deeply infolded bark around frost cracks or other deformities are almost impossible to kill by girdling. Mortality may also be reduced if the girdled trees are supported by substances transferred to them through root grafts by uninjured trees of the same species. For these reasons, girdling should not be expected to result in complete or immediate killing of all trees that are so treated.

Girdling can be done with axes or by cutting parallel, horizontal grooves through the bark several centimeters apart with chain saws or other cutting devices. Ordinary power saws are quite effective on large trees, especially because the blades can sometimes be poked into the stems deeply enough to sever overgrown inclusions of bark and cambium. Devices consisting merely of saw-chains with handles on both ends can be used on smaller trees.

The most common and effective method of girdling with axes is **double-hacking**. In this method a horizontal line of chips is removed by striking two downward blows; the second is made about 6 cm *above* the other so that the chip may be pried entirely out of the cut with a twist of the axe handle. With this technique it is easy to verify that no strands of inner bark and cambium remain.

In **single-hacking** or **frilling,** a single line of overlapping axe cuts is made through the bark. This is seldom effective by itself because the cuts usually heal. However, if herbicides are applied to the cuts this becomes one of the best alternatives to purely mechanical girdling; the cutting can be done swiftly and the chemical generally kills any unsevered tissues.

Notching involves cutting notched rings around the tree through the bark and a centimeter or more into the wood. This usually causes the quickest kill and the most basal sprouting because it is the method that most nearly halts the upward movement of water and nutrients. Unless the functional sapwood is deep, it has almost the same physiological effect as cutting the tree down entirely. It is cheap when done with motorized cutting devices. However, it is the most expensive, laborious, and dangerous way of axe-girdling because there are so many upward strokes; it can be used to sever ingrown folds of bark with axes but has little else to recommend it.

In **peeling,** strips of bark at least 20 cm wide are stripped off after continuous cuts are made at the top and bottom of the strips. Unless the bark is too thick or tough, this can be done quite effectively at those times when the cambium is highly active and the bark is "loose," as in spring and early summer in temperate climates. If the bark is "tight," however, it is not feasible. Peeling is the best technique from the standpoint of interrupting the phloem without halting the upward flow of stored substances from the roots. The reason for removing such wide strips is that the cambium sometimes can survive and regrow unless there is thorough exposure to desiccation.

Sometimes water solutions of highly phytotoxic and readily translocated herbicides are painted on surfaces exposed by peel-girdling to cause the bark to be shed from the entire stems of merchantable trees so that it will be left in the woods rather than removed in the mill.

The time and cost of girdling, whether mechanical or chemical, depends mainly on the amount of stem circumference treated so it can be predicted as some function of the sum of the diameters of the trees to be killed.

Girdling is much less effective than herbicides in preventing sprouting. This may be an advantage if one wishes to increase the supply of food for browsing herbivores. If sprouting is not wanted, it means that girdling is best reserved for nonsprouting species or stems too large to sprout. For trees of sawlog size girdling can be as effective as injecting herbicides. It can be done effectively on trees as small as 10 cm D.B.H., although herbicides are better for medium-sized trees. Below that diameter, cutting is easier, cheaper, and safer but herbicidal treatments are usually superior to either.

APPLICATION OF RELEASE OPERATIONS

Cleaning is usually associated with comparatively intensive silvicultural practice but the liberation of young stands from the remnants of old ones is a practice most characteristic of the early stages of new programs of long-term silviculture. Previous high-grading or use of land for grazing or agriculture can cause the development of stands cluttered with wolf trees or other undesirable specimens that encumber smaller and better trees. If the overtopped trees can respond vigorously and speedily to form stands of better quality, liberation is the quickest and cheapest silviculture method of making a silk purse out of a sow's ear (Fig. 5-7).

FIGURE 5-7 Stand of loblolly and shortleaf pine in Louisiana, approximately 5 years after liberation by the girdling of an overstory of oak. Most of the young pines were small seedlings at the time of release. (*Photograph by U. S. Forest Service.*)

The kind of trees eliminated in liberation operations have large crowns so that many small trees are released by killing a few big ones. For such reasons, liberation commands high priority in the early stages of intensifying silvicultural programs. If the practice comes to seem archaic, that is probably a sign of success. Liberation operations are commonly thought of as means of releasing conifers from hardwoods but they can play a very important role in refurbishing hardwood stands after their long and typical histories of high-grading.

The time between liberation and the reaping of benefits is shortest if the trees that are released are rather tall, although it is entirely possible for young, vigorous seedlings that have been released to overtake much older saplings that have suffered major reductions in live crown ratio.

In liberation operations, care should be taken not to replace high shade with more detrimental crown competition in the crown stratum occupied by the desirable seedlings or saplings. This can happen if the trees that are eliminated sprout profusely of if understory weed species profit more from the release than their desirable associates.

When cleaning could be done only by cutting, it was characteristic only of highly intensive silviculture. Aerial spraying with selective herbicides has made it feasible to contemplate release operations even for natural stands that may receive no other silvicultural treatment until the final harvest cuttings. However, the release of individual plants still belongs only with intensive practice.

Herbicides have increased the power to control species composition on many of those inherently good sites where so many species can grow that the better ones are mostly overwhelmed by the poor. In fact, there is a tendency to regard rather dry, mediocre sites as "best" simply because certain conifers or other desirable drought-resistant species can be maintained on them without excessive hindrance from many undesirable ones. Now the sites that are "best" in this sense may be found higher on the scale of true biological productivity. This has the important effect of extending the area on which highly productive and useful softwoods can be grown. However, selective placement of herbicides can also play an important role in developing good hardwood stands on good soils, especially by early elimination of malformed trees.

While improvements can be made in the composition of stands at almost any stage of a rotation, it is usually best to try to make them with early cleanings. When a stand is young it is much more plastic than after the sapling stage. The longer that undesirable trees dominate the more will be the loss suffered by suppressed, desirable ones. As time passes the more likely it is that any released stand will be irregular and full of gaps. Planted stands, in particular, represent such a large investment that it is folly not to conduct cleanings in them if necessary.

It is generally easier, although less necessary, to conduct the kind of release that accelerates plant succession than that which retards or reverses it. For example, it is much harder to free eastern white pine from red maple and oaks than from pioneers such as gray birch. The pioneers can usually be counted upon to die after several decades if one is ready to wait that long; they often cast such light shade that they do not entirely arrest the growth of more shade-tolerant species beneath them.

Unless they can be done by selective broadcast aerial spraying, cleanings are so costly that it is important to reduce the need for them as much as possible *before* stands are established. Efforts should be made throughout the rotation to forestall problems before they arise. Thinning should not, for example, unwittingly be made so heavy that they allow establishment of advance growth of undesirable species. The simplest means of eliminating an undesirable species is to get rid of the seed source; cleaning is the hardest way.

The silvicultural operations that cause new stands to become established naturally or artificially can seldom be conducted so perfectly that composition can be governed entirely by site preparation or skillful treatment of the previous stand. Consequently it must be anticipated that there will be many cases in which releasing operations will be desirable and will set the stage for subsequent rewarding efforts.

BIBLIOGRAPHY

Ashton, F. M., and A. S. Crafts. 1981. *Modes of action of herbicides*. 2nd ed. Wiley, New York. 525 pp.

Balmer, W. E., K. A. Utz, and O. G. Langdon. 1978. Financial returns from cultural work in natural loblolly pine stands. *South. J. Appl. For.* **2**:111–117.

Bovey, R. W., and A. L. Young. 1980. *The science of 2,4,5-T and associated phenoxy herbicides*. Wiley, New York. 462 pp.

Buchel, K. H. 1983. *Chemistry of pesticides*. Wiley, New York. 518 pp.

Garner, W. Y., and J. Harvey, Jr. (Eds.) 1984. *Chemical and biological controls in forestry*. Amer. Chem. Soc. Symp. Series 238. 406 pp.

Gratkowski, H. 1975. Silvicultural use of herbicides in Pacific Northwest forests. *USFS Gen. Tech. Rept.* PNW-37. 44 pp.

Gratkowski, H. 1978. Herbicides for shrub and weed control in western Oregon. *USFS Gen. Tech. Rept.* PNW-77. 48 pp.

Holt, H. A., and B. F. Fischer (Eds.) 1981. *Weed control in forest management*. Purdue Univ., Dept. of For. and Nat. Resources. 305 pp.

Klingman, G. C., F. M. Ashton, and L. J. Noordhoff. 1982. *Weed science, principles and practices*. 2nd ed. Wiley, New York. 449 pp.

Newton, M., and F. B. Knight. 1981. *Handbook of weed and insect control chemicals for forest resource managers*. Timber Press, Beaverton, Oregon. 213 pp.

Stewart, R. E., L. L. Gross, and B. H. Honkala. 1984. Effects of competing vegetation on forest trees: a bibliography with abstracts. *USFS Gen. Tech. Rept.* WO-43. 1 vol.

U.S. Forest Service. 1981. Forest management chemicals. *USDA, Agr. Hbk.* 585. 512 pp.

U.S. Forest Service. 1984. Pesticide background statements. Vol. 1. Herbicides. *USDA Agr. Hbk.* 633. 850 pp.

Walstad, J. D., and F. N. Dost. 1984. Health risks of herbicides in forestry: a review of the scientific record. *Oregon State Univ. For. Res. Lab., Spl. Pub. 10.* 60 pp.

Weed Science Society of America. 1984. *Herbicide handbook*. 5th ed. 515 pp.

CHAPTER 6

Improvement and Salvage Cuttings; Control of Operations

IMPROVEMENT CUTTINGS

Improvement cuttings are made in stands *past the sapling stage* for the purpose of improving composition and quality by removing trees of undesirable species, form, or condition from the main canopy. The unsatisfactory conditions corrected by improvement cuttings are generally those that might have been avoided if cleanings and liberation cuttings had been made earlier in the life of the stands.

Improvement cuttings may be applied in stands of almost any combination of species and age classes. Those applied to even-aged stands are nearly identical with selection thinnings; in both kinds of cutting, dominant trees are removed to favor better trees in the subordinate crown classes. If the species removed from an even-aged aggregation are essentially the same as those favored, the operation may be regarded as selection thinning and as improvement cutting if they differ. Improvement cuttings are more easily recognized when applied to stands of irregular age distribution. A cutting designed to free good trees, which have grown beyond the sapling stage, from the competition of *older*, undesirable trees is clearly an improvement cutting, provided that there is no definite effort to make way for new reproduction.

Improvement cuttings are often conducted simultaneously with true thinnings or reproduction cuttings. Improvement cuttings are preliminary operations designed only to set the stage for systematic thinnings or reproduction cuttings. They are not carried out continuously throughout a rotation and should not be regarded as permanent substitutes for the standard methods of cutting.

In spite of the difficulty of defining improvement cuttings, the situations in which they are needed are easily recognized in the field (Fig. 6-1) and their application is simple. It is easy to choose the trees to be removed.

The trees designated for elimination in improvement cuttings may be

FIGURE 6-1   A stand of bottomland hardwoods in Mississippi with defective trees, the remnants of an age class older than the main stand, marked to be killed or harvested for fuelwood in an improvement cutting. (*Photograph by U. S. Forest Service.*)

harvested in conventional fashion or merely killed and left standing. If they can be converted into useful products at a cost no greater than that of killing them, it is usually desirable to do so. The trees involved are generally of sufficient size to be suitable for cordwood or sawlogs. However, they are not necessarily of good enough quality and may be either too scattered for efficient utilization or too numerous for the markets to absorb. If there is no economically sound prospect of utilizing trees, they can be killed by the techniques described in the discussion of liberation cuttings (Chapter 5).

Where improvement cuttings are desirable there is no reason to delay them unless it be to let the trees grow to merchantable size. Neither is there much reason to restrict the rate of harvest to achieve sustained yield; to do so might imply that one wanted a continuing yield of poor trees. The only limitation on the rate of cutting of such trees is the amount that the market can absorb. The markets so often set the limit that it can be logical to cover a large amount of territory with light improvement cuttings taking out only the poorest trees. This can be better than squandering the available quota on heavier and more thorough improvement cuttings on more limited area.

Improvement cuttings should precede final harvest by enough years so that adequate benefit can be obtained from release of the better trees. One improvement cutting may suffice to put the stand into condition for subsequent thinning or regeneration cutting. About the only reason to do improvement cuttings in two or more stages would be some need to avoid making gaps so large that regeneration started prematurely.

If a stand has only a few trees likely to increase in value from added increment, it can be best to replace the whole stand by natural or artificial regeneration. If natural reproduction is to be sought in such cases, the best seed sources may well be the very same malformed dominants of desirable species that might have been removed in true improvement cuttings.

The importance of improvement cuttings should not be gauged by the amount of space devoted here to their discussion. They represent one of the most common practices employed in previously unmanaged or poorly managed stands. These are still common in the eastern United States, especially in hardwoods. There are, in fact, times and places where improvement cutting is temporarily so nearly synonymous with sound practice that it obscures the possibility that any other kind of cutting might become advisable.

SALVAGE CUTTINGS

Salvage cuttings are made for the primary purpose of removing trees that have been or are in imminent danger of being killed or damaged by injurious agencies other than competition between trees.

The kind of salvage cutting easiest to visualize is that aimed at capturing the highly perishable values in trees that are seriously damaged, dying, or already dead. A more sophisticated variant, sometimes called **presalvage cutting**, is that designed to anticipate damage by removing highly vulnerable trees. **Sanitation cuttings** involve the elimination of trees that have been attacked or appear in imminent danger of attack by dangerous insects and fungi in order to prevent these pests from spreading to other trees. Such cuttings are not necessarily confined to the removal of merchantable trees.

The removal of dead or dying trees from the subordinate crown classes constitutes thinning rather than salvage cutting, regardless of the nature of the agency ultimately responsible for mortality. The initiating cause of death of such trees is competition, and the agency that finally eliminates them is usually one of natural selection rather than a potential source of damage to the main stand.

Economic Role of Salvage Cutting

The recovery of timber values that might otherwise be lost is one important and expeditious silvicultural means of securing yields greater than those available from the managed forest. Damage from fungi, insects, fire, wind, and other agencies occurs almost continually. While there are ways of reducing the losses it is nearly impossible to prevent damage entirely.

The objective of salvage cuttings, as the name indicates, is to utilize the injured trees with the idea of minimizing the financial loss. They are not conducted unless the material taken out will at least pay for the expense of the operation, except in cases where real justification exists for true sanitation cuttings.

Even when the damaged material can be salvaged at a profit, the injuries that made the operation necessary ordinarily result in a loss of production. This is true especially when the damage occurs in the first half of the rotation. The loss is caused partly by deterioration of the injured trees before being salvaged, partly by reduction in stocking, and partly by the necessity of cutting the injured trees before they have reached optimum size. However, if the salvaged trees are parts of overmature stands that are not increasing in value, there is a subsequent gain of production to the extent that the vacancies are claimed by younger and more vigorous trees of desirable species.

The immediate financial loss depends largely on the extent and distribution of damage. If trees die sporadically and at widely scattered places in a stand, they may become a total loss because of the impracticability of harvesting them. The loss may be small if the amount of damage is not great and is concentrated in time and space. When catastrophic losses have occurred over a wide area, the returns from salvage cutting are often reduced by the necessity of selling the products on a glutted market. The costs of logging are generally higher in salvage cutting than in operations where the trees to be removed have been chosen by intention rather than by accident.

Simple Salvage Cutting

Ordinarily it is no problem to select the trees to be taken out in salvage cuttings. Sometimes, however, the mortality caused by the attack of a damaging agency does not take place immediately. This is particularly true where surface fires have occurred because the main cause of mortality is the girdling that results from killing of cambial tissues. As with other kinds of girdling, the top of the tree may remain alive until the stored materials in the roots are exhausted. It is usually a year or more before the majority of the mortality has occurred. By this time, those trees that were killed immediately have often deteriorated seriously. It is, therefore, advantageous to have some means of anticipating mortality before it has actually occurred. The predictions must be based on outward evidence of injury to the crown, roots, or stem.

The severity of a salvage cutting depends entirely on the proportion of the stand occupied by the damaged trees. If isolated trees remain undamaged it is best to harvest them if they are likely to succumb eventually or if it will become difficult to remove them after they are surrounded by young saplings. Occasionally they can be left for seed trees or for further growth, but only if such action is clearly justified.

Regeneration after salvage cuttings is often a difficult problem, especially if the residual stand is too young or too poorly distributed to provide a reliable source of seed. It is commonly necessary to resort to artificial regeneration. Unfortunately, prompt salvage cutting is frequently premature from the standpoint of obtaining natural reproduction because the damaged trees may bear large crops of seed before they die. As a consequence, the forester may be faced with the alternatives of serious loss through deterioration of damaged timber or heavy expense for planting. This dilemma can sometimes be solved by the temporary reservation of some of the less damaged trees as a source of seed.

Salvage cuttings should normally be completed as soon as possible after mortality or injury has occurred. Dead trees generally start to deteriorate rapidly during the first growing season after death, so it is usually advisable to get the trees out of the woods before insects and fungi have become active in the spring.

Priorities in Salvage

Unfortunately there are many situations in which the amount of salvageable material is so great that it cannot be removed within a few months or a year, even if all other operations are suspended. Under such conditions it is highly desirable to know how long the dead or damaged trees are likely to remain sufficiently sound to be worth salvaging. Entomological and pathological investigations have made this information available for a number of different species (Boyce, 1961). With this type of knowledge it is possible to conduct large salvage operations in a systematic and efficient manner.

The amount of time allowable varies widely depending on the circumstances. The sapwood of virtually all species is highly perishable but the heartwood of the most durable may remain sound for many years. Ordinarily dead trees of small diameter become valueless long before large trees. Deterioration usually proceeds more rapidly on good sites than on poor. Differences in the rate of decay of various species are also significant. The basic objective should be to schedule salvage operations in different places in an order such that the value of timber saved will be at a maximum. This does not necessarily mean that the most valuable material should be salvaged first.

The harvesting of damaged timber is often more difficult and expensive than cutting in undamaged stands. This is especially true if large groups of trees have been broken off or uprooted by wind. In conducting salvage operations it is important to remember that the money lost through inefficient logging can easily exceed the potential values wasted through decay of unsalvaged timber. The extremes of both haste and procrastination should, therefore, be studiously avoided in salvage cutting.

It must be anticipated that salvage cuttings will be necessary from time to time in any forest. In stands that are extraordinarily susceptible to injury, the time and place of harvesting operations may be dictated largely by damaging agencies. It is also a good policy to expect some mortality after even the most well-conducted partial cuttings. If full advantage is to be taken of opportunities for salvage of merchantable material, it is essential that all parts of a forest be made accessible by a good system of roads. The gains in production from salvage cutting and thinning can help justify the road network.

Presalvage Cutting

Silvicultural practices are always guided to some extent by the intent of harvesting trees that are especially vulnerable to loss. Such anticipatory "presalvage" cutting has sometimes played an important role in putting decrepit old-growth stands under management, as on some forests in the western United States. The goal of developing the full range of age classes required for true sustained yield often requires that old, overmature stands be retained for decades until they are scheduled for replacement. The losses of high-value timber can be quite substantial during the waiting period, but these can be salvaged or forestalled by salvage and presalvage cuttings. The opportunity to engage in such races with the damaging agencies depends mainly on whether means of access to the stands can be developed.

Once access is gained to forests of this kind, the forthright replacement of the most decrepit stands should be vigorously prosecuted by means of true reproduction cuttings. The temptation to deal first with the closed stands that have not yet started to break up should be restrained and attention turned as much as possible to those that are already deteriorating. The purest kind of presalvage cuttings are applicable to an intermediate category of stands that cannot be scheduled for early replacement but are already subject to significant losses of scattered trees.

The same approach is desirable in stands, young or old, in which attacks by damaging agencies are so chronic that silviculture is ruled by them. Usually more complete command of the situation can ultimately be taken by shifting to species or age classes less vulnerable to damage. Sometimes there is no better solution than the indefinite continuation of regularly scheduled salvage cuttings as the basic program of silviculture. One of the most outstanding examples of this distressing state of affairs is to be found in those badly eroded areas of the Piedmont Plateau where shortleaf pine is badly affected by the little-leaf disease and replaceable only with species that also have disadvantages.

The rational conduct of presalvage cuttings depends on the identification of those trees that are likely to be lost and the estimation of the length of time that they may be expected to endure. This is a more difficult problem than determining how soon obviously damaged trees will die or how rapidly dead trees will deteriorate.

If the main source of anticipated damage is wind, ice, or other climatic agency, the mechanical structure of the trees and their position within the stand are the important criteria. Trees with asymmetrical crowns or previous injury from frozen precipitation are prone to additional damage from this source. In managed forests losses to fire are usually kept in check by normal means of fire control or by true salvage cutting when these fail. However, the removal of trees that have been chipped for naval stores is an example of presalvage cutting done in anticipation of the serious damage such trees might suffer in event of fire or beetle outbreaks.

When biotic agencies or physiological factors such as those related to site factors are the causes of anticipated losses, the vegetative vigor of the trees is usually the criterion employed in presalvage cutting. Trees of relatively high vigor are less vulnerable if attacked because they are likely to endure the amount of loss of vital tissues that would kill trees of low vigor. The relatively abundant pitch flow from trees of good vigor even actively resists attack by *Dendroctonus* bark beetles, which dictate much presalvage cutting. The presalvage of trees of low vigor is rendered all the more logical because they grow little in volume or value.

A number of tree classifications have been developed for the guidance of presalvage cutting (Hedden, Barras, and Coster, 1981). One of the best-known examples of such a classification is that developed by Keen for the interior ponderosa pine type of California and the Northwest (Miller and Keen, 1960). Vulnerability to loss from bark beetles is so closely correlated with growth rate and the various stages of the life cycle of ponderosa pine that this classification is suitable for selecting trees for cutting even where risk of beetle attack is not the main consideration.

Keen's classification (Fig. 6-2) is based on the two major factors of age and crown vigor. There are four age classes and four vigor classes within each age class, making a total of sixteen classes. The four age classes are termed: 1—young, 2—immature, 3—mature, and 4—overmature; they should be thought of as grouped by relative maturity rather than by any definite ranges of age. Actual age limits for the groups vary in different parts of the ponderosa pine region. Color and type of bark, total height, shape of top, characteristics of branches, and diameter are the chief external indications of maturity. The four crown vigor classes are: A—full vigor, B—good to fair vigor, C—fair to poor vigor, and D—very poor vigor. The size of the crown (length, width, and circumference), its density, and the shape of its top indicate the crown vigor and consequently the inherent capacity of the tree to grow and endure exposure to beetles.

Keen's classification was originally designed to provide a means of evaluating the risk that trees left after selection cutting would be attacked by beetles during a cutting cycle of approximately 30 years. While it is still useful for guidance of many kinds of partial cutting and as an example of a

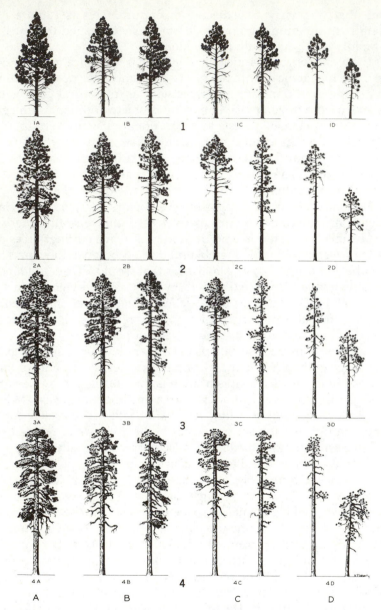

FIGURE 6-2 Keen's tree classification for ponderosa pine. The four age classes range from 1, the youngest, to 4, the oldest; the four crown vigor classes, from A, *the most vigorous, to* D, the poorest. (*Sketch by U. S. Forest Service.*)

tree classification, simpler means are now used to determine which trees are in imminent danger of death. This is because it is now possible to make much lighter presalvage and salvage cuttings at much more frequent intervals.

One potential advantage of presalvage, salvage, and improvement cuttings is that they may, in limited degree, reduce the proportion of genotypes that carry genes conferring vulnerability to the agencies causing the damage.

In the conduct of such cuttings it is important to consider the effects, good or bad, that they have in thinning the stands and making way for regeneration. It is not simply a case of finding trees to salvage.

Sanitation Cutting

These differ from other forms of salvage cuttings only to the extent that they are combined with or represent precautions to reduce the spread of damaging organisms to the residual stand. Sanitation cuttings may also be undertaken in anticipation of attack in attempts to forestall the establishment of damaging organisms entirely. They can be and usually are combined with salvage or presalvage cuttings; any cutting is a sanitation cutting to whatever extent it eliminates trees that are present or prospective sources of infection for insects or fungi that might attack other trees.

Sanitation cuttings are not effective and not worth conducting unless the characteristics of the stands and the organisms are such that the removal of susceptible trees will actually interrupt the life cycle of the organisms sufficiently to reduce their spread to other trees. The elimination of *Ribes* shrubs from stands of five-needled pines can be a moderately effective kind of sanitation cutting (or cleaning). However, the removal of pine already infected with the blister-rust fungus is not effective sanitation because the spores that carry *Cronartium ribicola* from pine to *Ribes* can travel so far that no practical advantage is gained.

The light partial cuttings often applied to deal with *Dendroctonus* beetles in ponderosa pine in the western interior are termed "sanitation–salvage cuttings" because they reduce the number of the very kind of feeble trees that sustain the beetle population.

It should not be tacitly assumed that the vigor of trees is the sole criterion of their susceptibility to attack, even though it may be a good indicator of their capacity to endure damage. Some economic pests of the forest, notably insects, multiply most rapidly on vigorous hosts just as grazing animals prefer forage from healthy plants. Fortunately the vast majority of insects attracted to thrifty trees are well-adapted parasites that do not endanger their own existence by killing their hosts; these insects rarely become cause for concern and may even escape notice. However, there are some important exceptions. In other instances, the presence of certain especially attractive species, regardless of their vigor, may determine whether a stand is threatened by pests that can harm more than one species.

The spruce budworm, one of the most dangerous insects of the North American forest, is an excellent example of an insect that is not merely a predator of the weak and also has a marked preference among species. The most typical form of this versatile insect is the scourge of the eastern portion of the spruce–fir forests, where massive outbreaks develop at intervals of several decades. It is poorly named because it is actually very dependent on and damaging to balsam fir. Spruces tend to be attacked when the supply of fir foliage has dwindled. The trees that are most susceptible to attack, but not especially high in vulnerability to loss, are large balsam firs. These bear abundant staminate flower buds which are highly nutritious for young budworms. Susceptibility to attack is determined more by the proportion of mature balsam fir in whole forests than by the characteristics of individual trees or stands. It can be reduced to a limited extent by replacing large tracts of mature forests with an intermingled arrangement of stands in which the susceptible mature stands are diluted among younger and less susceptible stands. Vulnerability to the losses that are likely to occur after a stand is attacked can be somewhat reduced by presalvage of firs and spruces with poor live crown ratios.

There are other defoliating insects that are governed more by the species composition of the forest than by the vigor of the trees. However, the wholesale changes in stand composition often necessary to reduce their depredations involve so many other considerations and procedures that they are not looked upon as sanitation cuttings. A sanitation cutting is usually thought of as an operation in which a few trees are cut in a stand almost exclusively with the object of safeguarding the much larger number that remain.

Sometimes sanitation cuttings must be associated with special measures to provide additional assurance that the damaging organisms will not spread to the residual stand. If the pests are exclusively parasitic on living tissues, it is sufficient to kill the infested trees and leave them in the woods. Sometimes it may help to hasten death by injecting organic arsenical herbicides or other pesticides into trees; this may not only shorten the time when pests can continue to feed on the trees but there may also be direct killing of bark beetles or blue-stain fungi.

If the insects or fungi involved are capable of multiplying as saprophytes in dead material, it may be best to utilize the wood even at a loss, provided that its transportation does not spread the infestation. This approach can be used to increased advantage against bark beetles if "trap" logs or trees are left temporarily to attract them and then hauled to mill or log pond before the emergence of the entrapped brood. If trees are killed with organic arsenical herbicides, the beetles may be attracted to them and then killed within the trees; if this is successful, the dead trees and beetles can be left in the woods. It may also help to burn slash or treat stumps with insecticides in ways that directly kill the beetles.

Those species of heart-rotting fungi that continue to produce spores from conks on fallen trees are not easy to control if infected logs must be left in the woods. However, the felling of infected trees does reduce the distance over which the spores can travel and the accelerated disintegration of the wood shortens the period of danger.

There are some agencies of damage against which sanitation cuttings are, for practical purposes, so ineffective as to be useless. Fungi that inhabit the soil and damage the roots of trees are not likely to be halted even by sanitation cuttings that are carried to the extent of removing the stumps. Some organisms spread so rapidly or over such long distances that sanitation cutting may be a meaningless gesture as far as the effective protection of the stand is concerned. It is for reasons of this sort that simple salvage cutting is more common than sanitation cutting.

Sometimes the salvage of entire stands may be looked upon as a desirable measure of sanitation even if the damaged stands by themselves are not worth the effort. Stands that have been badly damaged by fire, wind, or similar agencies frequently support the development of large populations of bark beetles that can cause serious injury to adjacent stands. The expense of salvaging such foci of infection may be amply rewarded by reduction of losses in adjacent stands.

In any kind of salvage or sanitation cutting in which the effects of insects or fungi are involved, there is no substitute for a thorough understanding of the life histories of the organisms involved. The details of a procedure useful in dealing with one kind of injury may be hopelessly ineffective or needlessly intensive as a means of reducing losses from other seemingly similar agencies of destruction.

CONTROL OF OPERATIONS

The success of silviculture heavily depends on the proper execution of the harvesting operations that are ordinarily used to carry them into effect. Cutting is the chief tool by which the forest is controlled; this is true even when the main objectives are the management of resources other than timber, such as water, forage, or wildlife.

The most important problems are encountered in the application of the numerous kinds of partial cutting that are so commonly essential to the efficient use of the capacity of the forest to produce wood and other benefits. These problems are changed in kind but not in magnitude when whole stands are cut in single operations.

The harvesting of trees for wood products is, from the silvicultural viewpoint, an indispensible means of controlling forest vegetation and, more than incidentally, an economic end in itself. The felling and removal of trees is the most costly operation that goes on in the forest. In most instances, the objective is to proceed so that the value of the products substantially exceeds the costs of harvesting and subsequent operations.

However, the opportunity to recoup some of the removal costs through conversion to wood products can make it possible to grow forests for aesthetic or similar reasons at endurable rather than ruinous expense.

If a tree is to be cut and used, there is nothing that can be done about the fact that its products must be moved over an essentially fixed distance from where it grew to the point of use. The best that can be done about this cost is to try to avoid incurring it while the tree is any smaller than it must be.

The most favorable balance between costs and returns in harvesting a stand are more likely to be achieved not by cutting all the trees but by removing the largest and best. Divergence between the long and short-term viewpoints arises mainly over the kind of trees cut and left but not so much over the total volume cut and left. Not uncommonly large trees of good quality, which are most attractive for immediate harvest, are of sufficient vigor that they will actually bring a better return if left to grow. Conversely there is a temptation to leave small or poor trees of kinds that are attractive for neither present harvest nor future growth but just impede the growth of anything that might be better.

From the silvicultural standpoint it is best to have all operations connected with harvesting, from the tree at least to the point where forest roads leave the forest, under the direction of the forester. The prevalent practice of putting harvesting responsibilities in the hands of purchasers or contractors, whose primary interest is necessarily in cheap logging, is not ideal.

Designation of Trees and Stands for Cutting

The trees or stands that are to be cut, left, or otherwise treated in harvesting or any other silvicultural operation can be designated by (1) specification in words, (2) physical marking, or (3) some combination of the two.

Cutting instructions specified in words are normally used under circumstances when it is either unnecessary or not economically feasible to make sophisticated distinctions between trees. The simplest specifications consist of nothing more than designation of the diameter limits or species to which cutting is to be confined. This approach usually gives results that are poor to mediocre from the silvicultural standpoint. The more the structure of a stand varies from point to point, the poorer are the results of diameter-limit cutting. It often leads to skipping dense patches of slender trees that ought to be thinned and overcutting where the trees are sparse and large.

Better results can be achieved by laying out more definite rules for cutting or by marking small plots to demonstrate the type of cutting desired. This method works best with well-trained personnel who understand the objectives and have no reason, economic or otherwise, to depart from the instructions. Advantage should be taken of opportunities to train woods personnel so that the simpler silvicultural operations can be conducted in this way.

The use of such general specifications must be associated with frequent and careful instructions, particularly at the beginning of each operation.

Tree Marking

A much closer degree of control, with less dependence on inspections, can be obtained by marking the trees that are to be removed or those to be left. The workers are responsible only for removing the designated trees with due care for the site and vegetation to be left. The timber marker bears a high proportion of the responsibility for the planning of both the logging and the future stand.

Ordinarily the trees to be cut are marked, because they are normally less numerous than those to be left. However, if the majority of the trees are to be cut, it is better to mark those that are to be left; this should be done without wounding the living portion of their bark. Trees to be left should be marked in the case of very heavy thinnings or those kinds of reproduction cutting in which only a few trees are reserved. This has the incidental advantage of focusing the attention of the timber marker on the residual growing stock. However, the cost of marking is great enough that it is always best to put the marks on whichever category of trees is in the minority. Where an area is to be cut clear, it is sufficient to mark the boundaries of the area to be so treated.

Sometimes it is possible to combine verbal specifications with partial marking. For example, there are circumstances in which it may be useful to reserve all trees below a certain diameter limit and those above that limit that are specifically marked. One may also specify the removal of all merchantable trees other than marked individuals of certain stated species. It may also be desirable to encircle certain patches of a stand for clearcutting but to limit cutting to marked trees outside those patches. The parsimony of such means of economizing on the marking effort must be balanced against the risk of causing confusion.

Mechanics of Marking

Many of the same marking procedures used for designating timber to be cut for utilization are also used to control pruning, release cutting, and a variety of other silvicultural treatments.

The marks themselves should be readily visible, difficult to counterfeit, and durable enough to last through the period of the projected operation. In selecting tools for doing the marking, the ease, rapidity, and expense of the work are important considerations. Among the implements used are axes, timber scribes, and paint guns.

Usually two marks are placed on each tree, one at a height where it is easily visible and another on a root swelling or so low on the trunk as to appear on the side of the cut stump. The mark on the stump enables one to determine whether the felling has been conducted according to the marking. The lower mark may be omitted when the felling crews can be depended upon to adhere to the marking.

Various schemes involving numbers and shapes of marks can be devised to stipulate such things as the direction in which a tree is to be felled, the product for which it is to be prepared, whether it is to be felled or killed, how high it should be pruned, or whether its utilization is at the discretion of the logger. Occasionally it may be desirable to put distinctive marks, as permanent as possible, on crop trees to expedite consistent treatment in marking for subsequent cuttings. Any such special marking schemes should be so devised as to eliminate the possibility of misinterpretation. The colors of paint chosen for marking trees and for such other purposes as designating property boundaries should follow some standard system.

Marking axes and timber scribes are used mainly where the amount of timber to be marked is too small to justify the advance preparations and transportation of materials necessary for the use of paint guns. Axes are also useful where much of the timber is likely to be rotten because they can be used to "sound" the trees to detect rot. Marking axes are sometimes fitted with special dies so that distinctive initials can be stamped on the blazes.

It is usually expeditious to use paint and paint guns rather than cutting tools because the chief cost of marking is that of time rather than of materials. The use of paint guns substantially reduces the time spent in making the marks and obviates the necessity of walking right up to each tree. No wounds are left and the tools are safer than axes or scribes. The only real problems lie in the cost and bother of dealing with paint and clogged paint guns.

The best way to mark a stand is to proceed back and forth across it systematically, completing one long, narrow strip on each trip. The marks should be placed so as to face the next unmarked strip, unless they are used to designate the direction of felling. If the strips are sufficiently narrow, one may even be able to mark one strip and simultaneously note the characteristics of the back sides of trees to be considered on the next strip. Sometimes it is necessary to tally the trees that are to be cut and, less often, those that are to be left. Tallying can be costly enough that careful consideration should be given to the value of the data gathered and to the question of whether efficient schemes of sampling can be devised to provide necessary information at lower cost.

Costs and Benefits of Marking

The cost of timber marking is usually, but not necessarily logically, figured as a charge per unit of product designated for harvest. It varies widely depending upon the size of the timber, the data gathered during the marking, the amount cut per acre, topography, density of under-growth, travel time, cost of materials, and the rate of pay of the personnel. It may be negligible where patches of large old-growth timber are being designated for clearcutting or it may be a major portion of the stumpage value when small pulpwood trees are marked.

The immediate benefits are the only ones properly charged against the cost of marking the timber that is removed. These values lie in things such as the information gathered about the quantity and quality of material being cut and the important opportunity to lay out efficient patterns for felling and extraction of the trees.

The essentially silvicultural benefits come mainly from increases in *future* production gained by exercise of discrimination among individual trees and conscious creation of vacancies suitable for establishment of re-production. The cost of these benefits is more logically charged against the acre or hectare, and the future yield thereof, than against the timber re-moved. Justification rarely exists for deciding that all partial cutting will or will not be controlled by tree marking. The seemingly costly marking of trees of low value in a young stand may actually yield more future benefit than marking in an old stand of valuable trees.

Guidance of Timber Markers

Professional foresters who have attained positions of considerable re-sponsibility are usually too expensive for routine timber marking. Ironi-cally they do less and less of it but become more and more concerned about how it is done. They become instructors and supervisors of timber mark-ing; instead of doing it themselves they may be able to do little more than prescribe the marking of a stand and inspect the results. This situation is an argument for trying to develop homogeneous stands, thus avoiding need-lessly complicated schemes of silviculture. Even-aged stands can be admin-istered more expeditiously than uneven-aged and the actual expense of marking them is lower because the work is more concentrated in space.

A forester who supervises timber marking must have personal profi-ciency and, above all, a clear idea of the ultimate objectives sought in the treatment of each kind of stand. These must be developed in the light of thorough understanding of local economic conditions, logging, and the ecology of all species in the stand. These ideas must be conveyed to the staff in such a manner that they can be depended upon to apply their judgment to attainment of the objectives. If the training is wisely done, the people who do the actual marking become more proficient than the super-vising forester, who should be prudent enough to recognize the fact.

Good judgment about timber marking is developed by observant ex-perience in a given locality. The best test of the wisdom of a particular marking project comes a number of years afterwards when the vegetation has adjusted itself to the conditions created by the cutting. The pressures of forest administration are so great that it is too easy to concentrate on marking for this year's cutting and forget to go back to inspect the results of the cutting of several years ago. It is also easy to become so obsessed with the imaginary virtues of an established marking policy as to be blind to any shortcomings that may be manifest in the results of its past application. The forester should take advantage of every opportunity to improve skill in marking timber for cutting and then observing the results as dispassionately as possible.

Within a given forest management organization or for a given forest it is usually necessary to have some means of standardizing practice among individuals and for various areas of the same forest type. There must be some sort of continuity and uniformity in the treatment of individual stands, management of growing stock, development of stand structure, and pattern of logging. These prescriptions or marking rules often describe the kinds of trees to be cut or to be fostered as candidates for the final crop. The use of tree classifications, sometimes with pictures depicting crown shapes, bark characteristics (Fig. 6-3), or other indicators of tree vigor, may be helpful. It is often especially important to specify the kind of stand that is to be left after partial cutting. Such rules should not only set forth the details of the marking itself but should also indicate what the objectives of stand manipulation are.

Useful as such rules may be, arbitrary insistence on conformity to them can inhibit the development of skill and good judgment on the part of the individual. The extent to which this is true depends on the intelligence and knowledge of the personnel, the nature of the rules, and the manner in which they are enforced. Furthermore, rules by themselves are no substitute for the use of demonstration plots, training sessions, and similar means of education and exchange of information among personnel.

Control of Waste and Damage in Logging

Some waste and destruction is inevitable in any logging operation and the work is so arduous that it is too easy to become inured to the damage. Constant effort is necessary to minimize it and to distinguish between the inevitable and the preventable.

Waste of presently merchantable material comes from (1) leaving unnecessarily high stumps, (2) failure to utilize merchantable portions of felled trees, especially in the tops, (3) breakage or other damage of stems in felling or skidding, and (4) failure to utilize suitable material in dead, dying, or partly defective trees. It may also be important to direct each part of a tree to its best use. Discriminating judgment must be exercised in determining standards of merchantability because these may vary with market fluctuations or even from the most to the least accessible parts of stands. These kinds of losses are simply evidence of inefficiency in logging and are best controlled by appropriate supervision and properly planned conduct of harvesting operations. Logging supervisors can normally be

FIGURE 6-3 Bark characteristics indicative of different rates of growth of southern red oak (from Burkle and Guttenberg, 1952), showing patterns typical of many species. As growth slows the individual bark plates become larger and the fissures between them deeper. A very fast-growing tree would have smooth bark composed of minute plates. Since the bark grows outward as the wood grows inward from the cambium, the outermost bark indicates the conditions of some years ago. The stem shown above are not necessarily of the same diameter. (*Photographs by U. S. Forest Service.*)

expected to cooperate in reducing waste of merchantable wood because it is in their interest to do so.

Greater administrative effort is necessary to curtail damage to desirable residual vegetation, to the soil, and to other values in which loggers have no immediate economic interest.

Protecting Vegetation

Destruction of residual vegetation is a most vital silvicultural concern in those kinds of cutting in which either trees or advance regeneration are left as sources of future production (Hamilton, 1976; Nyland *et al.*, 1976). If they interfere with such production, it may even be desirable that they be destroyed. Most damage to residual vegetation comes from the cutting, breakage, or crushing of trees that impede the felling and removal of the trees designated for cutting. The wounding of standing trees during felling and skidding is another important source of injury, especially because of the wood-rotting fungi that often invade exposed wood. Young trees are also cut for skids, corduroy roads, and other uses incidental to the harvesting operation. Supplies of such material can often be secured, with little if any added cost, from inferior species, tops, or cull material.

The greatest destruction of growing stock is likely to occur in stands of irregular form, where trees of various ages from small seedlings on up are intermingled. Greatest damage is likely to occur to large seedlings and small saplings that are tall enough to impede operations but not durable enough to resist breakage. Seedlings less than 2 feet in height are comparatively supple and are, therefore, more likely to escape destruction. Heavy cutting causes more damage than light.

It is difficult to be specific about the extent of damage that can be accepted without endangering future production in partially cut stands. Although losses are usually expressed in terms of the percentage of stems destroyed or injured, the most significant point is their distribution. For example, if each milacre in need of reproduction has one seedling the area may be fully stocked even though 70 percent of the total number were destroyed. On the other hand, a loss of 10 percent of the total might be so concentrated as to leave large openings entirely without trees. If a thinning is being made in a stand at a stage of development when the optimum spacing of the residual trees is 4 m, skidding equipment that leaves a swath of destruction 3 m wide can be used to some advantage with no damage of long-term consequence to the stand. However, it must be noted that root damage and soil compaction may make this path of destruction effectively wider than it superficially appears.

Damage to reserved trees or other desirable residual vegetation includes not only breakage but also the wounding of the bark, cambium, or even the wood of stems and roots. Pushing a tree into a leaning position so that it forms undesirable reaction wood is sometimes more damaging than killing it because it may become useless yet continues to usurp growing space. The same may be true of a tree so badly wounded or broken that it develops heart-rot or becomes malformed. The commonest damage to trees of pole size and larger is trunk wounding; this is most likely to occur when the cambium is actively dividing and the bark is knocked off easily (Shigo and Larson, 1969).

Damage from felling can be reduced by refraining from dropping trees upon advance regeneration or against trees that are to be left standing. If damage to young growth cannot be entirely avoided, the trees should be felled in such a pattern that the damage will be well distributed and have the effect of thinning the residual stand. Much difficulty can be forestalled by giving attention to the matter at the time of marking as well as during the operation itself.

In marking timber, it is important to consider each tree not only from the standpoint of its productive potentialities, but also with regard to the problems that might be encountered in felling it or one of its neighbors. *No tree should be marked for cutting unless a way can be seen, or created, to fell it without inconvenience or excessive damage to the residual stand.* It may then be necessary to supervise the operation closely or train the felling crews to ensure that they follow the plan embodied in the pattern of marking. Even then some unanticipated losses may occur. In such cases, provision should be made for salvaging any unmarked trees that are seriously damaged in the course of logging.

Little additional damage is likely to occur during limbing and bucking of felled trees, if the crews refrain from swamping out any young trees that do not actually impede the operation or constitute a source of danger to personnel. In logging heavy timber it is sometimes advantageous to finish the operation in two stages, removing the logs produced in the first stage before cutting the remaining marked trees. This practice has the favorable effect of avoiding congestion and confusion, thereby reducing damage to the residual stand, breakage of felled trees, and the amount of labor required for both limbing and bucking.

Protecting Soil and Water

Precautions should be taken to avoid the deposition of logging debris in streams or bodies of water. The resulting addition of organic matter may impair the quality of the water and harm the life in it by increasing the biological oxygen demand or through other effects of a chemical nature.

The greatest risk of damage to residual vegetation and virtually all of that of injury to the soil and downstream interests comes not from the felling of trees but from the subsequent transportation of logs and other products.

Physical injury to the soil is of consequence regardless of whether there is any residual vegetation to preserve. If it is serious its effects may last longer than damage to vegetation because trees regrow faster than the soil repairs itself. Cutting by itself rarely causes erosion or any other form of damage to the soil. Removal of the unincorporated organic matter, if not too extensive horizontally, may even be beneficial in baring mineral soil for establishment of light-seeded species, if they are desirable. However, deep gouging, rutting, compaction, or overturning of the mineral soil can lead to serious erosion, especially on steep slopes, on fine-textured soils, and in dry or cold regions where revegetation is slow. These effects impair the productivity of the areas involved but the resulting siltation of streams and reduction of water quality are often even more serious.

Damage may occur not only during skidding but also from construction of roads and landings. On very steep terrain heavy cuttings may bare as much as 15 to 25 percent of the soil in haul roads and skid trails. The area that is rather permanently removed from production by haul roads varies significantly depending on the kind of skidding equipment used; it is very high if most of the movement of logs within the cutting area is done after logs are loaded onto trucks. The disturbance of the soil on skid trails is mostly temporary and so much less serious than that on haul roads for trucks that it is desirable to accomplish as much movement of logs as economically possible in the skidding phase. Because of this it may be desirable to use fast or powerful skidders rather than those with a small economic operating radius; while the former may damage skid trails more, their use does not necessitate putting so much area into even more destructive truck roads.

Erosion along roads and skid trails can be reduced by locating them so as to avoid steep grades and places where extensive cutting and filling will be necessary. On steep slopes, skid trails that converge downward should be avoided as much as possible. The location of roads close to and parallel with streams increases siltation. Proper attention to drainage of roads not only reduces erosion but prolongs their usefulness. This applies not only to roads in current use but also to those that are not used between logging operations. If the drainage systems of temporarily abandoned roads cannot be cleared of debris periodically, they should be opened enough at the end of each period of use that they will function without attention.

It is frequently advantageous to suspend log transportation during periods when roads are muddy because the risk of creating conditions favorable to erosion is then at a maximum. Precautions of this kind often have the effect of reducing logging costs by preventing the damage and delay that occur when equipment gets stuck in the mud. Some readily accessible stands should be reserved for cutting in muddy weather.

The damage, waste, and unsightly appearance associated with road construction can be reduced if the work is done as a carefully coordinated job in advance of the main logging operation (Pearce and Stenzel, 1972). It is best that all trees and saplings in the way of the road are cut before any earth moving is done. Merchantable material should be converted to logs and put aside for removal when the road reaches the storage points. Burning or other disposition of unmerchantable material should be one of the first steps. It is often very advantageous to form the bed and side-ditches of a road some months before the final surfacing material is applied and the road is put to use. This gives the bed time to settle, become firmer, and thus less subject to the rutting and other difficulties that aggravate erosion and damage to equipment. The crucial consideration is planning the disposition of water. A forest road is really a water-control device designed to provide a cohesive but well-drained surface strong enough to support rolling vehicles.

Effect of Different Kinds of Equipment

Cable yarding with stationary engines is the kind of log transportation that does the least damage to the soil. There may be some gouging and scraping action by dragging logs but the heavy machines do not move over the soil and compact it. Some modes of cable yarding involve so much sideways shifting of the cables that almost all of any residual trees are knocked over; these are compatible only with clearcutting. On the other hand, with sky-line systems in which cables are suspended *above* the stand canopy it is sometimes possible to lift logs out of the stands vertically in thinning operations. With such equipment it is necessary to move the logs over well-defined corridors so any partial cutting must be modified accordingly.

There are some kinds of very light cable yarding equipment that move material *under* the crown canopy, usually to bunch logs for other skidding devices. This kind of cable yarding is very suitable for thinning operations. One significant advantage of cable yarding is that the machines do not burn fuel merely to pull themselves over the ground. In tractive skidding the power source expends many times more energy moving itself than in pulling its load.

Helicopter logging is very costly but it reduces road building as well as damage to soil and the residual stand. Balloon logging is, with respect to these problems, much like sky-line yarding except that the cableway is supported with a balloon rather than a tower. The use of these aerial systems in old-growth timber leaves no permanent transportation system. This can be a blessing for watershed management but a curse for any other form of management that requires good access to stands.

Skidding done with draft animals generally causes the least damage to residual trees and to the soil within the stand. However, because animals tire, skidding distances must be short and this usually increases the amount of area in roads, which are the chief source of soil and water damage. The main problems with animals are that they are slow and require much care.

Where heavy tractive machinery is used there is the choice between skidding out material as (1) logs bucked into short lengths, (2) whole delimbed stems, and (3) as whole trees minus the roots. The longer the lengths pulled out of the woods the greater the risk of damage to the residual stand and the greater the need to arrange for movement along straight lines. When whole trees are dragged out of the woods there is little gouging of the soil but there can be a sweeping of the litter that may or may not be desirable. The removal of small branches and foliage can have serious consequences for the nutrient capital of the site. The practice of doing the delimbing at the log landings and leaving their vicinity choked with debris just impairs the productivity of even more land.

Crawler tractors are much easier on the soil than rubber-tired devices because their weight is spread over much more area. Unfortunately they are slow and the maintenance of the metal tracks is costly. With any such machines it is silviculturally desirable that they not be any heavier or larger than is necessary. It can help to use one device for assembling batches of material to be hauled out of the stand with another kind of device, especially when it is advantageous to collect small stems into bundles. Certain kinds of feller-bunchers, especially those with long arms that can reach out between the trees and bring in trees from several meters away, can reduce the amount of ground that needs to be run over in partial cutting operations.

Many difficulties can be avoided if both timber markers and loggers plan for the felling and removal of trees. It usually helps to arrange for logs to be moved out in straight lines. If it is necessary to pull them around curves, it must be anticipated that any adjacent trees will be damaged; they can either be protected with buffers or simply harvested as a final step of the operation. Trees should not be felled parallel with skid roads or in such other orientations that the logs from them will have to be pivoted around good trees to move them to the roads. If trees are felled in a herringbone pattern in relation to the road, they can be pulled out with a minimum of turning. If they can be felled so that their tops point toward where they are to be moved, that reduces the total ground-skidding distance; however, it is often necessary to move the logs out butt first.

It takes cooperation and understanding between foresters and loggers to do good logging. There must be adequate incentives for the loggers. The cheapest possible logging is inevitably poor logging and ultimately invites cumbersome public regulation. Almost any kind of equipment can produce good logging jobs if supervisors make it clear that good work is expected. The presumption that logging damage is inevitable becomes a self-fulfilling prophecy. It helps tremendously if all parties concerned plan and think through what they do at each step, whether it be deciding which trees to cut or how to move a skidder around in the woods to collect a load of logs.

BIBLIOGRAPHY

Boyce, J. S. 1961. *Forest pathology*. 3rd ed. McGraw–Hill, New York, 572 pp.

Burkle, J. L., and S. Guttenberg. 1952. Marking guides for oaks and yellow-poplar in the southern uplands. *Southeastern For. Exp. Sta. Occasional Paper* 125. 27 pp.

Hamilton, G. J. (Ed.) 1976. Aspects of thinning. *U.K. For. Comm. Bul.* 55. 138 pp.

Hedden, R. L., S. J. Barras, and J. E. Coster (Eds.) 1981. Hazard-rating systems in forest insect pest management: symposium proceedings. *USFS Gen. Tech. Rept.* WO-27. 169 pp.

Miller, J. M., and F. P. Keen. 1960. Biology and control of the western pine beetle. *USDA Misc. Pub.* 800. 381 pp.

Nyland, R. D., *et al.* 1976. Logging and its effects in northern hardwoods. *State Univ. of N. Y., College of Env. Sci. & For., AFRI Res. Rept.* 31. 134 pp.

Pearce, J. K., and G. Stenzel. 1972. *Logging and pulpwood production.* Ronald, New York. 453 pp.

Shigo, A. L., and E. vH. Larson. 1969. A photo guide to the patterns of discoloration and decay in living northern hardwoods. *USFS Res. Paper* NE-127. 100 pp.

PART TWO

Regeneration

CHAPTER 7

Ecology of Regeneration; Direct Seeding

Even trees do not live forever; the time comes when they must be, or naturally are, replaced by new ones. The world also has a chronic oversupply of treeless areas where reforestation is needed to remedy the effects of previous misuse or natural accidents. Young vegetation is so plastic that what is done during the period of regeneration or establishment determines most of the future development of trees and stands. The events or treatments of the first few weeks or months can govern the future characteristics more than even the most strenuous subsequent tending. Many of the successes or failures of silvicultural treatment are preordained during stand establishment. Physicians bury their worst mistakes but those of foresters can occupy the landscape in public view for decades. The ultimate act of regenerating a stand is so crucial that it should be kept in view throughout the whole rotation.

Ecological Role of Natural Disturbances

At all stages of stand development, but especially in the earliest ones, it must be kept in mind that one is regulating all of the vegetation, or, for that matter, the animals as well, and not just the trees. True forests grow only where the climate is more conducive than average to the development of land vegetation. New vegetation generally appears only after some lethal disturbance has eliminated all of some of the preexisting plants. This is because woody perennials, once established, usually command the growing space so tenaciously that only the death of some of them can make vacancies large enough for new ones to start and make rapid growth.

Trees evolved many millions of years before man. Each species represents some sort of adaptation to a kind of lethal disturbance that occurred in nature. The natural vegetation of any locality is a repertory of species, each adapted to colonize some microenvironment or ecological niche that might commonly have become vacant in nature and is at least barely favorable to plant growth. Evolution has produced lichens that can grow on bare rocks and epiphytes adapted to crevices high on the boles of standing trees. It is through this collective versatility that vegetation is able to approach full occupancy of the growing space. It is seldom a question of whether there will or will not be vegetation but only of what kind.

These rules are not repealed for introduced, exotic vegetation. No exotic can be successful unless its new habitat is at least as favorable as its natural one. For example, the annual plants of most agriculture are herbaceous pioneers that colonize drastically disturbed sites around the world. During the growing season the microenvironment of a wheatfield in Manitoba differs little from that of one in Iraq. Exotic trees differ mainly in that they do not thrive on such extreme disturbance and must be adapted to year-round climatic conditions over many decades.

Silvicultural Simulation of Constructive Disturbances

If one wants to create a certain kind of forest the best point of departure is consideration of the natural disturbances that bring it into being in nature. Once this is known, it is possible to conjure up ways, sometimes quite artificial, that simulate the appropriate disturbance.

In determining what disturbances to simulate and in assessing the success of the technique devised, it is necessary to remember that large trees all start small. One can tell remarkably little about regeneration simply by observing large trees and old stands even though it can be very useful to deduce as much as one can from them. The literature of silviculture and forest ecology may be helpful in particular cases.

Often the quickest and best way to get new answers or test the validity of old ones is to crawl around on hands and knees examining the seemingly chaotic vegetation that develops during the first few years after regenerative disturbances. If one sees an abundance of desirable seedlings from a road it often means that there are too many.

Kinds of Natural Regenerative Disturbances

Natural vegetation ultimately arises after any of many kinds of lethal disturbances. No species is adapted to all kinds so it is necessary to know which appears after what.

In most parts of the world the most common natural disturbance is fire (Mooney et al., 1981), which was kindled by lightning and volcanic eruptions long before mankind put it to use and abuse. Fire, by itself, generally kills small trees more effectively than large ones; in other words, it tends to kill old stands from the bottom upward. The species that it favors are generally of two different categories. One of these is perennials that sprout from roots or the bases of fire-girdled stems. These sprouting species include many angiosperm trees and shrubs as well as a few conifers such as pitch pine and coast redwood (Boe, 1975). Frequent repetition of fires in forest climates usually perpetuates stands of sprouting shrubs or perennial grasses.

The other category of species favored by fire comprises those that germinate from seeds, usually small seeds, of wind-dispersed species adapted to germinate on mineral-soil surfaces bared by fires. This category of post-fire seeders includes most of the pines and so many other important commercial species that it is necessary to keep in mind that it does not include them all. The importance of fire-followers in silviculture merely testifies to the fact that most forests are associated with climates that are seasonally conducive to fire.

Severe Disturbances

Erosional geological events constitute the most severe kind of natural disturbance. In ecological terms, the only true primary succession starts with landslides, the melting of glaciers, or the formation of new land by water, wind, or vulcanism. Man-caused erosion or earth-moving can also expose soil parent materials that are free of organic matter and deficient in nitrogen or other nutrients.

The true pioneer vegetation adapted to colonize such vacancies is more likely to be herbaceous than woody plants but there are some tree species, such as the true poplars, alders, and certain river sand-bar species, that can do so without some initial herbaceous stage. Ordinarily trees do not start on such barren surfaces until some other vegetation has begun to build up the organic matter or provide enough shade for tree seedlings to endure microclimatic extremes.

In silviculture it is seldom wise to simulate such drastic disturbances deliberately. However, foresters are often called upon to afforest parent material that has been exposed by erosion, road building, strip-mining, or similar events. This very difficult form of silviculture works best if one has a very clear idea of how the natural pioneer vegetation develops on similar sites. The natural regeneration sources must be seeds or spores spread by wind, water, or animals.

The next most severe natural disturbance is the very hot fire that burns large amounts of dead fuel created by blow-downs or pest outbreaks. Lesser kinds of forest fires, even crown fires in living stands, are fueled mainly by the unincorporated organic matter of the forest floor but usually do not consume all of that. However, if there are large quantities of dead, combustible material on the ground, fires can burn so hot that they can be almost as lethal as a landslide. The main difference is that most of the incorporated organic matter remains intact.

Fires of this severity usually eliminate most sprouting species thus leaving the site nearly vacant for establishment of new vegetation from seed. The subsequent ecological successions are usually thought of as secondary ones because the remaining organic matter enables some species to start earlier than they would otherwise.

The significance of severe fires for silviculture is their similarity to much more artificial disturbances. Cultivation for agriculture and certain kinds of silvicultural site preparation also leave only bare soil with incorporated organic matter and without woody perennials capable of sprouting. It is noteworthy that many of the illustrations of discrete, orderly, sequential stages of plant succession used in the United States come from the natural revegetation of agricultural lands abandoned during a century-long epoch now coming to a close. The pure conifer stands that represent a stage in such old-field succession are so productive that they will be perpetuated as a silvicultural legacy of that epoch. The silviculture that simulates these cases depends on either natural seeding or planting of species adapted to colonize vacant areas.

Much of the foregoing paints a fiery picture of what can be called "scorched-earth silviculture." This term is not intended to be pejorative because so many forests were and are regenerated in nature by fire. The tree species involved usually grow rapidly as individuals probably because early attainment of seed-bearing status is crucial to their perpetuation and the fires are apt to be frequent. There are important economic advantages in having trees that grow rapidly in diameter and height, regardless of how well they produce in terms of the separate consideration of annual growth per hectare. Consequently these species are very important in silviculture, especially that in which large initial investments are made in planting.

Releasing Disturbances

The other general category of regenerative disturbance is that in which such agencies as wind or pests of large trees kill forests from the top downward but spare most plants of the lower strata. The species that are adapted to such circumstances are those with foliage that is constructed and displayed in ways that make it tolerant of shade. Their seedlings are usually not adapted to exposed microclimates and their juvenile growth is slow.

These species can endure for many years as advance growth beneath old stands and retain the capacity to initiate rapid height growth whenever some lethal event releases them (Gordon, 1973). Some endure as stunted seedlings and saplings; others, as perennial rootstocks that survive while their tops grow up and are repeatedly killed back. Once released, most of these species maintain height growth for long periods. This characteristic, combined with the ability to maintain deep canopies of shade-tolerant foliage, can lead to long-sustained periods of high per-hectare production. However, such species tend to be limited to sites and regions that are continuously moist enough to reduce exposure to fire or drought stress. At risk of over-generalization, they can be thought of as "advance-growth-dependent" species best regenerated naturally under some sort of protective cover.

Regeneration Mechanisms

As far as sources of regeneration are concerned, there is the basic difference between sexual regeneration from seed and asexual from various modes of vegetative sprouting. All tree species can regenerate from seed but some are so dependent on vegetative sprouting that the capacity of sexual reproduction seems to be retained only as a nearly vestigial capacity for further evolutionary adaptation.

With regeneration from seed, distinction can be made between new seed carried in from outside the stand being regenerated and that which is "stored" on the site (1) on the trees (as in unopened serotinous cones), (2) in the forest floor as ungerminated seeds, or (3) as various kinds of advanced growth. Another kind of regeneration that may be "stored" on site is that from vegetative sprouts that can arise from (1) basal portions of stems ("stump-sprouts," Fig. 7-1), (2) root systems ("root-suckers"), or (3) rooted branch ends ("layers").

FIGURE 7-1 Not all forests grow on dry land nor do all regenerate only from seeds. It may be anticipated that a good tree will develop from one of these 2-year-old stump sprouts of water tupelo in a very wet part of a flood-plain or bottomland forest in Mississippi. (*Photograph by U. S. Forest Service.*)

Seeds from outside a stand can be classified according to the agency, such as wind, water, gravity, or animal life, that normally transports them. Artificial regeneration can be from the sowing of seeds on the site, or from the planting of seedlings or vegetative cuttings started in nursery or greenhouse.

Actually there are so many ways in which new trees can start naturally or be deliberately established that only broad categories can be distinguished. There are a very large number of ways in which plants can be dispersed and established. If one attempts to set up detailed classifications of the sets of ecological adaptations of species it is soon found that each species of a given locality seems to have its own separate pigeonhole. Similar sets of adaptations can usually be found among representatives from geographically separated regional flora. However, one species seldom exhibits exactly the same adaptations throughout its natural range. The only real solution to the problem is to find out as much as possible about the regeneration requirements of each species in each locality.

Adjustment of Growing Space

Successful regeneration of any sort, natural or artificial, can occur only if the right amount and kind of vacant growing space becomes available for the establishment and subsequent growth of the desired species. The ideal objective is to create vacancies that are not merely favorable to the desired species but are more favorable to them than to any others. This objective would be achieved if one created an environment in which the species one wanted would start and then grow faster in height than any competing vegetation. If this perfect goal could always be attained, subsequent releasing operations could be avoided (Fig. 7-2).

In silvicultural practice, the appropriate kinds of regeneration environment are created partly by regulating the spatial pattern in which the previous stand is removed. Most of the names of methods of regeneration cutting discussed in later chapters refer to these patterns.

However, the pattern of tree removal falls very far short of being the only factor that controls regeneration. It is also frequently necessary to regulate the lesser vegetation that can either compete with or favor the desired species. This includes not only the vegetation that is already established but that which may appear in response to the regenerative disturbance. If it is undesirable, it cannot be wished away merely by single-minded concentration on the desired tree species.

Sometimes regeneration, especially that from seed, is regulated by various treatments of the soil and the forest floor. The physical characteristics of the surfaces on which germination takes place exert a very powerful control on seedling establishment. Minute differences can cause remarkable variations in the species of plants that appear.

The treatments that can be used to regulate accessory vegetation and soil surface conditions include the various forms of release cutting, site

FIGURE 7-2 In regenerating forests it is necessary to anticipate the development of *all* the vegetation and not just that of the desired species. The first picture shows a Douglas-fir seedling barely visible in front of a stake during the year after a cutting in a forest on an excellent site on the Oregon Coast. The second picture shows the same tree three years later and almost submerged by red alders and thimbleberry bushes that were not apparent earlier. (*Photographs by U. S. Forest Service.*)

preparation, and prescribed burning discussed in other chapters. Most of the rest of this chapter is concerned with regeneration from seed because that is the kind of regeneration for which regulation of the microenvironment is most important. Regeneration by planting and from vegetative sprouting evades most of the crucial problems of the germination stages; these forms of regeneration are not immune to microenvironmental effects but are not so tightly regulated by them.

The sizes of the vacancies created in the growing space should be neither so large that the desired seedlings are overwhelmed by new undesirable vegetation nor so small that they are overtopped by preexisting vegetation expanding from the sides. Excessively heavy cutting or overly drastic site preparation can call forth too much unwanted pioneer vegetation.

Openings can also be so small that they create no real vacancies in the soil growing space. If intraspecific root grafts exist, the root systems of the trees that were cut may simply be taken over by those of adjacent uncut trees of the same species. Even more commonly, the roots and crowns of adjacent trees expand into the vacancies faster than the newly established trees can claim them. This can be deceptive because roots, which do not have to provide for their own mechanical support, often expand into vacancies faster than the more visible crowns. This kind of effect is most spectacular on sites where soil moisture is so limiting that a stand may fully occupy the soil without ever achieving full closure of the crown canopy.

The sizes and shapes of regeneration vacancies may be limited by any requirements for a supply of seed from adjacent trees.

It is not absolutely necessary that vacancies be created all at once because they can be enlarged in a series of operations. This sequence is often useful for species that can become established only under some sort of shelter yet will not commence rapid growth without release.

Natural regeneration is easy if any tree species that appears is acceptable and the site is not marginal for natural tree growth. It becomes more difficult when one seeks some chosen species or wants very prompt reestablishment. One of the keys to controlling the process is understanding of the adjustable microenvironmental phenomena that govern where given species will appear.

The Environment of the Microsite

New plants germinate and live through the most dangerous weeks of life within tiny worlds not more than several centimeters in any dimension (Fig. 7-3). The chief things that matter initially are the environmental conditions of these small **microsites**. If these conditions are favorable, it makes no difference whether the spots are in the middle of a huge clearcut area or are the result of thinning. Seedlings do not read silviculture books but

FIGURE 7-3 A newly germinated pine seedling with the seed still encasing the ends of the cotyledons. (*Photograrah by U. S. Forest Service.*)

respond to water, light, carbon dioxide, chemical nutrients, and biotic influences.

As far as initial establishment is concerned, a surprisingly small number of suitable microsites may suffice to regenerate a large area. The establishment of 2000 desirable seedlings per hectare theoretically requires only that 2000 suitable spots are distributed over the hectare. If each microsite occupied 10 cm^2 this would be only 0.005% of the total area. Of course, this presumes that one germinable seed landed on each spot and was overlooked by mice and birds. It is worth noting that the ultimate outcome would also depend on the question of what other vegetation might possibly appear on the other 99.995% of the area. Ordinarily the early mortality of seedlings is so very large and the distribution so uneven that it takes more seedlings to achieve satisfactory natural regeneration. On the other hand, it is possible to have conditions so favorable that tens or hundreds of thousands of seedlings survive on a hectare and grow into badly stagnated thicket stands.

The crucial environmental characteristics of seedling microsites are very different from those that will govern the development of the tree after its top and roots have extended a few centimeters above and below the soil–air interface.

Water Loss

Within the uppermost layers of the soil and organic material water is lost to the atmosphere by direct evaporation. Such loss can proceed swiftly if the surfaces are exposed to large amounts of direct solar radiation. The loose materials of forest floor litter generally lose water so fast that they are seldom hospitable to the roots of plants. The uppermost layers of mineral soil or finely divided organic matter are somewhat more favorable but are also subject to water loss by direct evaporation. This comes about from the effects of surface tension in the very slender water columns within these materials. As water is lost to evaporation from the top ends of these columns more is pulled upward to be lost in its turn. The finer the materials the deeper is this capillary fringe. This kind of water loss differs from the transpiration of water by plants, which take up water from the soil through the roots. The important point about direct evaporation is that it explains the loss of water from the uppermost levels of the soil even when these layers contain no roots.

Fortunately the strata from which water can be lost by direct evaporation are seldom more than several centimeters thick in total. If the roots of a new seedling do not quickly penetrate below this layer or are not repeatedly rescued by rain, they die (Kozlowski, 1949). Shade and mulching effects can greatly slow down the direct evaporational losses but cannot add any new water.

Surface Temperature and Energy Transfers

The same general kind of situation prevails in the first few centimeters of air above the forest floor (Oke, 1978). Here frictional effects greatly impede the turbulent transfer by which most vertical movement of heated air, water vapor, carbon dioxide, and other airborne substances takes place. The sluggish movement thus induced greatly restricts the upward transport of heated air by day and the downward movement thereof that offsets radiational cooling losses of heat by night.

In this connection, it is well to note that much of the energy from the sun comes through the atmosphere almost as if there were nothing there to block it. The ozone of the ionosphere does absorb most of the life-threatening, shortwave, ultraviolet radiation. Otherwise clouds, water vapor, and leaves are the only significant things that can shield the soil surface from solar radiation. The atmosphere is heated from below by heated land and water surfaces just as air is heated when it passes over a hot stove.

The survival of new unshaded seedlings depends heavily on the interaction of the various physical processes by which this stupendous load of solar energy is dissipated. If heat accumulates, the surface temperatures soar; if it is lost too fast by night, the seedlings may freeze. The processes that take heat energy away from the absorbing surfaces are reflection, convection, conduction, evaporation, and outgoing re-radiation. While the important concern is usually the prevention of extremes of temperature, it is sometimes necessary to provide for temperatures high enough for germination.

Shading, that is, the reflection or absorption of solar energy by leaves or other objects, is the most important physical process subject to silvicultural manipulation. Opaque objects obviously divert the radiation but leaves reflect, absorb, and transmit. The chlorophyll of leaves not only absorbs visible blue and red-orange light used in photosynthesis but also reflects substantial amounts of green and, more importantly, invisible infrared light. A high proportion of the solar heating effect comes from the longwave infrared radiation.

Large amounts of radiation are reflected, without ever being absorbed, by the surface materials themselves. Most soil and dead organic materials do not vary much in their reflectivity or albedo; therefore, not much can be done to foster seedlings by trying to modify surface reflectivity. Charcoal from forest fires, in spite of its blackness, is only slightly less reflective than unburned organic matter or exposed mineral soil almost without regard for their color. The wetting of surface materials slightly reduces their reflectivity.

For practical purposes, the only way in which one can use reflectivity to control forest regeneration is to take advantage of the reflectivity of chlorophyll and absolutely opaque rocks and wood. This does not always have to be by shade; moss and other green vegetation growing around new seedlings can cool them by adding to the reflectivity of the green seedling leaves themselves.

The most important way by which heat moves vertically through air is turbulent transfer, which, with respect to heat, is called convection. This process involves the chaotic movement of large clumps of air molecules driven by the downward transfer of wind energy from the atmosphere hundreds of meters above. Conduction, the movement of heat by collision of single molecules, transfers heat through the air very slowly because there are so few molecules to collide. In fact, if the movement of heat upward from the surface depended only on air conduction, the temperature about 5 meters above would be warmest in midwinter and the midsummer surface temperatures would approach the boiling point of water.

The low heat conductivity of air is the cause of a phenomenon that can make the stratum immediately above the surface as dangerous for seedling tops as that below is to the roots. The air in the first few millimeters is so tightly held by friction that convection has scant effect on it. The air molecules right next to solid materials can only vibrate and thus conduct heat; they cannot swirl and join in convection. While this effect diminishes very rapidly with height, it makes the first few centimeters a bottleneck for heat transfer. If the surface temperatures rise above 50°C, as they readily can on some kinds of surface, the succulent stems of small seedlings can be girdled by heat injury.

Aside from providing shade, the best way to forestall severe surface heating is to stimulate downward conduction of heat by the soil. The denser the surface material, the more will be the amount of heat conducted and less extreme will be the surface temperature. Leaf litter, with its very large amount of included air, is about as low in conductivity as any naturally occurring substance. Some of the highest temperatures naturally induced by the sun on the face of the earth may be those of flat litter surfaces composed of small conifer needles; these can be as high as 75°C. It is because of this and several other factors that the burning, physical displacement, or other modification of the unincorporated organic matter plays such a critical but manageable role in regeneration.

The finely divided humus of the H-layer of organic matter does have good heat conductivity. Except for coarse sands, most bare mineral soil surfaces conduct enough heat downward to prevent fatally high surface temperatures. The heat conductivity of soil actually depends mostly on the relative amounts of air and water in it. Water is a comparatively good conductor, so the more water and the less air the greater the conductivity.

Water has other important effects in stabilizing the microclimatic extremes of surfaces. Because of its remarkably high specific heat, it can absorb tremendous quantities of heat yet its own temperature increases slowly. Its latent heat of evaporation is so great that large quantities of heat are absorbed, without change of temperature, when it changes from the liquid to the vapor phase. If the water vapor is then swept aloft, it carries large amounts of heat energy with it. Moreover, the latent heat of condensation is so low that large amounts of heat must be removed to convert it from a liquid at the freezing point to ice at the same temperature. In other words, its benefits to plants can be physical as well as physiological. However, the soaking of surface litter by rain postpones the extreme heating of such material by only about a half hour once the direct rays of the noonday sun hit it.

The remaining heat transfer process important for plant life is radiation. This goes on all the time and would do so even if there were no air. Much of the heat absorbed by the earth's surface materials is lost to outer space as infrared radiation of wavelengths even longer than that which comes from the sun. The warmer the material the more it radiates. It does reduce surface heating effects by day but there is not much that can be done to alter what happens then.

The same surface materials that absorb most of the solar heat by day also lose the most by radiation at night and thus can become much colder than the air a short distance above. This creates risk of frost damage to plants, especially succulent seedlings. The sluggish movement of air in the boundary layer aggravates the problem even more than with heat injury during the day. During daytime heating, the heated clumps of air wrested from the surface layer are less dense than the cooler air around them so they are convected aloft rapidly. At night any warmer air that is pushed downward to replace very cold air at the surface must penetrate a denser medium. If the air is humid enough to have a dew point higher than 0°C, the condensation of water vapor into dew usually slows the cooling of plant tissues enough to prevent frost damage.

The most practical silvicultural method for preventing excessive radiational cooling is interposition of overhead cover. If the quanta of heat energy that radiate from a surface hit something above, there is a strong statistical probability that some of them will bounce back to the original radiating surface and reheat it. There is a subtle difference between this kind of protective effect and that of daytime shading from direct solar radiation. The sun's rays come in at slanting angles, except at noontime in the tropics. The most effective route by which outgoing radiation can escape is straight up toward the zenith. This means that the side shade that helps in daytime may not help as much at night. In general, the wider the angle of the inverted cone through which radiation can escape unimpeded from any point, the greater will be the radiational cooling of that point.

Species Adaptations and Their Silvicultural Significance

The chief purpose of this detailed account of surface microclimate lies in the fact that the success of seedling regeneration and the kind of species obtained can be regulated to some extent by deliberate manipulation of overhead cover and the characteristics of the surface materials. It helps to note that the physical environmental factors that control the small, new seedling may not operate the same after the roots have penetrated below the capillary fringe and the tops have grown up into the more turbulent air strata.

Once the seedling is tall enough, and at all subsequent stages, it is not threatened as much by extremes of temperature. The risk of frost damage persists longer than that of true heat injury. There still have to be all kinds of adaptations to restrict excessive water loss, but there is usually enough turbulent convection to take care of other microclimatic problems.

A very crucial race against time which the seedling must make is that of extending its roots downward faster than the loss of water through direct evaporation from the capillary fringe can overtake them. After this penetration is achieved, the roots move permanently into soil strata in which water loss is governed by gravity and transpiration but not by direct evaporation. Only the transpiration of established plants can move water upward from these strata. The less established vegetation there is to transpire the better will be the supply of water beneath the first few centimeters.

The general result of these phenomena is that the ecological and silvicultural ground rules that govern the lives of new seedlings are quite different from those that apply later. If one reduces the amount of competing vegetation, as by thinning, the supply of water for established trees is *increased*. On the other hand, reduction of vegetative cover during the early and most crucial stages of regeneration *reduces* the water supply for new seedlings; it also threatens them with microclimatic extremes that are of little consequence later. If this distinction is overlooked, one can be very puzzled by seemingly paradoxical results.

It might seem from all this that no tree seedling could ever survive in the forest even though many obviously have for many millions of years. Forests have regenerated because the trees and other plants involved are collectively capable of colonizing virtually any kind of microsite with seedlings. Each species is adapted to a range of microsites; this range is wider for some than for others but none are adapted to the full range. If one wants to favor any given species, this can, given enough knowledge, be done by deliberately creating a suitable number of appropriate microsites.

While there are exceptions, the seedlings of most tree species cannot colonize severely exposed microsites without the protective cover of some sort of hardy pioneer herbaceous vegetation. Sometimes this herbaceous vegetation is so ephemeral that it is taken for granted or is overlooked because it soon vanishes leaving only the small trees that it once sheltered. The seeds and seedlings of many tree species have adaptations to take advantage of such cover.

The cotyledons and juvenile foliage of new tree seedlings are often built or arranged in ways that make them more shade-tolerant or efficient in photosynthesis than those of the adult stages. The seeds of many have enough stored material that they are more dependent on water to mobilize these reserves than on their own initial photosynthesis. The supply of carbon dioxide for photosynthesis is quite large because of close proximity to the respiring organisms that feed on soil organic matter. The supply of light becomes much more crucial later on, but even then the requirements of different species vary tremendously.

The availability of chemical soil nutrients usually does not limit the growth of new seedlings; the seeds themselves supply some and more nutrients are usually also released by the decay of the plants that were killed to make the regeneration vacancy in the first place.

There are great variations in the means and rates of penetration of seedling roots downward to the relatively stable moisture supplies available at depth. The supplies themselves have different characteristics. Most species germinate at a certain time of year when the soil is well supplied with water. If the ensuing period is rainless, as in the western United States, or subject to drought from high evapotranspiration, as in parts of the South, everything depends on whether the seedling roots can extend downward more rapidly than the "drying front" that moves downward from the surface. Most of the species involved are adapted to produce roots that grow 15–50 centimeters deep the first year. If the necessary building material cannot be provided from the seeds, there must be enough light for adequate photosynthesis or the seedlings simply die of drought in the shade.

If there is plenty of continuing rain, the seedlings are often adapted to survive with much shallower root systems. However, in this connection, it is well to note that moistening of the soil from above can proceed only to the extent that there is enough water to bring the uppermost soil strata to field capacity (the amount of water that can be held against gravity). This means that there is a "wetting front" as well as a "drying front" and both move downward.

Because of differences in ability of roots to penetrate and for many other reasons, many of the microsite differences that can be used to discriminate between species depend on the nature of surface materials and their alteration. Close examination reveals many such differences but most are poorly documented. Much depends on how deeply fire or physical disturbance has exposed various soil layers.

In general, the only species that can effectively establish themselves on thick leaf litter are those, such as oaks (Carvell, 1979) , that have very large seeds. However, this is not because they can germinate on top of such surfaces but because they were buried beneath them by rodents or more falling leaves and because they have enough stored materials to grow back up through the leaves. Oak acorns and other large seeds usually cannot successfully germinate on bare soil surfaces. Their new, blunt roots simply roll them around over the surface without penetrating; they need to be buried to hold them in place and keep them moist. If smaller-seeded species appear to have established themselves on thick litter, close examination usually shows that it was on some localized spot where wind or something such as a little ant mound made the litter very thin.

Most small-seeded species can germinate only on dense media such as exposed mineral soil. This is mainly because there must be close contact between the seed and the moisture-supplying medium. Even then, the germination is seldom good unless the action of rain, frost, or other agencies has caused the seeds to be slightly buried.

With these dense seedbed media it can make much difference how deeply they are exposed and what sort of materials thus forms the surface. One of the most ideal seedbeds for some very small seeds, such as those of birches, is the finely divided black humus of the H-layer above the true mineral soil. This has appropriate density coupled with high concentration of chemical nutrients that the small seeds did not bring with themselves.

The uppermost strata of mineral soil are favorable for many species with small- or medium-sized seeds. For reasons that are not altogether clear, the number of species that can become established seem to dwindle the more deeply the soil is gouged. Sometimes they become limited to sedges or plants other than trees. It is rare that there is any useful purpose in extending seedbed preparation any deeper than the surface of the mineral soil except to bury the seeds slightly or to alter the drainage characteristics of the soil.

Regulation of Regenerative Canopy Openings

In very small openings, the development of extremes of surface temperature is impeded by side shade; in large openings, the wind causes enough turbulent transfer of heat to restrict the diurnal range of surface temperature. There is evidence that the greatest extremes of temperature occur when the diameter of an opening is 1½ times the height of the surrounding trees (Geiger, 1965). Presumably this is the situation in which the combined effect of side shade and ventilation is the least.

When an opening is more than 2–3 times the height of the surrounding trees, the environmental conditions at the center are about the same as those that would prevail in any very much larger opening (Minckler and Woerheide, 1965). Some important things that would differ would be the effectiveness of seed dispersal from the adjoining trees and effects that had to do with the travel of animals between the opening and the cover of the taller trees.

Microclimatic differences between spots can also be systematized in terms of the distinction shown in Fig 7-4 between conditions found (1) in unshaded, *open* areas, (2) in the side shade and partial protection of uncut timber along the *edges* of openings, and (3) under *full* shade. The edge zone is exposed to the scattered, diffuse light from the sky but not to any harmful effects of direct sunlight. Although the total amount of light is distinctly reduced, diffuse radiation is relatively rich in the blue wavelengths most effective in photosynthesis. This transitional zone extends, for practical purposes, not only for a fluctuating distance outward from the edge of an uncut stand but also a short distance inward beneath standing trees. The illumination under full, overhead shade consists mainly of whatever direct sunlight penetrates through interstices in the crowns and the filtered light, rich only in the photosynthetically useless green and infrared wavelengths that are transmitted through the foliage. The differences between these

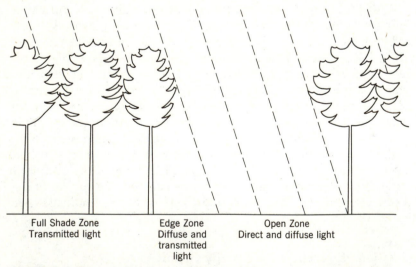

Full Shade Zone	Edge Zone	Open Zone
Transmitted light	Diffuse and transmitted light	Direct and diffuse light

FIGURE 7-4   The zonation of solar radiation which, in combination with alteration of surface conditions, controls most of the alterable microclimatic factors significant in regulating regeneration from seed.

zones correspond roughly to differences in degree of root competition, except that sunlit edges of openings are subject to nearly full light and undiminished root competition. Figure 7-4 does not include depiction of another pattern in which forest cover can be opened to create vacancies for regeneration; all it shows is full crown closure in the adjoining trees. A diffuse cover of scattered trees, such as might be left in what is called uniform shelterwood cutting, would theoretically allow more ventilation; the extent of the effect of increased ventilation in narrowing the extremes of surface temperature is not known. This kind of cutting is the best way of increasing the area of the edge zone. It virtually eliminates the full shade zone but puts varying amounts of surface into the open zone.

Some useful ideas about controlling regeneration from seed can be deduced by observing the patterns of species and their height growth in different parts of forest openings. These things can be observed and diagnosed most simply in circular openings, as shown in Fig. 7-5; the same things can be observed in openings of any shape but are not as simply analyzed.

In a circular opening, those effects associated with the slanting rays of direct solar radiation will be arranged in *crescentic* patterns. Those that are controlled by root competition of adjoining vegetation, diffuse solar radiation, air movements, and outgoing radiation (frost and dew) should have

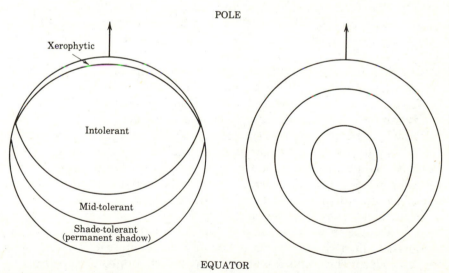

FIGURE 7-5 Crescentic (*left*) and concentric (*right*) patterns of horizontal variation in the effects of microclimatic factors in circular openings at some mid-latitude. The kinds of species favored by different kinds of shading and exposure are indicated in the crescentic pattern. Concentric patterns are more likely to appear during the development of established vegetation than to affect species composition. Recognition of these two kinds of microclimatic variation may help in deducing the factors governing regeneration and determining how to control it.

concentric arrangement. Except for the effects of root competition, crescentic patterns are, because of the power of direct sunlight, much more common than concentric ones. However, these patterns will be altered by the degree and aspect of slope and by latitude. Since there is no environmental factor more readily predictable than the sun, the extent of side shade can be calculated (Brown and Merritt, 1970; Harrington, 1984; and Halverson and Smith, 1979).

On poleward slopes all shade zones are broadened and the steeper the slope the more they are broadened; on equatorward slopes these effects are the opposite. At high latitudes the various shade zones become very broad; diffuse solar radiation may even become fully as important as the direct. In the tropics most such patterns become concentric and side-shade is effective only in very narrow bands.

A case most common at midlatitudes in the Northern Hemisphere is depicted in Fig. 7-5. Ordinarily the only place where new seedlings of species that are tolerant of shade and intolerant of exposure can appear is in a crescent along and under the southern edge. More shade-intolerant and exposure-resistant species find their optima for establishment and height growth in crescents arranged successively northward. At the very northern edge there is often a peculiar zone, almost like a little desert, in which the combination of unimpeded direct sunlight and root competition of taller trees can prevent much of anything from regenerating.

These patterns are greatly modified by differences in seedbed conditions that can vary considerably from spot to spot. Close-by shade of stumps, rocks, and woody debris can also be as effective as that of more distant tall trees. Furthermore, after exposure-tolerant seedlings or saplings are established, seedlings of exposure-intolerant species can start beneath them. However, their height growth is likely to be slow until whatever time the trees above them are eliminated.

The patterns can also be obscured and confused in nature by preexisting advance growth and vegetative regeneration, that is, by plants already present when the opening was created. Observations of these patterns can help greatly in deducing the kinds of microsites that one would aim to create over larger areas to produce stands of the species composition desired. The desired spatial pattern will usually be something different from the circular openings that are ideal for the diagnostic purpose.

From these considerations it can be deduced that even small forest openings cannot be regarded as uniform habitats for regeneration. If the regeneration in a quarter-hectare opening is all lumped together and counted collectively without regard for spatial variations as much useful information is buried as is revealed.

Essential Steps in Natural Regeneration

Successful natural reproduction depends on the completion of a long sequence of events (Wenger and Trousdell, 1957); failure of any single link in the chain is fatal.

Seed Supply

The first and most obvious prerequisite is an adequate supply of seed. No tree or group of trees is a dependable source unless it is sufficiently old and vigorous to produce seed. Seed bearers should, furthermore, be located so that wind or other agencies will ensure pollination and properly distribute the seeds over the area to be regenerated. Regardless of how carefully the seed bearers are chosen and fostered, it must be remembered that most species do not annually produce the abundant crops of seed necessary for satisfactory regeneration. This characteristic makes it difficult to carry out reproduction cuttings with equal chance of success each year. The limited crops of seed that are produced almost every year give rise to little or no reproduction. The origin of many stands can be traced to unusually good "seed years" that were followed by satisfactory conditions for germination and establishment of seedlings (Fig. 7-6).

Good seed years occur at intervals that are better thought of as sporadic rather than predictably periodic. If there are any real cycles of seed production they are complicated and have yet to be verified or explained. The first essential is an ill-defined state of physiological readiness for flowering; the supply of nitrogen compounds for protein building may be one of the crucial factors. Even after flowers or strobili appear, it is common for adverse weather or infestations of insects and fungi to intervene and prevent pollination or maturation of flowers and fruit. The depredations of insects are a common cause of irregularity of seed crops. Sometimes good seed crops follow years of total failure that may have caused collapse of populations of specialized seed predators. In general, the more favorable the conditions of soil and climate for plant growth, the more frequent are good crops of seed.

Seed Dispersal

In spite of remarkable adaptations for dispersal over long distances, the seeds of only a few species are distributed in adequate numbers beyond a distance of a few times the height of the seed bearers. With many wind-disseminated species, particularly the conifers, seeding is not uniform in all directions because dispersal is most effective when dry winds are blowing. With species disseminated by birds, mammals, gravity, or flowing water, it is important that the forester proceed with clear understanding of the factors involved.

Some adaptations are quite bizarre. For example, the seeds of some tropical mangrove species germinate on the trees and the fallen seedlings can then float upright over miles of ocean to take root on some distant shoreline.

Both before and after the seeds fall from the trees they are likely to be eaten by insects, birds, and rodents. Although seed predators are held in check by natural controls that may be strengthened by direct or indirect artificial measures, the most practical means of combating the menace is to

FIGURE 7-6 Ponderosa pine regeneration under difficult climatic circumstances in northern Arizona. The two pictures show the same spot just after a group-selection cutting in 1909 and again 1938. The saplings of the second picture did not become established until abnormally favorable circumstances took place in 1919. Unusual May rains probably overcame the competitive effect of the grass cover. (*Photographs by U. S. Forest Service.*)

ensure that the supply of seed is sufficient both to feed the predators and to regenerate the forest. Sometimes the most important predators are also the most effective agents of dissemination; if it were not for squirrels and other rodents the heavy seeds of oak, hickory, and walnut would not be carried farther than gravity would take them. Edible fruits are often adaptations for seed dispersal by animals, especially birds and tropical fruit-eating bats.

Germination

Successful germination depends largely on the rainfall and the nature of the spot where the seeds are deposited. Moisture is the most critical and variable factor. The embryos in excessively dry seeds either die or remain dormant; if the seedbed is too wet the seeds will rot without germinating or they will suffer from deficiency of oxygen. The kinds of dormancy encountered in artificial seeding control the timing of germination. Processes that break dormancy are normally completed in seedbeds that are otherwise favorable for development of the seedling, and are rarely critical in *natural* regeneration.

In general, the seeds of wind-disseminated species germinate best on or slightly beneath the surfaces of seedbeds that tend to remain moist; bare mineral soil is usually the best seedbed for such species. Seedbeds of horizontally oriented materials, such as coniferous litter, tend to be unfavorable because they inhibit penetration of seeds into moist substrata; vertically arranged materials, like grass, are less resistant. Large seeds, such as oak acorns, usually do not germinate unless buried beneath litter or mineral soil, because they must be completely surrounded by a moist medium if they are to survive. Delays in germination are dangerous because they lengthen the period of exposure to birds and rodents. Losses from germination failure, which are usually caused initially by desiccation of seedbeds, are frequently more serious than losses of germinated seedlings.

Early Survival

New seedlings are most vulnerable during the first few weeks of their existence, while their stems are still green and succulent. Heat injury resulting from extremely high temperatures on surfaces exposed to direct solar radiation takes a heavy toll, particularly among conifers. Tender seedlings in shaded situations are subject to damping-off caused by a wide variety of weakly parasitic fungi that are normally saprophytic and incapable of attacking larger seedlings. Cutworms and other insect larvae are particularly active during this period; most of the species involved are omnivorous feeders that will eat anything green and succulent. The better known forest insects and fungi generally attack trees that have passed the succulent stage. Seedlings that germinate late and remain succulent in the fall are commonly killed by frost. The dangerous succulent period ends when the outer cortical tissues of the stem become dry and straw colored, a process referred to as "hardening." Thereafter the attrition of various damaging agencies continues, but the period of catastrophic losses is generally at an end.

Drought mortality is distinctly different from heat injury and can take place after the succulent stage. It will not occur if seedling roots extend themselves rapidly enough to maintain contact with portions of the soil where water is available. The seedling must manufacture enough carbohydrate to enable the roots to grow downward faster than the deepening of the stratum in which water is unavailable. This is one reason why shaded seedlings are more likely to die of drought than those growing in the open.

DIRECT SEEDING

Success in natural regeneration usually depends on whether a very small proportion of a large quantity of seed lands in favorable spots and is overlooked by all the animals that feed on seeds. This random process could ordinarily be duplicated artificially only by broadcast distribution of hundreds of thousands of seeds per acre; it would be much more costly and not often as successful as planting. In establishing stands by direct seeding, the odds against success must be reduced by (1) control of seed-eating animals and (2) distinctly favorable conditions of site and seedbed. Artificial means are now at hand for providing this combination of effects. Ordinarily success also depends on whether there is sufficient rain after sowing to keep the uppermost layer of soil adequately moistened throughout the period of germination and the succulent stage.

Control of Seed-Eating Animals

Almost any site that supports a forest or is capable of doing so also supports a large population of small mammals and is also combed over periodically by birds seeking food. These seed predators are rarely obvious to casual inspection, especially during the middle of the day. However, superficial digging in the litter and soil often reveals so many rodent tunnels that one is led to wonder how any seeds would go uneaten. These populations vary widely in time and space; variations between seasons and between years are large and not impossible to predict. In general, areas of rather barren soil support lower populations of small mammals than those covered with vegetation or litter; they also offer less shelter to seed-eating birds. In direct seeding it is well to take advantage of and even to create times and places of low animal population, but dependable results usually require more direct measures to ward off these animals.

Site preparation to reduce the amount of cover over the soil is very important in improving germination and survival of seedlings but is not especially effective in controlling the seed predators. Screens and traps are amply effective but so expensive that they are really useful only for various investigative purposes. Rodents have a good sense of smell and are rarely outwitted by covering the seeds. Even the wholesale poisoning of rodents on regeneration areas is a rather inefficient and questionable technique, although it can be made to work. If the population of seed predators is reduced this merely creates an ecological vacuum; breeding or reinvasion can restore the population even during the period that seeds remain edible.

During the 1950s direct seeding became practical because of the development of chemicals that effectively repelled small rodents and birds (Abbott, 1965; Derr and Mann, 1971). An ideal repellant is a compound that makes the seeds distasteful to the seed predators but does not necessarily kill them. If the seed-eaters of a seeded area can be "educated" not to feed on the seeds, their presence tends to forestall invasion by their "uneducated" brethren. Unfortunately many of the repellants were compounds that had suitable effects at the low dosages in which they were applied but were either dangerously toxic at higher dosages or excessively persistent.

Endrin, for example, is an effective rodent repellant but is basically a chlorinated hydrocarbon insecticide with all of the persistence of DDT coupled with very much higher mammalian toxicity. Its use is now banned in many localities and the development of suitable replacements is unfinished business.

Two chemicals are available that repel but do not kill birds. One of these is Arasan (thiram or tetramethyl thiram disulfide). It was developed as a fungicide for seed treatment and is thus beneficial in reducing damping-off and other difficulties with fungi; it is sometimes effective by itself as a rodent repellant. Sublimed synthetic anthraquinone is less irritating to the skin and mucous membranes and should be substituted for Arasan if sowing is to be done by hand. The seeds are first coated with special latex or asphalt adhesives and then, within 1 or 2 minutes, mixed with the repellant chemicals in a rotating drum. A small amount of aluminum or graphite powder may then be added to make the seeds flow more easily.

Some species, such as birches, spruces, eucalypts, and western red cedar, have seeds small enough that they are not especially subject to rodent predation. Sometimes these can be sown without use of repellants. However, the lack of generally acceptable rodent repellants has greatly curtailed use of direct seeding. Much of the subsequent discussion of the practice must be regarded as information stored against the time when new and suitable rodent repellants are developed.

Effect and Modification of Site Factors

Success in direct seeding also depends on rendering the microsite as favorable as possible and ensuring prompt germination. The seeds should be placed in contact with mineral soil and, if possible, covered to the greatest depth consistent with successful germination. Moisture must be almost continuously available at or close to the surface of the mineral soil until the seedling roots have penetrated to a stable moisture supply. The amount of water several inches below the surface that might support the damaged root system of a newly planted tree is not necessarily enough. Consequently there are some climates and soils that are too dry for successful direct seeding.

The most favorable soils are those well supplied with moisture because of their topographic position, although poor aeration in excessively wet spots can also cause failure. Moist but well-drained soils that are supplied with water by seepage from higher ground are also very favorable. Soils of extremely coarse or extremely fine texture are less favorable than those of loamy texture.

As far as the germination and initial establishment of seedlings is concerned, a light cover of vegetation over bare mineral soil is almost ideal. It allows the seeds to come in contact with the mineral soil yet it shields the surface from direct sunlight thus reducing heat injury and direct evaporation from the soil. Although desiccation of most of the soil layer results from removal of water by plants, that of the surface and about the first half inch of soil is caused more by direct evaporation than by transpiration. Therefore, the vegetation that restricts the supply of water to a well-rooted seedling can actually increase the amount of water available during germination and the crucial days thereafter.

Sometimes a thin cover of herbaceous annuals or grass can improve the success of direct seeding by shielding seeds and seedlings from birds, heat, or frost. The ideal seedbed for longleaf pine is one year's "rough" or growth of grass after accumulated litter has been eliminated by prescribed burning. This species germinates in the autumn so frost is a factor. Such annuals as mustard and rye have been useful when sown in mixture with conifer seeds on bare mineral soil. They do not reappear during the second year and may serve to exclude undesirable plants as well as provide shelter during the first year.

The shade of woody plants is beneficial in the early stages, but its ultimate effect depends on the extent to which it competes with established seedlings for light, water, and nutrients. The best shade is that cast by dead materials, such as stumps, logs, or light slash.

Very low cover may impede foraging by birds but almost all the other effects of cover are likely to favor seed predators. Vegetative cover tends to reduce mechanical erosion of repellant materials from seeds. Sometimes this makes it possible to sow in the fall and allow stratification to proceed naturally. The sowing of open areas often must be restricted to the period immediately before germination because the repellant coatings deteriorate too much if exposed.

On most sites, the outcome of direct seeding still depends on the vagaries of rainfall. The careful selection and preparation of sites can only mitigate the baneful effect of one or two rainless weeks during the period of germination and initial development. If climatic records indicate that failures ought to be anticipated for example, in 1 year out of every 4, the true cost of each successful operation should be regarded as increased by 25 per cent. There are regions where the rainfall is so undependable that direct seeding may have to be restricted to sites where the soils remain moist during drought periods.

The chances of success are significantly increased if the seed is treated so that it will be capable of prompt and vigorous germination. This shortens the period of exposure to predators as well as that during which unfavorable weather can be critical. The same seed of mediocre quality that germinates well in the nursery may fail under the more rigorous conditions of the wild. Therefore, fresh seed is distinctly preferable to that which has been in cold storage for several years. If the seed is dormant and cannot be after-ripened naturally by fall sowing, it must be stratified. Such stratification must precede treatment with repellants and sowing should follow within a day or two because stratified seed is very perishable. Finally, results are greatly improved if the seed is buried at optimum depth on favorable microsites.

Relative Merits of Direct Seeding and Planting

There is much more risk of poor survival with direct seeding than with planting. New seedlings that germinate and grow in the field have scant protection from the numerous lethal agencies that can be controlled in the nursery. Trees established by direct seeding grow no more rapidly than natural seedlings, so they suffer more than planted ones from competing vegetation. Furthermore, there is no opportunity to shorten rotations as there is in planting.

Direct seeding is inherently cheaper than planting because it involves less labor and equipment; investments in nurseries and the overhead charges involved in their operation are avoided. Large areas can be seeded more quickly, on shorter notice, and with fewer organizational problems. The only preliminary step is seed collection, although such large quantities are required that the normal problems of procurement are aggravated. Seed of demonstrably superior genetic quality is rarely cheap or abundant enough to permit its use in direct seeding.

The roots of trees established by seeding develop naturally and are not subject to the deformities that are suspected of making planted trees susceptible to windthrow and root rot. Species that develop taproots or distinctly shallow, lateral root systems grow best if their roots develop naturally.

Direct seeding is possible on soils where planting is not feasible because of stones, stumps, or other obstructions, *but only if the sites are otherwise favorable.*

If dense stocking is desirable, it is more economically secured by direct seeding than by planting. However, the risk of localized understocking and overstocking is far greater. Stands established by direct seeding are likely to require more subsequent treatment, such as refilling, release cutting and precommercial thinning, than planted stands.

Direct seeding can often be conducted over longer periods than planting and during colder or wetter weather. The main limitations on timing are those imposed by excessively dry weather and any need for minimizing the period of exposure to seed-eating animals and for avoiding unseasonable germination. Since germination often occurs after older seedlings have broken dormancy, the best time for direct seeding may actually come shortly after the regular spring planting season thus enabling continuation of activity even where planting is the standard method of artificial regeneration.

Broadcast Seeding

The simplest type of direct seeding consists of scattering seeds uniformly over the area to be restocked. This is likely to be a waste of seed unless the conditions of the seedbed surfaces and competing vegetation have been made appropriately receptive. In the common case in which direct seeding is being used to regenerate some species adapted to follow fire, it is ordinarily essential to expose much mineral soil and eliminate most preexisting vegetation. Accidental or intentional burning often suffices. Mechanical site preparation for broadcast seeding is justified mainly as a means of removing vegetation that will ultimately compete with the new crop after it is established.

Broadcast sowing is very rapid and is most often used when there is necessity of covering large areas quickly. Its most serious drawback is the lack of any provision for covering the seeds. For this reason it is best done during moist weather.

The seeding rates vary considerably depending on the favorability of site and anticipated weather. Recommendations ordinarily call for 10,000 to 25,000 viable seeds per acre, provided that these seeds are treated with appropriate repellants. However, the seeding rate should be carefully adjusted from year to year on the basis of quantitative observations of the results of previous seeding projects.

Most broadcast seeding is done from the air (Panel on Aerial Seeding, 1981). The use of aircraft is usually feasible only if several hundred acres or more can be done in a single project. Uniformity of seeding is possible only if the distributing equipment is closely calibrated and if moving flagmen are stationed on the ground. Fixed-wing aircraft do a faster and cheaper job than helicopters but are best adapted to seeding over gentle terrain where the areas to be seeded are also large and uniform. Helicopters are most suitable if the terrain is rugged or if the areas to be seeded are intermingled with those on which seeding is unnecessary or unlikely to succeed. The cost of aerial seeding itself is small compared with that of the seeds, logistical organization, and any associated site preparation. Broadcast seeding can also be done on the ground at rates up to 8 hectares per day with crank-operated "cyclone" seeders like those used for sowing grain crops. The cost is about the same as for aerial seeding but it takes much longer to cover a given tract.

Strip and Spot Seeding

Failure in direct seeding is least likely if the seeds are sown in spots or strips that are specially prepared or selected. The sowing itself is more expensive than broadcast seeding but cheaper than planting. Strip and spot sowing are more economical of seed than broadcast sowing and can be done successfully in a much wider range of times and places.

If the terrain is suitable and there is no necessity for attempting to eliminate all the vegetation that will ultimately compete with the new crop, the mechanical preparation of the soil is most efficiently accomplished by furrowing or disking in strips. Although the seed is sometimes broadcast on the strips, it is usually much better to take full advantage of the preparatory work and apply the seeds so that they are covered with soil or pressed into it. If the site preparation has left air pockets in the soil, it may be necessary to lightly recompact the soil in the spots or strips to be sown.

Tractor-drawn equipment has been or can be devised to do light soil preparation and to sow and cover seeds one-by-one all in one pass of the equipment. Such machines can cover 10 to 25 hectares per day on gentle terrain at total cost roughly half that of planting. In preparing spots or strips for the seeds, it is important to manipulate soil surfaces in ways that will prevent soil or litter from being moved by water, ice, or wind in ways that undermine or bury germinating seedlings. Shallow furrows can be plowed to anchor blowing leaves and ridges can be constructed to deal with poor drainage, but it is otherwise better to keep the soil surfaces as level as possible to avoid erosional effects.

Spot seeding is limited to operations that are too small to justify use of tractors or where steep terrain and obstructions prevent their use. The spots are prepared with hand tools or merely by kicking the debris to the side.

It is difficult to arrive at the proper relationship between the spacing of seed-spots and the number of seeds applied to each. If one seed is sown on each spot, many spots must be prepared and seeded to allow for all the failures; if enough are sown on each spot to ensure that each has one established seedling, some spots will be choked with seedlings. The ordinary compromise is that of sowing several seeds each on spots spaced at intervals closer than would be used in conventional planting; it must be anticipated that the stocking of the stand will be erratic and that it might become desirable to thin overcrowded spots and patches.

The seeds can be placed by hand, although any repellants that are used may be poisonous enough to dictate use of rubberized gloves. Various special seeding devices resembling corn planters are much more convenient; they require less contact with the seed and enable the sowing of up to 5 acres per person-day. The only other tools especially designed for spot seeding are the dibbles used for planting acorns and other large nuts. These are bars shaped to punch holes into the soil obliquely so that the seeds can be planted on their sides.

Application of Direct Seeding

Broadcast seeding from the air was more common in the 1960's in North America than it has been subsequently. Planting has come into greater favor mostly because it provides more economical use of the costly seed produced in tree improvement programs.

Where direct seeding works, it can enable swift reforestation of various kinds of barren, devastated areas, especially when done from the air. It can also provide a useful supplement to planting programs, which almost always fall behind schedule because of their cumbersome logistics.

Direct seeding is often one of the best ways for regenerating species with very small seeds, provided that there are enough microsites with bare mineral soil suitable for their germination. The supply of seed must also be abundant and cheap. This technique is often used for Australian eucalypts, which have seeds so small (300-800 per gram) that they are coated with clay to facilitate application and ward off insects.

One advantage of direct seeding over planting is that the roots of the trees grow in natural, normal arrangement. This advantage is seldom appreciated because the roots are hidden and abnormalities usually affect long-term development rather than short-term survival. However, there are some species, ordinarily those that are taprooted or have comparatively large seeds, that do not thrive if transplanted from nurseries.

One of these is the longleaf pine of the South (Mann, 1970) (Fig. 7-7). The stemless, taprooted seedlings are nearly impossible to plant successfully because the buds are so likely to become buried or excessively exposed by the slightest erosion around them. The discouraging tendency of the seedlings to linger in the grass for years without growing in height has also led to the common practice of reforesting extensive areas of old longleaf lands with planted slash and loblolly pine. Direct seeding of longleaf pine is much more successful than planting; it can be done by broadcast seeding, preferably on one year roughs, or by drill sowing on prepared strips. In fact, the reduction of grass competition on prepared strips sometimes causes speedy emergence from the grass stage.

Heavy-seeded hardwoods, such as oak, walnut, and hickory, have large taproots that are easily damaged in planting. The best artificial regeneration of these species is obtained by direct seeding, although the control of rodents is still a serious problem. Screens and other physical barriers are usually necessary; the use of repellants on large nuts has yet to be perfected.

Direct seeding is potentially useful as a low-cost means of establishing advance growth of shade-tolerant, exposure-intolerant species beneath full

FIGURE 7-7 A stand of longleaf pine in Louisiana established by direct seeding on disked strips. The seedlings are in the middle of the fourth growing season and have been sprayed once with a fungicide for control of brown-spot disease. (*Photograph by U. S. Forest Service.*)

or partial cover of older trees. The slow juvenile height growth of most such species so delays return on planting investment that cheap initial establishment is advantageous.

Planting is ordinarily a more effective means of artificial regeneration, especially if it enables genetic improvement. Direct seeding is employed mostly for quick reforestation of large, barren areas, on unusual terrain that is physically difficult to plant but is otherwise favorable, and for rather peculiar species that do not withstand transplanting well. The potentialities of this low-cost technique could be exploited better if suitable rodent repellants were developed. The most serious inherent shortcoming is the need for frequent rain after sowing of seeds.

BIBLIOGRAPHY

Abbott, H. G. (Ed.) 1965. *Direct seeding in the Northeast.* Mass. Agr. Exp. Sta., Amherst. 127 pp.

Benzie, J. W. 1977. Manager's handbook for jack pine in the North Central States. *USFS Gen. Tech. Rept.* NC-32. 18 pp.

Boe, K. N. 1975. Natural seedlings and sprouts after regeneration cuttings in old-growth redwood. *USFS Res. Paper* PSW-111. 5 pp.

Brown, K. M., and C. Merritt. 1970. A shadow pattern simulation model for forest openings. *Purdue Univ. Agr. Res. Bul.* 868. 11 pp.

Carvell, K. L. 1979. Factors affecting the abundance, vigor, and growth response of understory oak seedlings. *In: Regenerating oaks in upland hardwood forests.* Purdue Univ. W. Lafayette, Ind. pp. 23–26.

Clark, F. B., and S. G. Boyce. 1964. Yellow-poplar seed remains viable in the forest litter. *J. For.* **62**:564–567.

Derr, H. J., and W. F. Mann, Jr. 1971. Direct seeding pines in the South. *USDA, Agr. Hbk.* 391. 68 pp.

Finley, J., R. S. Cochran, and J. R. Grace (Eds.) 1983. Regenerating hardwood stands. *Proc., 1983 Pennsylvania State Forestry Issues Conference.* 241 pp.

Geiger, R. 1965. *The climate near the ground.* 3rd ed. Harvard Univ. Press, Cambridge, Mass. 611 pp.

Gordon, D. T. 1973. Released advanced growth of white and red fir: growth, damage, mortality. *USFS Res. Paper* PSW-95. 12 pp.

Halverson, H. G., and J. L. Smith. 1979. Solar radiation as a forest management tool. *USFS Gen. Tech. Rept.* PSW-33. 13 pp.

Harrington, J. R. 1984. Solar radiation in a clear-cut strip—a computer algorithm. *Agric. and For. Meteorology* 33:23–40.

Harris, A. S. 1967. Natural reforestation on a mile-square clearcut in southeast Alaska. *USFS Res. Paper* PNW-52. 16 pp.

Kozlowski, T. T. 1949. Light and water in relation to growth and competition of Piedmont forest tree species. *Ecol. Monogr.* **19**:207–231.

Logan, K. T. 1966. Growth of tree seedlings as affected by light intensity. II. Red pine, white pine, jack pine, and eastern larch. *Can. Dept. For. Publ.* 1160. 19 pp.

Lopushinsky, W. 1969. Stomatal closure in conifer seedlings in response to leaf moisture stress. *Bot. Gaz.* **130**:258–263.

Mann, W. F., Jr. 1970. Direct-seeding longleaf pine. *USFS Res. Paper* SO-57. 26 pp.

Marquis, D. A. 1966. Germination and growth of paper birch and yellow birch in simulated strip cuttings. *USFS Res. Paper* NH-54. 19 pp.

McDonald, P. M. 1976. Forest regeneration and seedling growth from five cutting methods in north central California. *USFS Res. Paper* PSW-115. 10 pp.

Minckler, L. S., and J. D. Woerheide. 1965. Reproduction of hardwoods 10 years after cutting as affected by site and opening size. *J. For.* **63**:103–107.

Mooney, H. A., et al. (Eds.) 1981. Fire regimes and ecosystem properties. *USFS Gen. Tech. Rept.* WO-26. 394 pp.

Noble, D. L., and R. R. Alexander. 1977. Environmental factors affecting regeneration of Engelmann spruce in the central Rocky Mountains. *For. Sci.* **23**:420–429.

Oke, T. R. 1978. *Boundary layer climates.* Methuen, London. 372 pp.

Panel on Aerial Seeding. 1981. *Sowing forests from the air.* National Academy Press, Washington, D.C. 35 pp.

Ronco, F. 1970. Influence of high light intensity on survival of planted Engelmann spruce. *For. Sci.* **16**:331–339.

Sander, I. L., and F. B. Clark. 1971. Reproduction of upland hardwood forests in the Central States. *USDA, Agr. Hbk.* 405. 25 pp.

Tappeiner, J. C., and J. A. Helms. 1971. Natural regeneration of Douglas-fir and white fir on exposed sites in the Sierra Nevada of California. *Amer. Midland Nat.* **86**:358–370.

U. S. Forest Service. 1965. Silvics of forest trees of the United States. *USDA, Agr. Hbk.* 271. 762 pp.

U. S. Forest Service. 1983. Silvicultural systems for the major forest types of the United States. *USDA, Agr. Hbk.* 445. 191 pp.

Wenger, K. F., and K. B. Trousdell. 1957. Natural regeneration of loblolly pine in the South Atlantic Coastal Plain. *USDA Production Rept.* 13. 78 pp.

Willson, M. F. 1983. *Plant reproductive strategy.* Wiley–Interscience, New York. 282 pp.

CHAPTER 8

Preparation and Treatment of the Site

In considering the regeneration of stands emphasis is traditionally and logically placed on the pattern by which the preexisting stand is removed. Most of the remaining chapters of this book are, in fact devoted to these patterns of cutting. Such emphasis on harvest cutting should not obscure the fact that regeneration may also depend on accessory measures to dispose of debris, reduce the competition of unharvested vegetation and prepare the soil for the new trees. Sometimes it is as logical to prepare the ground for forest crops as for agricultural crops. However, it is necessary to remember that most tree species are of later successional status than most agricultural crop species, which are mainly herbaceous pioneers adapted to colonize severely disturbed soils.

Acceptable regeneration can often be obtained without deliberate site preparation, but it is important to distinguish between total absence of such treatment and the unintentional kinds resulting from logging and slash disposal or prescribed burning undertaken for fuel reduction. This is not to say that the effects of logging, slash disposal, and prescribed burning always facilitate regeneration. If reliance is being placed on established advance regeneration, it may be logical that the harvesting of the previous crop be accomplished with as little disturbance as possible. The risk of harm to the soil must also be considered.

Site preparation, whether deliberate or unintentional, may be more crucial in the establishment of regeneration than the method of reproduction cutting. Many species can be reproduced by several different methods of cutting but by only one general program of site preparation dictated by the characteristics of species and site. The important objective is to prescribe and create environmental conditions conducive to the establishment and growth of the desired species. Efforts can be wasted or damage done if techniques of site preparation are either overdone or automatically applied as ends in themselves.

Some of the techniques have overlapping objectives and can also be intended for objectives other than growing trees. Most of the treatments are applied during the period of establishment, but some are started well in advance of harvest cutting or applied occasionally throughout the rotation.

The more important of the treatments, other than those covered in previous chapters, may be imperfectly divided as follows:

1. Disposal of logging slash
2. Treatment of the forest floor and competing vegetation:
   a. prescribed burning
   b. mechanical treatment
3. Treatment of the mineral soil
   a. fertilization
   b. drainage and irrigation

SLASH DISPOSAL

Effects of Slash and Its Treatment

The appearance of debris left by harvesting operations is so offensive that it is not easy to be entirely objective about determining the extent of disposal. Slash can be simultaneously harmful and beneficial; its treatment can be very expensive and the resulting benefits are mostly rather indirect. Consequently, the problems created by slash and other organic materials must be thought of as an integrated whole in terms of their effect on the productivity, utility, and safety of specific stands and site (Cramer, 1974; Martin and Dell, 1978; Kraemer and Hermann, 1979).

Slash in Relation to Forest Fires

Most slash disposal is still applied primarily to reduce the potential fuel for forest fires (Brown and Davis, 1973; Chandler *et al.*, 1983; Pyne, 1984). Slash is a fire hazard because it represents an unusually large volume of fuel; it is often so distributed that it dangerously impedes construction of fire lines. Therefore, the debris left after trees are cut is potential fuel that would not be there but for the cutting.

The greatest menace exists during the short period in which the foliage and small branchlets remain on the slash; they are readily ignited and burn rapidly. The larger materials are not easily kindled and do not normally burn rapidly, although they give off large quantities of heat. However, when conditions are favorable to very hot fires the size of the units of fuel is no longer a factor limiting the rate at which a fire will spread. Therefore, during bad fire weather, fires can burn rapidly in large concentrations of slash and may "blow up" into well-nigh uncontrollable conflagrations. The main objective of slash disposal for fire control is the prevention of such catastrophes.

Fires on cutover areas almost invariably start and spread in the litter of the old forest floor. An area can be rendered temporarily fireproof through the elimination of this blanket of fuel, but the effect lasts only until the first crop of herbaceous vegetation dries out. Therefore, no method of treating the potential fuels of forest fires is a substitute for a good system of fire control. The prevalence of fires on heavily cutover areas is the result not of the presence of slash but of the desiccation of the exposed forest floor.

Policies of slash disposal have been heavily influenced by the popular misconception that the danger of bad fires on cutover lands can be eliminated by destruction of logging debris. Actually the menace of slash can be diminished not only through reduction of its amount but also through any measures that break up its continuity, protect it from sun and wind, or decrease the risk of ignition.

Effect of Slash on Reproduction

In addition to being a hazard and impediment in fire control, logging debris often hinders the establishment of reproduction. This harmful influence is caused principally by the heavy shade and injurious mechanical effect of dense concentrations of slash. In such places advance regeneration is buried or crushed and the establishment of new seedlings is prevented by the shade. Slash composed of green branches is more detrimental than that of dead branches because of the heavier shade.

The magnitude of the harmful effect depends on the proportion of the area covered as well as the thickness and density of the slash. Thick, dense layers of slash, until broken up by decay, prevent the establishment of reproduction. Sometimes the first plants to appear on the sites of old slash piles are undesirable grasses, herbs, and shrubs that may usurp the area for long periods. Thin, loose layers of slash, on the other hand, may be of real benefit to young seedlings by protecting them from extremes of temperature, desiccation, grazing animals, and the competition of intolerant vegetation. In fact, the complete removal of all potential slash in whole-tree logging can sometimes cause the death of the small crucial advance growth of exposure-intolerant species such as the true firs.

Slash disposal is very often done to reduce physical impediments to hand or machine planting.

Management of Slash, Litter, and Soil

Logging debris and forest-floor litter can be viewed as storehouses of carbohydrate energy and inorganic nutrients that can be manipulated in various ways. They can be left to decay naturally, burned, or removed from the site.

Decisions about how to treat such material depend on the extent to which unincorporated organic matter accumulates on the forest floor. Under conditions that do not favor decomposition, organic matter may accumulate and tie up nutrients to such an extent that it may take a hot fire every century or two to rejuvenate the system. This situation can exist in the fire-ruled kinds of boreal forests in parts of Canada. At the other extreme is the true tropical rain forest where litter does not accumulate on the forest floor and almost all of the nutrient capital is continually cycling through the living vegetation. There one should try to conserve every scrap of dead organic matter because the organic part of the system is the only important nutrient reservoir.

Most forests lie between these extremes and are places where the mineral soil is the chief storing place of chemical nutrients. However, they vary enough that one should know where the various stores of nutrients and energy lie at the time of any treatment.

If these organic materials are allowed to decay naturally, most of the nutrients are ultimately returned to the soil and living vegetation. In the meantime, they are apt to remain unavailable to the vegetation. Substantial amounts of nitrogen remain bound away in the body proteins of the micro-organisms responsible for the final stages of decay. The effects of immobilization of nutrients can be detrimental on infertile soil or in climates where thick layers of organic matter normally accumulate beneath the forest.

If these kinds of dead organic matter decay in place, some of the energy stored in them goes to nourish the large and small soil organisms that churn it and are chiefly responsible for maintaining its good physical properties. The concomitant incorporation of organic matter in the mineral soil is important in maintaining the capacity of the soil to hold water, oxygen, and nutrients. Slash covering tends to decelerate the rate at which snow melts. Slash and litter also restrict direct evaporation of water from the soil, as well as prevent rain-splash erosion and frost heaving. However, those effects that maintain the porosity of the mineral soil are the ones chiefly responsible for preventing soil erosion.

Deposition of slash over actively eroding areas is not as effective as casual consideration might suggest. The way to prevent water erosion is to get the water to infiltrate so that it does not run over the surface picking up organic and mineral fragments. Infiltrated water usually emerges later in clear-water springs; clean water thus comes through the "dirt."

Effects of Burning on Nutrients and Soil

If dead organic materials are burned, their stored energy goes mostly to heat the air and the stored chemical nutrients are released. Some nitrogen compounds are volatilized and lost into the atmosphere. Most of the nutrient elements that are of essentially mineral origin are returned to the soil in more readily available form than before. Sometimes the increased amount of chemical nutrient in the mineral soil stimulates the nonsymbiotic nitrogen-fixers enough that the supply of available nitrogen becomes greater than before. It is possible for some chemical nutrients, especially nitrate nitrogen and potassium, to be made mobile enough by burning to accelerate loss by leaching and surface runoff. In most situations, however, burning either improves the chemical properties of the soil by accelerating nutrient cycling or does them little harm.

The effects of burning on the physical properties of the soil are either minor or deleterious. They are usually minor because most fires do not burn up all the incorporated organic matter and enough remains to continue most beneficial effects. Severe heating of the mineral soil takes place only where large concentrations of debris or logs burn for an hour or more; this incinerates incorporated organic matter and can cause severe soil damage, but usually only in small spots where the soil is baked red.

There are some soils that acquire **hydrophobic** properties, that is, resistance to wetting, because the soil particles get coated with waxy substances. These abound on the leaves of xerophyllic vegetation such as that of the California chaparral.

Fire can damage forest soils but only under unusual circumstances such as combinations of hot fires, steep slopes, and compact soils. The effects are more commonly small or even beneficial. The removal of any organic matter from a site inevitably takes away some nutrients and stored energy. Forest systems can replace the energy easily, so that kind of loss becomes serious only if impoverished animals or people divert almost all of it away from the soil-improving organisms. The nutrient losses are of greatest consequence because they are harder to replace. Aside from the special case of nitrogen, the available inorganic nutrients of the soil are replaced by (1) decomposition of rock minerals in the soil and (2) atmospheric fallout of dust, sea-salts, and the products of other forms of air contamination.

The crucial nitrogen compounds come and go from the huge but remarkably inert reservoir of nitrogen gas in the atmosphere; virtually none are of mineral origin. Most useful nitrogen compounds are captured from the air by nitrogen-fixing microorganisms, but they can also be fixed by lightning discharges and high-temperature combustion. While available nitrogen compounds are almost always in short supply and a cause for concern, they are more easily replaced by nature than the nutrients of truly mineral origin.

Effects of Removing Organic Materials

If nothing but stemwood larger than about 10 cm in diameter is removed, nutrient depletion is seldom likely to be of consequence. While this large wood is a major part of the energy storage by forests, the nutrients in wood are greatly diluted in the products fixed from water and carbon dioxide. If only large wood is removed, the nutrients lost with it are usually somewhat more or somewhat less than the amounts made newly available by natural processes while that amount of wood grows. Although this statement is ordinarily true, there is no inherent reason why it should be. It would become dangerously false in those cases where rock weathering could not be much of a source of newly available nutrients. This might be true of peat bogs, sands composed largely of silicon dioxide, and the strongly leached soils of tropical rain forests, where most of the nutrients are in the vegetation already.

In the most common condition, most of the nutrients are concentrated in the leaves, twigs, rootlets, bark, and especially the litter layers of the forest. Often there is also a large reservoir in the mineral soil, but none of the treatments now under consideration affect that directly. If these materials are left on the ground (or burned in place) at times of harvest, no soil damage is likely. If they are removed for utilization or pushed too far to the side, varying degrees of nutrient depletion can result.

The annual removal of the litter for fuel or agricultural mulches, common where land resources are overstrained, is one of the most damaging things that can be done to forests and their soil. It leads not only to nutrient depletion but also to serious erosion because the food supply of the soil-improving organisms is diverted away from them.

Some of the same kinds of problems could result from very close utilization of forest production by whole-tree chipping for fuel, pulp, and animal food. These purposes do not require removing the litter and there is little use in removing green leaves except for their food value. The extent of any potential problems depends on how much of which plant structures are removed and how frequently. Need for replacement of nutrient losses by fertilization may be anticipated, although at rates far less than those common in agriculture.

With any form of close utilization it is well to try to leave the leaves, twigs, and small roots where they grew. Their high content of water and poor fiber characteristics make them more valuable for their nutrient content than for fuel or pulp. Their inclusion in chipping processes is often only an attempt to avoid the high cost of delimbing. Of course, it is also theoretically possible to bring some of the nutrients back to the forest in the form of ash from burned wastes.

It is important to note that timber-production silviculture is very much less depletive than most forms of agriculture and grazing. Furthermore, fire and, for that matter, herbicides are kinder to the soil than almost any mechanical treatment. These observations are contrary to popular intuition but they are true so long as forest systems are understood and not pushed beyond their limits (Stone, 1973).

Effect of Burning Forest Fuels on the Air

The most important problems with smoke from forest fuels come from unburned particles that make it a source of dirt and restriction to visibility (Southern Forest Fire Laboratory, 1977). Therefore, the drier the fuel and the quicker and more complete the combustion the better it is for the quality of the air. Furthermore, the conditions conducive to such good combustion are usually ones in which smoke columns rise quickly so that the pollutants are soon dispersed thinly in the atmosphere. The more rapidly the air temperature decreases with height the better is this kind of vertical dilution. If there is a temperature inversion, that is, a situation in which warm air has settled atop cooler air, the smoke will accumulate beneath an otherwise invisible ceiling formed by the warm air.

Unfortunately, the atmospheric conditions that dilute the smoke most rapidly are the same ones most conducive to dangerously rapid spread of fires. Therefore, compromises have to be made and these reduce the number of days when silvicultural burning is possible. Greatest care is needed where the spreading of fire would do great harm or where the spreading of smoke would cause dangerous restrictions of visibility along highways or at airports. Sometimes special permits are required for such burning. It is at all times necessary to coordinate the operations with the best available knowledge about how the weather conditions will affect behavior of fires and smoke from them.

Small amounts of noxious gases, mostly carbon monoxide, are produced by burning forest fuels but they are so quickly diluted that there is little evidence of harm caused by them. As is the case with the dirt from unburned particles, the poorer the combustion the greater is the production of noxious gases. In this respect, cool, smouldering fires are more harmful than hot, vigorous ones. While the dangers do not seem great, it is best to caution woods workers not to breathe much smoke.

Most of the smoke generated from forest fuels comes from wildfires or land-clearing fires. Many forests grow in seasonally dry climates in which it is not a question of whether the forest fuels will burn but of when and under what conditions. The pollution effects and other kinds of harm will be least if the burning is conducted at deliberately prescribed times rather than determined by accident or human incendiaries.

Both burning and decay of organic substances from the forest and elsewhere change the carbon of those substances into atmospheric carbon dioxide. There is concern that continuing increases in the amount of this gas in the atmosphere will block so much outgoing radiation as to cause significant climatic warming. Unfortunately it does not appear that the increased carbon dioxide goes entirely toward making the world's vegetation grow faster because the amounts in the atmosphere seem to increase. There is a net transfer of carbon to the atmosphere when forests are destroyed and not replaced with new ones. However, forests that are actively accumulating wood remove carbon dioxide from the air. Wood put to structural use continues to sequester carbon. Silviculture, especially that aimed at timber production, is not a cause of this problem but is really one of the very best cures available.

Slash in Relation to Insects and Fungi

The insects and fungi that feed on logging debris are more beneficial than harmful because they are primarily responsible for the disintegration and decay of unburned slash. The vast majority of them are scavengers or saprophytes which do not attack living trees.

The few species of bark beetles and heart-rotting fungi which can spread from slash to living trees are mostly found in the cull logs, stumps, and large branches that are rarely eliminated in conventional slash disposal. The injurious fungi that proliferate in large pieces of slash are, for example, those which had already infected the living trees. Some of them may produce fruiting bodies more abundantly after cutting than in living trees. The treatment of slash to control insects and fungi is, therefore, best accomplished by such measures as close utilization, application of insecticides, and indirect methods of combating the proliferation and spread of harmful organisms. Ordinary slash-burning does not necessarily have any effect on the situation and may even aggravate it by killing unmerchantable, standing trees.

The most important aspect of the influence of insects and fungi on slash is the rate at which they bring about disintegration. It may take anywhere from several years to several decades for slash to decompose, depending on climate and species. The question of how much time is required has an important bearing on whether or how to treat the slash and also on the planning of subsequent operations.

Hardwood slash, for example, tends to remain moist and decays so rapidly that it is seldom necessary to burn it. Slash generally decays more swiftly on good sites than on those that are extremely wet or dry. Partial shade is conducive to decomposition.

Slash in Relation to Harvesting Operations

Efficiency in log transportation depends to a large extent on the success with which the slash is concentrated during felling. The interests of both logging and silviculture are usually best reconciled by concentrating the slash in a large number of small compact piles or in long, narrow strips. The consolidation of slash into a few very large piles is as detrimental to silvicultural purposes as the diffuse scattering of debris is to efficient transportation of logs. There are occasions when the slash from one cutting may remain long enough to impede subsequent operations.

The amount of logging residue and the diameters of its components depend on the extent of utilization of the felled trees. Large, untrimmed tree-tops are especially detrimental. They cover much space and their loosely arranged twigs and foliage allow fires to travel rapidly. Increased intensity of utilization reduces the amount of large debris; moreover, it almost automatically ensures that any fine fuels that remain are left more compact and closer to the ground so that they burn more slowly and rot faster.

Slash in Relation to Aesthetics and Wildlife

Logging debris is so unsightly that it is desirable to dispose of it or refrain from cutting in recreation areas and along public ways. Because of this and also to help with fire control, laws often require slash disposal within specified distances from highways, railroads, habitations, and adjoining properties.

Slash should not be deposited in waterways because its decomposition may reduce the oxygen content of the water below the level required for many species of fish. Heavy cutting along the margins of small streams may also increase the temperature too much for maintenance of desirable fish populations. On the other hand, it is possible to overdo these precautions. Some log dams are needed to preserve gravel beds for fish-spawning and some debris is needed to provide insect food and cation-exchange sites in the water. Distilled water at temperatures slightly above freezing may be full of oxygen but not of life. Slash often provides excellent cover, nesting sites, and insect food for birds and small mammals so it can be managed for their benefit.

Methods and Application of Slash Disposal

Tremendous advances in fire control and more complete wood utilization have caused slash disposal to be one American forestry practice that becomes less important with the passage of time. When it is done, it is increasingly done to facilitate planting or natural regeneration.

Slash disposal for fuel reduction is common practice in the West, which has long, rainless summers and old-growth forests with many rotten trees. Most of it is done by cheap broadcast burning. In this region and others, efforts are often limited to eliminating slash in narrow bands along routes of travel or at intervals across cutover areas in order to break up concentrations and provide places for fire-line construction in event of need.

In most other parts of North America the disposal of slash for fire control has generally been less necessary. In the northern coniferous forests limited kinds of slash disposal are sometimes desirable. In the southern pine forests slash rots so quickly that it rarely requires treatment and the basic purpose is often more effectively accomplished by prescribed burning of the forest floor anyhow. In the eastern hardwood forests and in the Northeast slash also rots quickly enough that disposal is required only along roads and in similar places.

Broadcast Burning of Slash

In broadcast burning, the slash on clearcut areas is burned as it lies within prepared fire lines. Practically all the remaining vegetation, except for that of sprouting species, is destroyed. This precludes reliance on advance reproduction but eliminates much of any undesirable vegetation that is present. The extent of exposure of mineral soil is actually rather variable depending on the moisture content of the forest floor at time of burning. Usually an ample amount of mineral soil is exposed; the areas where fires burn with such sustained heat as to damage the physical properties of the soil are rarely large enough to be of much significance. The sites are left in reasonably good condition for hand planting, direct seeding, or natural seeding from adjacent stands.

Broadcast burning is often associated with the clearcutting of old-growth stands in the West (Fig. 8-1). It can also be applied where scattered seed trees have been reserved provided that fuel is removed from beneath the trees.

Unless some source of wind-dispersed seed is close by, broadcast burning is usually compatible only with artificial regeneration. It is desirable to set the fires quickly under just the right conditions and in patterns that will cause the fires to burn away from the edges. The technique can be expedited by dropping incendiary devices, first developed for use in Australian eucalypt forests, from aircraft.

Spot Burning

If the fires can be depended upon not to spread dangerously, it may suffice to limit broadcast burning to concentrations of slash in spots or patches. Sometimes this is a way of conserving seeds that have already fallen or of reducing any baneful effects of slash burning.

FIGURE 8-1 Broadcast burning in preparation for planting after the clearcutting of an overmature stand of the western white pine type, Deception Creek Experimental forest, Idaho. Note the large volume of defective grand fir and western hemlock felled before the burning. If this had been left standing, the resulting dead snags might have become ignited in wildfires and spread burning embers far and wide. (*Photograph by U. S. Forest Service.*)

Burning of Piled Slash

Slash disposal associated with partial cutting usually involves the burning of piled slash. Where much of this sort of work is necessary it is now ordinarily done by pushing the slash into piles with bulldozers or similar equipment for subsequent burning. Use of such equipment has the important incidental advantage of reducing the competing vegetation and exposing the mineral soil on the treated areas. This effect makes an important contribution to the success of natural regeneration or planting where competition from brush is a serious problem after partial cuttings as in the mixed conifer types of the Sierra Nevada (Fig. 8-2).

The older methods of expensive hand piling have little silvicultural effect other than freeing advance regeneration and exposing some mineral soil. As far as the techniques are concerned, distinction is drawn between **progressive burning** in which the slash is laid on piles as the burning progresses and **piling and burning** in which the piles are constructed well in advance of the time when it becomes safe enough to burn them.

FIGURE 8-2 Dense, 6-year-old natural regeneration of ponderosa pine resulting from very intensive site preparation at Blacks Mountain Experimental Forest on the eastern slope of the Sierra Nevada. Most of the slash was piled mechanically in windrows along the edge of the opening, which was created in a group-selection cutting. During the next good seed-year the area was disk-plowed and the rodent population was reduced by poisoning. (*Photograph by U. S. Forest Service.*)

Piling of slash is often useful for aesthetic and wildlife management purposes. If such work is done by hand, the cost can be kept in check by planning to move the material over only short distances to many small piles. However, the burning of small piles of slash can be very costly, partly because workers easily become mesmerized watching the fires.

Mechanical piling is well adapted to the rather open forests of the interior ponderosa pine type, provided that the terrain is not too steep or rocky. The opportunity to dispose of large material in which bark beetles might breed represents a highly important advantage as far as management of ponderosa pine is concerned. Mechanical piling has also been used after clearcutting in the lodgepole pine type, and it leads to fully as good fuel reduction and to better reproduction than broadcast burning.

Lopping and Scattering of Slash

Some objectives of slash disposal can be accomplished without burning. The fire hazard can be reduced simply by lopping the tops so that the severed branches lie closer to the ground. This is advantageous in freeing saplings and seedlings of advance growth that have been bent over by falling tree tops. A limited amount of this kind of work often obviates the necessity for planting or waiting for new natural reproduction and forestalls subsequent difficulties with malformed trees. Bent trees should be released as promptly as possible; they straighten up much better at the beginning of the growing season than if they are not given an opportunity to do so until growth has been going on for several weeks.

Although lopping is one of the cheaper forms of slash disposal, anything that involves scattering or moving slash can be rather expensive. Lopped slash is sometimes scattered in instances where any sort of burning would destroy too much advance reproduction. The deliberate scattering of slash provides an artificial means of seed dispersal for closed-cone conifers.

It is also possible to break up concentrations of slash or eliminate it in strips along routes of travel simply by moving it. None of these techniques of redistributing slash reduce the total amount of fuel. However, they are often more compatible with silvicultural objectives than methods involving burning, especially in situations where the risk of wild fire is not very great.

Direct Control of Bark Beetles in Slash

If the hazard of outbreaks of bark beetles is serious and the utilization cannot be close enough to eliminate the felled material in which they breed, it may be most expedient to spray the slash with insecticides. The best time for spraying the bark comes just before the normal time of egg laying. Other techniques of direct control are mentioned in the section on salvage cutting (Chapter 6).

Chipping and Yarding of Slash

Portable chipping devices are sometimes used where slash disposal is essential and equally expensive piling and burning is the only alternative or when burning cannot be done. Such devices are, of course, most useful when the chipped material can be utilized for pulp, fuel, or mulching road and trail surfaces.

Slash disposal problems can sometimes be mitigated by transporting large unmerchantable material along with merchantable logs to central landings. The resulting concentrations can be either burned at these points or stored where they will be accessible when market conditions will permit their utilization. This technique of "yarding unmerchantable material" is sometimes used in conjunction with the cable logging of old-growth timber in the West. Under the circumstances, there may be only one opportunity ever to move large, defective trees out of the woods.

TREATMENT OF THE FOREST FLOOR AND COMPETING VEGETATION

It has already been stated that the establishment and development of a new forest crop can take place only if sufficient growing space is made available by harvesting or killing all or part of the preceding crop. Similarly, any sort of regeneration from seed is affected by the condition of the seedbed. Basic requirements cannot always be met by judicious adjustment of the pattern of cutting, by intentional or unintentional disturbance in logging, or by treatments of undesirable vegetation that are delayed until after the reproduction is established. Deliberate measures such as prescribed burning and mechanical site preparation are sometimes necessary to create the appropriate environmental conditions for natural or artificial reproduction.

Seedbed Preparation

The preparation of seedbeds ordinarily involves treatment of the **forest floor,** which is the layer of unincorporated organic matter that lies on top of the mineral soil and is composed of fallen leaves, twigs, and other plant remains in various stages of decomposition. This material does not make a good seedbed for most small-seeded species. Mineral soil can be exposed by burning or mechanical scarification. Too much exposure can invite excessively dense regeneration or invasion of undesirable pioneer vegetation. Often the exposure of a small patch every meter or two is ideal and machines have been developed in Scandinavia for this very purpose.

Competing Vegetation

The harvesting of trees inevitably leaves some vacant growing space, at least temporarily. Even after a very heavy cutting, however, there may be more serious competition, existing or potential, from unwanted vegetation than might be inferred from outward appearances or the severity of the cutting. Unless the markets are good, there is certain to be a residue of trees that were not worth cutting. There may also be low vegetation consisting of grasses, herbaceous plants, shrubs, or advance growth of undesirable tree species. If no vegetation shows above ground, there may still be root-stocks of sprouting species. Finally, if little or no vegetation remains and if the site is reasonably favorable for plants, the way is open for invasion by whole armies of plants.

The most serious problems arise when the competing vegetation has existed for many years and is not worth harvesting at all. The low shrubs of brushfields and the grasses or other herbaceous growth of "open" lands do not cast as much shade as a closed forest but can cause even more root competition as far as seedlings are concerned. Consequently, preexisting vegetation is most likely to have to be controlled when it is so worthless that artificial regeneration is necessary (Stewart, Gross, and Honkala, 1984).

Grass and grass-like vegetation hamper the growth of trees more than might be inferred from their comparatively short stature (Larson and Schubert, 1969). Last year's grass, lying brown and battered on the ground at planting time, can be insidiously deceptive and should be assessed in terms of the height it attains. Tall, dense grass often competes with tree seedlings enough to reduce survival. Grass competition is almost always detrimental if there are serious seasonal moisture deficiencies. It is, for example, the chief cause of the "grass stage" of longleaf pine and of "check" in planted spruce, conditions in which seedlings grow little or not at all in height until, after some years, they are able to develop large enough root systems.

Not all of the effects of grass or other inhibiting vegetation are from competition for light, water, and nutrients. There is a growing body of evidence about **allelopathic effects,** which are chemical antagonisms between different species that enable one species to poison the progeny of other species or, sometimes, its own (Daniel and Schmidt, 1972; Rice, 1984).

The effects of competing vegetation can be useful as well as harmful. Eastern North America has vast areas of unnaturally pure stands of various pines and spruces that spontaneously reforested abandoned grassy fields and are testimony to the ability of grass to exclude many hardwood species (Fig. 8-3). Some of this was the result of selective grazing but much of it was probably from the competition and allelopathic effects of grass. These effects of grass can be great enough that planted stands of hardwoods may have to be cultivated like row-crops during the first year or two.

In the South, the annual and then perennial grasses that normally appear after burning or mechanical site preparation probably serve to inhibit hardwoods. This effect has been deliberately used to inhibit reinvasion by the native angiosperm forest in the Jari Valley of Amazonian Brazil after it was clearcut and replaced by planted Caribbean pine.

More generally the establishment of forest trees is often assisted by temporary protective effects of other vegetation, especially the herbaceous

FIGURE 8-3 An old-field stand of shortleaf pine invading abandoned agricultural land in the Arkansas Ozarks. Since the grass inhibits broadleaved species more than the conifers, the new pine stand will be unnaturally pure. (*Photograph by U. S. Forest Service.*)

kinds. Such cover may be essential in preventing damage by heat or frost when the tree seedlings are very young and succulent. The theoretically ideal kind of accessory vegetation would be that which protected but did not compete; mosses, which are not capable of pulling water and nutrients from the soil, sometimes come close to this ideal. The next best would be plants that died or became overtopped by young trees promptly after their protective effects were ended. Even this does not always help. A grass cover can, for example, cause frost damage to hardwoods after they emerge above it.

It must be fully anticipated that any lethal disturbance done deliberately or unintentionally in the process of forest regeneration will cause the appearance of some kinds of vegetation other than the species desired. Treatments that expose the mineral soil to sunlight inevitably call forth pioneer vegetation. The forester who carries out such treatments must have full knowledge of what this kind of vegetation will be and be ready to live with the consequences. Any treatments that eliminate preexisting vegetation usually fit best with the regeneration of species that are naturally adapted to follow fires. If the object is to regenerate shade-tolerant species not so adapted, such treatments can bring in so much undesirable vegetation as to be worse than useless.

The main thing is that one should know the kind of vegetation that will develop after any kind of treatment and conduct the chosen treatments in the light of this knowledge. If the preexisting vegetation is going to hamper the establishment of a new crop of trees it is usually best to eliminate as much of it as necessary before the regeneration step than to plan to temporize with it afterwards. Except for broadcast herbicide spraying, most kinds of selective weeding treatment are costly and cumbersome.

In the past, it has often been impractical to provide young trees with anything approaching complete freedom from competition. Herbicides and heavy mechanical equipment have now provided the power to kill the roots of competing vegetation. The spectacular increases in seedling growth that can be attained by such treatment have been observed in a wide variety of species and regions.

Techniques of Treatment

Prescribed burning and mechanical site preparation are techniques in which both the forest floor and competing vegetation can be treated simultaneously. Before considering these two methods in detail, attention is called to other means of accomplishing the objectives.

Some scarification of the mineral soil and reduction of competing vegetation is accomplished during logging and any disposal of slash that is undertaken. Although the resulting disturbance is often adequate it seldom exposes any substantial amount of mineral soil or eradicates much sprouting vegetation. Logging does not result in very complete scarification unless there is a deliberate attempt to skid almost every log over a different pathway on snow-free ground. The one kind of slash disposal most likely to achieve complete site preparation is that in which the slash is piled with bulldozers equipped with root rakes or similar equipment. Broadcast burning often eliminates most of the unincorporated organic matter, but does not necessarily reduce sprouting vegetation very much. Practically all the other methods of logging or slash disposal have limited and erratic effects.

There is no silvicultural treatment other than those mentioned that significantly interrupts the continuity of the forest floor, but there are a number of ways of attacking the competing vegetation. Most of these, such as cutting, girdling, and chemical treatment, were considered in Chapter 5. Cutting and girdling are, like fire, entirely adequate for species that do not sprout but are rather frustrating to employ against those that do. There are herbicides that can eliminate almost all vegetation from most kinds of sites but not always cheaply or with the kinds of species selectivity one might desire. Mechanical site preparation of the kind that involves ripping the roots of unwanted plants out of the ground is very effective but rather expensive and usually damaging to the soil. The combination of two or three methods of plant killing in sequence, but with each treatment of limited intensity, is often cheaper and less harmful than reliance on use of one treatment.

Grazing and browsing animals are sometimes selective enough in their feeding habits to cause moderate and temporary reduction of competing vegetation. Most of the grasses and other plants upon which they feed are capable of sprouting. Feeding that is heavy enough to reduce the competing vegetation substantially is often associated with effects harmful to tree seedlings. Nevertheless regulated herds of domestic animals can sometimes be used to control undesirable vegetation without suffering from malnutrition.

Prescribed Burning

Fire, like cutting, can be used both constructively and destructively in handling the forest. The practice of using regulated fires to reduce or eliminate the unincorporated organic matter of the forest floor or low, undesirable vegetation is called **prescribed** or **controlled burning.** The burning is conducted under such conditions that the size and intensity of the fires are no greater than necessary to achieve some clearly defined purpose of timber production, reduction of fire hazard, wildlife management, or improvement of grazing (Fahnestock, 1973; Chandler *et al.*, 1983).

These particular terms, the first of which is preferred, have been coined to distinguish the use of fire as a silvicultural tool from its application for purposes bearing little relationship to the maintenance of productive forests. Very similar types of burning have been traditionally employed in many localities, especially in the South, to keep the forest open enough for grazing or other uses.

For purposes of this discussion, prescribed burning is regarded as involving fires that are set to burn through fuels that naturally occur on the forest floor, usually under existing stands. The burning of slash involves hotter fires and much heavier concentrations of fuel so that it is a kind of treatment easily recognizable as being in a class by itself. However, it could be and often is regarded as a form of prescribed burning.

The Role of Fire in Nature and Silviculture

As has already been pointed out, the formative processes in the development of forest stands are disturbances that kill trees and make way for new ones. The characteristics of all stands are determined by the kind, frequency, and magnitude of disturbances that have affected the sites in the past. Climax communities are, in this sense, results of long series of small, light disturbances while pioneer stages are the product of catastrophe. Fire is one of the most common kinds of natural disturbance and many species, including some of the most valuable of the North American forest, represent natural adaptations of the vegetation to fire (Kozlowski and Ahlgren, 1974; Mooney *et al.*, 1981; Wright and Bailey, 1982; Pyne, 1984). Some are adapted to reproduce after severe fires with effects akin to those of broadcast burning and others to much gentler disturbances like those of prescribed burning.

Except for landslides, volcanic eruptions, and similar geologic events that set off true primary successions, the most severe natural disturbance of forests is the sequence of blow-down or insect kill followed by fire. Such fires have tremendous amounts of dead, dry fuel. The regeneration of some species appears to require their simulation. One of these is *Eucalyptus regans* of Tasmania, the species that grows almost as tall as coast redwood and does so faster; a lower stratum of non-eucalypt understory trees has to be felled to produce enough fuel to get fires of required intensity. The view has been advanced that some species have evolved to produce enough fuel to provide for their own pyrogenic renewal. Fire-fostered species may be divided into several categories depending on the nature of their adaptation to fire.

The *closed-cone* pines constitute the first group. They are best exemplified by jack pine and include lodgepole, Monterey, bishop, and sand pine. In these species regeneration occurs naturally after a crown fire has killed the old crop, exposed the mineral soil, and opened the cones. The second group is composed of species like the cherries and a number of undesirable shrubs, including certain species of *Ribes*, that have *hard-coated seeds* capable of surviving for long periods in the forest floor and springing up after fires. The third group consists of the large number of species that can reproduce from *sprouts;* this group is composed mainly of broad-leaved trees and shrubs but also includes a few conifers. The fourth category consists of *light-seeded* species that thrive on seedbeds of bare mineral soil exposed by fire but are not outstandingly resistant to fire.

The fourth group includes many valuable species such as the birches, Douglas-fir, eastern and western white pine, the spruces, sweet-gum, and yellow-poplar. Regeneration of these species after fires comes from seeds already present on the old trees or from those subsequently produced by trees that happened to survive because of their size or location. Those that are true pioneers form all or part of the main canopy almost from the time of their establishment. Those that are rather tolerant of shade are likely to become established simultaneously with faster-growing pioneer species and remain underneath until the death of the pioneers, which are usually short-lived. In other words, this fourth group is a heterogeneous one including a wide range of species that do not all respond to fire in exactly the same way.

In all four categories mentioned the species involved are primarily adapted to regeneration after catastrophic fires that destroy most of the trees of one generation and prepare the way for another. It would be unwise to duplicate these conflagrations because of waste and public danger. The broadcast burning of slash is sometimes used to simulate these effects but it is commonly necessary to resort to measures that do not necessarily include any burning at all.

Finally, there is a fifth category of species that have sufficiently *fire-resistant bark* to withstand burning at intervals throughout most of a single generation. This group consists of certain hard pines, notably those of the South. The most outstanding in longleaf pine, which thrives only as a result of periodic fires. The others are loblolly, pitch, shortleaf, slash, and ponderosa pine; there is a strong possibility that red pine also belongs in this group. Occasional surface fires were beneficial to the maintenance of these species in the original forest because they arrested natural succession, exposed favorable seedbeds, and prevented more destructive fires. Up to the present time prescribed burning has been extensively applied only to this group of fire-resistant trees. It has been practiced on a large scale mainly on the Atlantic Coastal Plain from New Jersey southward.

Many species have several kinds of adaptations to fire. Pitch pine, a species that grows on sites where it is constantly bedeviled by fire, is an outstanding example of such a species (Little and Somes, 1961). It develops the thick, fire-resistant bark characteristic of trees of the fifth group. Its cones have a tendency to be serotinous, like pines of the first category. Seedlings and saplings killed by fire sprout from the base, like species of the third group; older trees defoliated by fire usually sprout new crowns. Like all species of the fifth group it can be regarded as a particularly well-adapted member of the fourth.

Another way of viewing differing degrees of adaptation to fire is to note that a given locality may have species with adaptation to different frequencies of fire. In forested climates, sprouting perennial grasses often go with annual burning and sprouting shrubs with less frequent burning. Sprouting tree species are next in line if fires are too frequent for seed production. Serotinous-coned conifers or species with seeds that remain stored in the forest floor often come next. Coequal with these may be the pioneers with long-distance seed dispersal or the fire-resistant species that regenerate after light fires have burned beneath them. Next are usually those that regenerate after fires at long intervals, often under the protective cover of some simultaneously regenerating pioneer. The final members of the series are those that are simply so defeated by fire that they can return only after some intervening stages of succession. Not all members of such a series exist in a locality, but it is useful to know, for each locality, what species would be fostered by a given frequency of fire.

The fact that a species is adapted to fire does not necessarily mean that fire has a practical, safe, and feasible place in its silviculture. Various cutting practices, herbicidal and mechanical treatments, or other kinds of disturbance can be used to simulate the effects of fire. However, it is the most common of the regenerative disturbances of natural forests, its application is usually cheap, and its silvicultural role needs more use and understanding than it usually gets.

Purposes and Effects of Prescribed Burning

When properly done in appropriate situations, burning accomplishes a number of things.

1. The most common objective of prescribed burning is still **fuel reduction.** In many respects it is a more satisfactory method than conven-

tional slash disposal because it eliminates most of the readily inflammable fuels rather than just the debris left from logging. It is, however, important to note that prescribed burning does not render any area fireproof, except temporarily, and is, therefore, no substitute for a well-developed system of fire control.

The most important effect of prescribed burning in fuel reduction is the interruption of the horizontal and sometimes the vertical continuity of inflammable materials. The interruption of any vertical curtain of fuel is especially significant in the slash pine type of the Southeast and the oak–pine type of southern New Jersey. Areas of slash pine that have not been burned for a decade or more develop a tall understory of various inflammable shrubs that become draped with fallen pine needles. In fuels of this kind surface fires can rapidly develop into disastrous crown fires because there is a ready path for the flames to follow from the ground up into the crowns. In the oak–pine type of New Jersey the presence of a shrubby understory tends to create the same dangerous condition. In each of these regions, successful silviculture depends heavily on the prevention of crown fires and it has been shown that prescribed burning is the only dependable means of forestalling them.

2. Prescribed burning is also effective in **preparation of seedbeds** for regeneration of wind-disseminated species like the pines, which become established most readily on bare mineral soil. This effect is also beneficial in hampering reproduction of the heavy-seeded oaks that would displace many of the pines in the course of natural succession.

3. Prescribed burning is also a means of achieving **control of competing vegetation.** This often has the effect of arresting natural succession by killing understories representing stages later than the one desired. However, where the aim is to prevent invasion by hardwoods, as in stands of southern pines, burning must be done fairly often because only the seedlings are likely to be killed by fire; saplings will resprout and larger trees may only be scarred. If soil moisture is a seriously limiting factor, elimination of the understory may improve growth of the overstory substantially. Increases of as much as 25 percent have been observed (Zahner, 1955) on fine-textured soils in southern Arkansas, where the rainfall is much less than farther east.

Roots of perennial grasses are killed by prescribed burning only where there are large units of fuel, such as fallen snags or large chunks of wood, that ignite and burn for a long enough period to heat the soil in depth (Weaver, 1951). Both in nature and in practice, surface fires must occur quite frequently if they are to be very effective in keeping brush and other understory vegetation in check. Fire is really a rather cumbersome tool for the accomplishment of this purpose and it is now fortunate that it can now be supplemented or replaced by herbicides.

4. The most traditional use of fire in forests is for the **improvement of grazing** although it has commonly caused the destruction of forests. The best place for this kind of use in silviculture is in certain dry-site forests that have a characteristic understory of grasses. The amount and quality of this

kind of forage can be enhanced by periodic removal of litter and dead grass. Burning at the end of the dormant season accelerates the sprouting of green grass at the very time when the animals are most likely to be starving. The only nonsprouting North American tree species really compatible with frequent burning to favor grazing is longleaf pine and not even it will stand annual burnings. Many ponderosa pine forests have a typical grass understory and grass can be induced under stands of other American tree species by burning and grazing, but in these cases the burning has to be suspended during periods of tree regeneration. Attempts to combine grazing and timber production in closed stands that do not have a characteristic grass understory are seldom very good silviculture or animal husbandry.

5. Burning can also be used to stimulate herbaceous species or sprouting, woody perennials useful in the **management of wildlife** under some situations. One example of this is the use of fire to regenerate understory herbaceous legumes for bobwhite quail in southern pine forests.

6. Burning can be employed in **recreation management** to maintain a park-like appearance in stands that would otherwise develop understory jungles. Large areas of southern pine forests, which now have understory tangles of shrubs, small trees, and briars, were pleasanter places in the era of frequent burning to promote grazing. Openness in the lower strata also facilitates logging and other forestry operations.

7. Prescribed burning has sometimes been successfully used to achieve the effects of low **thinning** in sapling stands of ponderosa and southern pines. However, it is very difficult to strike the right balance between fires that kill too much and those that kill nothing.

8. The various effects of fire can probably be used more than they are for **control of pests,** although it aggravates problems with some. Most of the effects are subtle and indirect. One decisive role is in the direct control of the defoliating brown-spot fungus (*Scirrhia acicola*) of longleaf pine to be considered later. Burning also sometimes reduces problems with annosus root rot.

Potential Damage from Prescribed Burning

Fires started by incendiaries, accidents, or lightning cause so much damage to forests that it is not easy to reconcile prescribed burning with efforts to educate the public about fire prevention. Sometimes the fire-prevention advertising exaggerates the effects of fire and thus plants the seeds of future trouble.

Prescribed burning has its greatest usefulness, and is indeed used the most, in those very regions where difficulties with fire are the greatest, simply because valuable species that are naturally well adapted to fire are most likely to occur there. In many respects the use of fire in slash disposal and prescribed burning is actually a concession to the inevitable in localities where fires are discouragingly common and fuel reduction almost essential.

As already indicated only exceptional kinds of soils are harmed by fire. Prescribed burning is useful mainly in forest types where natural fires are common. If burning had any seriously harmful effects on the soils involved, they would probably already be ruined. Successful prescribed burning generally makes fires either less severe or less frequent than they were before the initiation of comprehensive programs of fire control and silvicultural management. The harmful effects of fire are much more likely to take the form of obvious and long-enduring effects on the vegetation than of subtle damage to the soil (Lutz, 1956).

The effects of prescribed burning on standing timber depend on the size of the trees and the extent to which their stems and crowns are heated by fires. Since the primary source of fuel lies on the ground, the damaging effects ordinarily take the form of complete or partial heat-girdling at the ground line. The hotter the fire the farther up the effects extend, especially on the leeward sides of the stems where heated air accumulates. If there is a large amount of fast-burning dry fuel on the ground, there is risk that enough heat will be generated to support burning or overheating of the foliage. No North American species other than longleaf pine can withstand burning until rough, thickened bark has developed on the lower parts of the stems; the main portion of the crown canopy must also be well above the height of the flames.

The extent of injury to any part of the tree depends on whether the living tissues are heated above the lethal threshold of 55°C and how long such a temperature is maintained. Therefore, any factor that hastens the transport of heated air out of the forest reduces the danger of damage. On level terrain it is, for this reason, better to conduct prescribed burning when there is a gentle breeze than in calm weather; however, on pronounced slopes sufficient updrafts develop to allow burning when there is no wind.

The initial temperature of the living tissues is also important in determining the highest temperature that they attain. The risk of injury is lowest when burning is done in the winter because it takes such a large amount of heat to raise the tissues to lethal temperatures. Head-fires, which travel with the wind, are often less damaging than back-fires, which burn against the wind, because the heat is carried upward more rapidly and high temperatures are not as long sustained close to the ground. However, head-fires are more likely to spread to the crowns and should thus be avoided where the crowns are likely to be damaged.

Prescribed burning is applicable only in stands composed of trees that have reached fire-resistant size. It is not easily used in uneven-aged stands because it damages reproduction. The only way in which controlled burning can be applied in uneven-aged stands is to have intervals between episodes of cutting and burning no shorter than the time required for reproduction to appear and develop resistance to fire.

The surest way to avoid damage to the forest from prescribed burning is to conduct it under the right conditions according to carefully developed plans. Like all silviculture operations, it is not an end in itself but a means of achieving some well-defined goal of management.

Methods of Prescribed Burning

The customary plan of action in conducting a prescribed burn is to isolate the area to be treated by means of plowed lines wide enough to contain the fire. Since the construction of these lines is costly, every reasonable advantage should be taken of preexisting barriers such as roads and swamps. The whole operation should be carefully mapped out with special regard for the selected conditions of wind and weather under which the burning is to be conducted (U. S. Forest Service, 1971). The fire lines should be plowed out in advance according to this plan, but not so early that they are covered over with fallen leaves before the burning is done.

The most common practice is to use back-fires (Fig. 8-4). Flanking or "quartering" fires, which are set in lines parallel to the wind, spread more rapidly, but are somewhat more likely to scorch the foliage of saplings or taller trees. Head-fires can be used when the forest floor is too moist to be burned by back-fires, thus increasing the number of days when treatment can be conducted. They are in some respects safer than back-fires because

FIGURE 8-4 Prescribed fire backing into the wind through needle litter and saw palmetto beneath a flatwoods stand of slash pine. (*Photograph by U. S. Forest Service.*)

they are more likely to go out than to escape from control if the wind shifts unexpectedly. The heat from them is carried away rapidly so that they are less likely to produce fire scars on the stems of trees than are back-fires. If head-fires can be used safely, they are cheaper than back-fires because they can cover an acre much more rapidly.

The area to be treated should be subdivided into units no larger than can be burned in a period of about 10 hours, although several units may be burned simultaneously. It is unwise to let a fire run for a longer period because of the possibility of unforeseen changes in wind and weather.

Weather forecasts and measurements of fire danger should be used to select times for burning that are both feasible and safe. The wind should be steady in direction and not erratic or too strong in speed. The moisture content of both fuel and soil is important not only for safety but also for determining whether the right intensity of burning will be secured. Fires that merely smoulder are costly to apply and cause much air pollution.

The burning is best planned and conducted by the same personnel involved in control of wildfires. Much can be learned about the behavior of wildfires from prescribed burning. There should be enough personnel and fire equipment at hand to deal with any fires that escape.

The cost of prescribed burning depends on the size and shape of the units burned, the inflammability of the fuels, the length of fire line needed, the size of the crew, and the shape of the terrain. If the units are more than 10 hectares, the costs are usually very low. The cost of fire lines is the chief variable. High cost may result if there is a heavy accumulation of fuel requiring a series of small fires or if there is so little fuel that the fires have to be relighted often. The costs of burning small tracts, areas with irregular boundaries, or narrow strips are rather high. However, if large areas can be treated in single operations, several burnings can be done for the cost of one broadcast spraying with herbicides. If the undesirable vegetation is small, such a series of burns may be almost as effective in controlling the vegetation and also reduces the amount of fuel very substantially.

Application of Prescribed Burning

The details of prescribed burning vary according to the species and objectives of treatment, the most important variable being the schedule of burning.

Longleaf pine is one important species that is ecologically almost totally dependent on frequent surface fires (Grelen, 1978). The seedlings germinate in late fall and thrive best beneath the one year's growth of grass developing after a previous winter burn. Until they form the first dormant terminal bud about 9 months later, they can be killed by fire. They stay in the peculiar grass stage, without growing in height, until the large tap-root becomes about one inch in diameter; a sudden spurt of height growth ensues with a brief period of moderate vulnerability to fire. During the grass stage, it is necessary to burn about every 3 years to keep the brown-spot disease in check. The spores that spread this disease can move so far by rain-splash that it is desirable to operate with even-aged stands and regeneration areas of about 100 hectares. Once the seedlings have become more than a meter tall, frequent burning remains desirable but the benefits become the same as with the other southern pines.

The most common purposes of prescribed burning under stands of southern pines are fuel reduction and the control of understory hardwoods that continually threaten to take over the stands (Fig. 8-5). The South has

FIGURE 8-5   The upper photograph shows an untreated 45-year-old stand of loblolly pine with a dense hardwood understory at the Santee experimental forest on the South Carolina coastal plain. The lower photograph was taken on the same spot 6 years later, just after the second of two partial cuttings and after four prescribed burning operations. (*Photograph by U. S. Forest Service.*)

many dry soils on which hardwoods grow fast during the juvenile stages and tend to overwhelm conifers that only later manifest their superior adaptation to such soils.

It may take rather frequent fires to keep hardwoods under control because only the seedlings are killed outright. Saplings usually resprout and can thus be killed back only to the ground. Larger trees have to be killed with herbicides or by girdling or cutting. One common sequence for eliminating degraded hardwood stands is (1) burning, (2) herbicide spraying of the resulting succulent sprouts, and (3) killing the large survivors by girdling or herbicide injection.

If large amounts of fuel have accumulated from long periods without fire, it may take several light winter fires to skim off enough successive layers to make it safe enough for fires that really favor regeneration. The fires that foster regeneration are often summer fires capable of exposing some mineral soil and killing small hardwoods (Fig. 8-6). Summer fires involve no risk of destroying any seeds that have fallen. Fires can be of many different combinations of timing and intensity; the choice of these can be a fine art.

Prescribed burning plays a very crucial role in prevention of crown fires in the slash pine forests of the deep South and the pitch pine barrens

FIGURE 8-6 A stand in the same place and treated similarly to that shown in Figure 8-5, but after an additional 5 years, with dense natural reproduction of pine replacing the original hardwood understory. The treatment here consisted of one winter fire followed by three successive annual summer burnings. (*Photograph by U. S. Forest Service.*)

of southern New Jersey. Although pitch pines can recover from crown fires by sprouting along the charred stems and branches, this impairs the form of subsequent growth of the trees badly. Crown fires in slash pine and most other tree species are lethal to trees and highly dangerous.

Fire has essentially the same role in the interior ponderosa pine forests as it does in the southern pines. On the best sites, fire prevents invasion by the relatively tolerant and less valuable Douglas-fir and true firs. There are also indications that light surface fires improve forage production and reduce the danger of serious fires.

In other kinds of American forests most silvicultural use of fire, except that for slash disposal, is still in an experimental stage. Fires burning beneath natural stands had an important effect in the maintenance of sugar and ponderosa pine, as well as the big-tree sequoia, in the mixed conifer forests of the Sierra Nevada (Kilgore and Taylor, 1979). These forests are as difficult to regenerate as they are magnificent. The situation is complicated by shrub species that are also favored by fire and by heavy fuel accumulations from decades of fire exclusion.

Many of the better forests of moist, cool, northern climates, especially the birch and coniferous types, owe their origin to the effects of fire (Vincent, 1965; Zasada *et al.*, 1977). However, most of the fires involved were of the catastrophic kind that occurred at long intervals rather than light, frequent ones. Therefore, the desirable effects of fire are normally achieved by broadcast burning of slash rather than by burning under the stands.

One characteristic that virtually all Australian eucalypts have in common is adaptation to fire (Hillis and Brown, 1978). In fact, Australian sites that are too moist for fires seldom have eucalypts. Bare mineral soil favors regeneration of this small-seeded genus. Many of the species sprout after crown fires; some can even renew their crowns by sprouting after crown fires. Many of the so-called ash group, usually of the moister sites, are adapted to regenerate after catastrophic fires, best simulated by clearcutting and slash burning. However, in some forests of drier sites prescribed burning is often conducted beneath the stands for fuel reduction and other purposes described in connection with the southern pines.

There is a significant pattern in the development of the application of prescribed burning in those localities where it is used. When management is started in a forest type that owes its better characteristics to repeated burning, it is most logical to attempt to conduct silviculture by methods that do not include prescribed burning. If this line of action produces a satisfactory result, there is no reason for resorting to the use of fire, which is at best a treacherous tool.

There are, nevertheless, some instances where such exclusion of fire created situations in which the problems multiplied progressively. The essential conditions for reproduction of the desirable trees were created less frequently, and the understories filled up with arborescent or shrubby species that were adapted to the new conditions and were often of little value. As fuels accumulated, wildfires that formerly caused little damage showed an increasing tendency to develop into conflagrations; oftentimes this development proceeded more rapidly than fire control could be improved to meet it. Frequently the point was reached were undesirable, sprouting vegetation could no longer be eliminated by fire and had to be left or attacked by more laborious methods. It is possible that this pattern of events will be recognized in more kinds of forests than is now the case.

Mechanical Treatment

The mechanical methods of site preparation include one or more of the processes described in the following paragraphs.

The **reduction of undesirable vegetation** can be done by uprooting large woody plants, chopping up smaller plants, or plowing under grasses and other herbaceous growth. The objective of this step is usually to disrupt the roots of the undesirable plants enough to kill them. Of course, mechanical equipment can be used simply to cut or break off the stems of woody plants, especially those that are incapable of sprouting. It is entirely possible that one might attempt to eliminate only the woody vegetation and leave any grasses or other low plants alone, especially if their competition is not likely to be significant and excludes more troublesome plants.

Practically all site preparation involves some **redistribution of dead vegetation.** This may range from concentrating uprooted trees into high windrows to the gentle scarification of thin litter beneath otherwise undisturbed vegetation. With large debris the objective is partly to remove obstacles to subsequent operations. It may also be desirable to hasten the destruction of debris; this can be done not only with fire but also by pushing it into compact piles or working it into the mineral soil so it will rot more quickly. With the gentler kinds of treatment there may be no objective other than **exposure of mineral soil.** Any method of mechanical site preparation exposes some of the mineral soil, often in the process of doing something more difficult.

If the site is subject to extremes of moisture conditions, activities occasionally include **reshaping** the soil surface. This usually is done by some sort of plowing or scraping to create low ridges in wet places or shallow trenches to collect water in dry areas. Sometimes **smoothing** of the surface with harrows may facilitate machine planting or other future activities.

The equipment used for the work varies widely in kind and size. Most of it requires tractor-powered devices. Simple scarification designed merely to expose mineral soil can be done with light equipment. Single or double-moldboard plows, including those designed for constructing fire lines, or special shaping devices, can be used for furrowing or bedding.

Bedding is the mounding up of low ridges or "beds" on flat, poorly drained soils such as those common on the Lower Coastal Plain of the southeastern United States (Fig 8-7). It has the effect of increasing the volume of soil that is sufficiently well supplied with oxygen and water to lengthen the period of physiological availability of water. This effect

FIGURE 8-7   A flat, poorly drained site in North Carolina bedded in preparation for planting loblolly pine. The raised beds or berms on which the trees are to be planted were cast up with a special plow and shaping device to provide ridges of aerated soil in which the roots can start their expansion.

speeds the growth of new stands at least until they close although some slight advantage might theoretically persist longer. Sometimes bedding is extended through force of habit to drier sites where it may do harm without any good.

Under extreme cases, the **loosening of compacted soils** may be done by plowing. Ordinarily the natural soil organisms of the forest keep the soil in a more porous and penetrable state than could be accomplished by mechanical churning. However, mechanical loosening does help if the hooves of herbivores or the weight of heavy machinery have made the soil very compact. The areas used for loading logs can be a particular problem. It has been found in Australia that it may be cheaper to do this with a small explosive charge for each planting hole than to move heavy plows to each landing site.

Very deep plowing is sometimes used to rupture sub-surface hardpans that have been formed by purely natural processes or as the result of unwise use of certain sensitive soils for grazing or agriculture. This sort of plowing is most common in very humid climates, such as those in parts of Britain and Ireland, where sphagnum bogs and hardpans can form, even on hillsides, because of the leaching associated with the podzolization process. If the accumulation of oxygen deficient water or merely resistance to root penetration restricts rooting depth, the breaking of impediments to downward movement of water helps growth. The reverse may be true if the hardpans conserve meager supplies of water. It is, as usual, well to proceed on the basis of knowledge of which growth factors are deficient in each case.

On very dry sites it may improve the survival and early growth of planted trees to plow out furrows or to construct contour terraces and depressions in ways that will collect water and uproot or bury competing vegetation. In some arid regions it is simply impossible to get trees to start without such artificial means of creating localized concentration of water. Terracing is a common practice in Israel and elsewhere in the Middle East

(Goor and Barney, 1968). It has also been successfully used in the northern Rocky Mountains although there has been some objection to the artificial appearance that it confers on the terrain.

Most site preparation is aimed chiefly at the destruction of competing vegetation. Plants of sapling size or smaller can be broken up and worked into the soil with heavy disk plows or rolling brush choppers (Fig. 8-8). It is often desirable to induce sprouting with an initial treatment and then follow it with a second to kill the new, succulent sprouts before they form buds. The second treatment can be another mechanical one or it can be prescribed burning or herbicide spraying.

Devices akin to large, rotary lawn mowers can be used to cut and shred small vegetation and woody debris preparatory to planting but they do not eliminate sprouting vegetation. Bulldozers or similar devices (Fig. 8-9) can be used against vegetation of any size from small brush to large trees. However, they must be fitted with special toothed blades or "root rakes" (Fig. 8-10) if they are to be very effective in uprooting anything without scraping away too much topsoil. The uprooting of large areas of sizable trees is sometimes done by using powerful tractors to drag battleship anchor chains over the ground. The huge links of these chains pinch around the bases of the trees and remain engaged until the trees are pulled

FIGURE 8-8 A bulldozer and two rolling brush cutters being used to destroy a cover of poor oak and grass preparatory to planting pine on a dry, sandy site on the Southeastern Coastal Plain. The two water-filled drums are hitched at an angle to one another and to the direction of travel to impart a shearing action to the sharp blades. (*Photograph by Caterpillar Tractor Co.*)

FIGURE 8-9 Clearing of oak on a dry upland site with a "K-G" blade. The sharpened blade is set at an angle so that it can be used to shear off the trees at or below the ground line; the upper bar pushes the trees over as they are being severed. The "stinger" at one end of the blade is shown splitting the stem of a large tree so that it can be cut off or pushed over in two passes of the machine. (*Photograph by Caterpillar Tractor Co.*)

over. Ordinarily pairs of tractors are used to drag one chain. The uprooted trees must then be pushed into windrows with bulldozers.

Mechanical site preparation of the kinds described is usually a preliminary to planting. If there has been any substantial overturning or upheaval of the mineral soil, it is necessary to delay planting for several months to allow the soil to settle. Otherwise there is risk that many of the roots of the seedlings will be planted in air pockets and die of drought. The smoothing of areas with heavy harrows is often aimed chiefly at making it possible to plant with machines. Although the competing vegetation is greatly reduced by such drastic treatment there are usually some sprouting rootstocks left in the soil. These techniques are often employed in the South to replace good pine stands. They are also advantageous in replacing degraded hardwood stands with pines; well-established hardwood stands are almost impossible to eliminate by prescribed burning.

In all treatments of this sort it is desirable to avoid the horizontal movement of organic materials and topsoil as much as possible. The churning or overturning of these materials in place is distinctly preferable to any sort of scraping action.

FIGURE 8-10   A root rake clearing a brush field for planting in the Sierra Nevada of California. (*Photograph by U. S. Forest Service.*)

Much of the nutrient capital of the forest is usually tied up in the litter and upper layers of the mineral soil. Most of the power of the forest system to resist erosion and protect water quality resides in the porous uppermost layer of mineral soil. There is simply no silvicultural practice that can impair the productive capacity of the forest or damage watersheds more than scraping off the surface materials. This is especially true if the nutrients are moved farther sideways than the future extent of the tree roots that might bring them back. The bulldozer should be regarded as an instrument with more potentialities for damage than fire, herbicides, or other forestry tools.

The practice of windrowing uprooted material is an especially questionable practice. It is impossible to do this without moving not only nutrients but also some mineral soil. This effect is often made apparent by the fact that planted trees (or weedy vegetation) grow fast in the windrows while the other trees may be stunted and chlorotic for a time (Fig. 8-11). The windrows sometimes waste growing space. Closer utilization may reduce the use of the practice but it is much better for the soil and site productivity to employ rolling brush cutters, plows, or other devices that work material into the soil without moving it sideways.

It is ironic that silvicultural efforts to emulate the clean cultivation of agriculture have developed at the same time that there are efforts to use

FIGURE 8-11 Windrowing of stumps and logging debris and the effect that the associated scraping action can have in moving nutrients sideways. Both pictures show site preparation for planting on the Atlantic Coastal plain in North Carolina. The upper one shows a very thorough job of concentrating the debris in a tight windrow. The lower picture shows tall loblolly pines growing in the nutrient concentration of such a windrow on the right but with stunted and somewhat chlorotic seedlings on the left in the zone from which materials had been pushed with a root-rake. (*Photographs by Yale University School of Forestry and Environmental Studies.*)

herbicides for "minimum-till" agriculture. Clean-cultivation agriculture is an inherently depletive process and the reason why silviculture is not is that it defends the soil against erosion by conserving rather than squandering soil organic matter.

Light scarification and the plowing of shallow furrows are usually cheap but measures that require powerful machinery can be very costly. Windrowing is one of the most expensive operations. If the investment in equipment is high the cost may vary considerably depending on how the depreciation of the machines is treated in the accounting. Efforts to reduce costs by clearing narrow lanes through tall undesirable vegetation do not always prove effective.

Mechanical site preparation can be useful, safe, and advantageous or it can be costly and harmful. Its use should be the result of thoughtful prescription and not a matter of unquestioned standard operating procedure.

It is important to note that the increase in growth that can be obtained in new stands because of the effects of reducing competing vegetation is limited to the period before the stand itself approaches full occupancy of the growing space. The early gains in growth are not lost but one should not extrapolate them beyond the time of stand closure.

IMPROVEMENT OF THE SITE

Fertilization, drainage, and irrigation are intensive treatments of the soil aimed more at improvement of site quality than at preparation for regeneration. The proper conduct of these treatments depends so heavily on knowledge of the complex chemical properties of soils and their moisture relationships that it is logical to refer to works on forest soils and other accounts, such as those of Pritchett (1979); Ballard and Gessel (1983), and Bowen and Nambiar (1984) for more complete information.

Fertilization

Few forest soils provide an optimum supply of the nutrient elements essential for the growth of trees. Sometimes marked deficiencies may exist because of improper land management in the past or merely because of inherently low natural fertility of the site. The nutrient elements most likely to be deficient are the NPK elements of most fertilizers, nitrogen, phosphorus, and potassium, in that order of frequency of deficiency.

It is often possible to make forest trees grow better by increasing the supply of nitrogen. The most common cases in which this does not happen are sites where the supply of water or phosphorus is the limiting factor instead. The most common nitrogen fertilizers used in forestry are urea compounds that yield positively charged ammonium ions, which can be absorbed in the cation exchange capacity of the soil and are easily taken up by plants. Urea–formaldehyde compounds have the advantage of releasing the ammonium slowly. While nitrate compounds can be used, the negatively charged nitrate ions are easily lost from the soil by leaching.

Nitrogen compounds in the forest system are constantly moving to and from the atmosphere or being lost to moving soil water. They can be very mobile and are lost because of leaching. They can also become unavailable to higher plants through being captured by decomposing organisms in the soil or by being locked up in undecomposed organic matter. This often happens where the soil climate is unfavorable to decomposition. It is sometimes possible to remedy nitrogen deficiencies by stimulating the symbiotic and non-symbiotic nitrogen-fixing organisms (Gordon and Wheeler, 1983). Fertilization with other elements sometimes stimulates the nitrogen-fixers as does the encouragement of legumes, alders, and other plants that support symbiotic nitrogen-fixing bacteria.

The cost of artificial fixation of nitrogen for fertilizer in terms of both money and energy is high. The effects of a given application are limited to several years, so repeated applications can be necessary. However, nitrogen is so crucial in building the proteinaceous biochemical machinery of life that spectacular effects can result from reducing the chronic deficiencies. Fertilization with nitrogen compounds alone has become moderately common in the humid parts of the Pacific Northwest (Miller and Fight, 1979) and in other forest areas where growing conditions are good and nitrogen may be the most important limitation.

Phosphorous deficiencies remediable by fertilization are common on poorly drained soils. In these and most other cases of phosphorous deficiency, the problem is that the acid condition of the soil or other factors cause too much phosphorous to be tied up in unavailable form in compounds with iron and aluminum. Phosphorous can be applied as ground phosphate rock or in more concentrated form. These are not very consumptive of energy in manufacture. Single applications have long-lasting effects.

The amounts of nitrogen and phosphorous available to plants must be well balanced. What might otherwise be the right amount of one can make a deficiency of the other more acute or even harmful. If a tree has more phosphorous to form the energy transfer machinery of photosynthesis, for example, it also needs more nitrogen to build the protein components of this machinery.

If there is opportunity for good nutrient recycling, single applications of potassium fertilizer in the forest seem to suffice for very long periods. Potassium is a very soluble and mobile compound. It is not only easily lost by leaching but even leaks out of green leaves. Deficiencies have been encountered on easily leached sandy soils previously subjected to highly extractive kinds of agricultural crop removal in New York State (Leaf and Leonard, 1973).

There are few cases in forestry in which actual benefit has come from fertilization with other elements. It is rather surprising that there is little evidence of improvements from adding calcium, magnesium, or manganese. Deficiencies of trace elements such as zinc, molybdenum, boron, cobalt, and copper are known mostly from a few districts in Australia.

Australia is an ancient and strongly leached continent where deficiencies of phosphorous and other elements have caused forest fertilization to be much more remarkably successful than in most parts of the world.

Forest fertilization is normally done from the air. It is usually desirable to avoid fertilizing the forest waters so as to avoid contributing to eutrophication. The materials to be spread are heavy enough that it helps to minimize flying distances. It is also important to be able to load the aircraft quickly.

In most cases it is best to restrict forest fertilization to the latter part of the rotation. A given amount of fertilizer seems to produce about the same amount of wood regardless of tree size and a given cubic volume put on large trees is worth more than the same volume put on small trees. Furthermore, the supply of available nutrients is generally greatest just after the destructive events associated with regeneration. Deficiencies are most likely to set in after the stands have filled all the growing space and more and more nutrients are getting tied up in living and dead organic materials on the site. Fertilization is done at the time of planting or regeneration only if the deficiencies are very serious, as on recently drained organic soils. Sometimes fertilization of young stands favors the competing vegetation more than the trees. There are other cases in which it is not needed and some in which it induces harmful imbalances between nutrient elements.

Drainage

Areas of stagnant, oxygen-deficient water can be almost like arid deserts. Paradoxically it is for the same reason, deficiency of available water (Kozlowski, 1984). The decomposing organisms attacking organic matter of swamps rob the water of the oxygen that the roots of higher plants must have to function. Therefore, trees growing in bogs can extend their roots downward only a few centimeters and the strata below are as impervious to roots as rock would be. This situation prevails only where water is slow-moving. Sites along freely flowing streams can seem equally wet but often are biologically the most productive of any of a given locality; in such places, the water is well aerated and freely available.

As is so frequently the case, it is important to put intuition aside and recognize that wet sites with ponded water are physiologically dry. At least it can be said that many tree species seem adapted to this phenomenon. Many species of literally dry sites (such as deep sands or thin soils) are also found in physiologically dry swamps. Some of the examples include red maple, black spruce, and some species of pine, although not all species of one habitat also grow in the other. The most common indicators of oxygen-deficient water, the world over, are sphagnum mosses. These mosses not only indicate bogs but they can, in very humid climates, even build them on hillsides because of their very large water-holding capacity.

Some of the previously described methods of using plows for bedding or puncturing hardpans accomplish some drainage, especially on sloping ground. However, water is viscous enough that such measures do not really move much off flat terrain. This requires ditches or canals as well as enough difference in elevation to provide places to which water can flow.

The most common method starts with the use of back-hoes or drag-lines to dig parallel primary canals; the fill placed between them becomes an access road (Fig. 8-12). Secondary ditches are constructed to lead to these usually at right angles and at intervals of 20–40 meters. In the process of site preparation for planting, the sites are often bedded with plows so that the trees can be started on low, well-drained ridges. Unless they have been enriched by nutrients flowing off adjacent uplands, the organic soils of such areas are deficient in nutrients, especially phosphorous and nitrogen. Fertilization is commonly needed and it usually takes much local experimentation to determine how much of what to add.

It must be anticipated that drainage will cause the newly aerated peat to decompose in such a manner that the surface will sink. Sometimes this proceeds to the point where the ditches have to be deepened. If there is then no longer enough difference in elevation for the water to move, the whole effort is defeated. The worst situation is with lands that have a thin layer of light fresh water over dense salty water. The salt water rises as the fresh water moves to the top while the surface subsides. On the other hand, if a treeless area is drained and afforested, the trees can transpire enough water to improve their own growth by removing excess water.

The largest forest drainage projects are those of peat bogs that have formed in former lakes in glaciated lands in Finland (Fig. 8-13), Scandinavia, and the Soviet Union. The techniques, which usually have to include fertilization of the sterile peats, are highly developed and based on detailed classifications of the different kinds of wet sites. So far there have been few attempts to drain similar bogs in Canada and the northern United States.

FIGURE 8-12 Drainage canal, with water-regulation device, designed to move water off a very flat site in North Carolina. The road is made of soil dug from the ditches on either side of it. The canal is adjacent to the bedded area shown in Fig. 8-7.

FIGURE 8-13 Scotch pine stands on drained peat land in Finland. The trees were absent or badly stunted before the drainage started 5 decades ago. The surface has subsided about a meter because of decomposition from the aeration of the drained peat. (*Photograph by Yale University School of Forestry and Environmental Studies.*)

Most American forest drainage is on the flat Lower Coastal Plain of the South. The most successful efforts have been with very flat but somewhat elevated areas distant from major rivers. These are wet mostly because of the flatness of the land and have comparatively thin layers of peat. If the drainage ways are kept open and phosphorous deficiencies are remedied, the drainage of these soils makes them so well supplied with aerated water that they become highly productive. The drainage of depressions in this region is more difficult and the results vary. Attempts to drain forest areas that are almost at sea level and underlain by salt water have been costly failures.

Irrigation

Except for seed-orchards, nurseries, and other sites of very intensive tree culture, there is little use of irrigation in forestry. Irrigation water is usually more valuable for agricultural use. There are a few places in the world where hybrid poplars and other species of alluvial flood plains are grown with supplemental irrigation water. The most common ways of increasing the water supply of forest trees are the various methods of reshaping the ground surface to concentrate surface runoff water on the roots of planted trees that were mentioned in connection with mechanical site preparation.

BIBLIOGRAPHY

Alexander, M. E., and F. G. Hawksworth. 1975. Wildland fires and dwarf mistletoes: a literature review of ecology and prescribed burning. *USFS Gen. Tech. Rept.* RM-14. 12 pp.

Ballard, R., and S. P. Gessel (Eds.) 1983. I.U.F.R.O. Symposium on Forest Site and Continuous Productivity. *USFS Gen. Tech. Rept.* PNW-163. 406 pp.

Bowen, G. D., and E. K. S. Nambiar. 1984. *Nutrition of plantation forests.* Academic, Orlando, Fla. 506 pp.

Brown, A. A., and K. P. Davis. 1973. *Forest fire: control and use.* 2nd ed. McGraw-Hill, New York. 686 pp.

Chandler, C., *et al.* 1983. *Fire in forestry.* 2 vols. Wiley, New York. 789 pp.

Cramer, O. P. (Ed.) 1974. Environmental effects of forest residues management in the Pacific Northwest: a state-of-knowledge compendium. *USFS, Pac. Northwest For. and Range Expt. Sta. Publ.* D-l-D-23. 537 pp.

Daniel, T. W., and J. Schmidt. 1972. Lethal and non-lethal effects of the organic horizons of forested soils on the germination of seeds from several associated conifer species of the Rocky Mountains. *Can. J. For. Res.* **2**:179–184.

Fahnestock, G. 1973. Use of fire in managing forest vegetation. *Trans., Amer. Soc. Agr. Eng.* **16**:410–413, 419.

Goor, A. Y., and C. W. Barney. 1968. *Forest tree planting in arid zones.* Ronald, New York. 504 pp.

Gordon, J. C., and C. T. Wheeler (Eds.) 1983. *Biological nitrogen fixation in forest ecosystems: foundations and applications.* Nijhoff/Junk, The Hague. 342 pp.

Grelen, H. E. 1978. May burns stimulate growth of longleaf pine seedlings. *USFS Res. Note* SO-234. 5 pp.

Hillis, W. E., and A. G. Brown (Eds.) 1978. *Eucalypts for wood production.* Australia, Commonwealth Sci. and Industrial Res. Org., Canberra. 434 pp.

Kilgore B. M., and D. Taylor. 1979. Fire history of a sequoia-mixed conifer forest. *Ecology* **60**:129–142.

Kozlowski, T. T. (Ed.) 1984. *Flooding and plant growth.* Academic, Orlando, Fla. 368 pp.

Kozlowski, T. T., and C. E. Ahlgren (Eds.) 1974. *Fire and ecosystems.* Academic, New York. 542 pp.

Kraemer, J. F., and R. K. Hermann. 1979. Broadcast burning: 25-year effects on forest soils in the western flanks of the Cascade Mountains. *For. Sci.* **25**:427-439.

Larson, M. M., and G. H. Schubert. 1969. Root competition between ponderosa pine seedlings and grass. *USFS Res. Paper* RM-54. 12 pp.

Leaf, A. L., and R. E. Leonard (Eds.) 1973. Forest fertilization. *USFS Gen. Tech. Rept.* NE-3. 246 pp.

Little, S., and H. A. Somes. 1961. Prescribed burning in the pine regions of southern New Jersey and Eastern Shore Maryland—a summary of present knowledge. *Northeastern For. Exp. Sta., Sta. Paper* 151. 21 pp.

Lutz, H. J. 1956. Ecological effects of forest fires in the interior of Alaska. *USDA Tech. Bull.* 1133. 121 pp.

Martin, R. E., and J. D. Dell. 1978. Planning for prescribed burning in the Inland Northwest. *USFS Gen. Tech. Rept.* PNW-76. 67 pp.

Miller, R. E., and R. D. Fight. 1979. Fertilizing Douglas-fir forests. *USFS Gen. Tech. Rept.* PNW-83. 29 pp.

Mooney, H. A., *et al.* (Eds.) 1981. Fire regimes and ecosystem properties. *USFS Gen. Tech. Rept.* WO-26. 394 pp.

Pritchett, W. L. 1979. *Properties and management of forest soils.* Wiley, New York. 500 pp.

Pyne, S. J. 1984. *Introduction to wildland fire; fire management in the United States.* Wiley, New York. 566 pp.

Reitveld, W. J. 1975. Phytotoxic grass residues reduce germination and initial root growth of ponderosa pine. *USFS Res. Paper* RM-153. 15 pp.

Rice, E. L. 1984. *Allelopathy.* 2nd ed. Academic, Orlando, Fla. 440 pp.

Southern Forest Fire Laboratory. 1977. Southern forest smoke management guidelines. *USFS Gen. Tech. Rept.* SE-10. 140 pp.

Stewart, R. E., L. L. Gross, and B. H. Honkala. 1984. Effects of competing vegetation on forest trees: a bibliography with abstracts. *USFS Gen. Tech. Rept.* WO-43. 1 vol.

Stone, E. 1973. The impact of timber harvest on soil and water. In: *Report of the President's Advisory Panel on Timber and the Environment.* Govt. Printing Off., Washington, D.C. Pp. 427–467.

Stone, E. L. (Ed.) 1984. *Forest soils & treatment impacts.* Proc., Sixth No. Amer. Forest Soils Conf. Univ. of Tennessee, Knoxville. 454 pp.

Tippin, T. (Ed.) 1978. *Symposium on principles of maintaining productivity on prepared sites.* USFS, Southern Area State and Private Forestry, Atlanta. 171 pp.

U. S. Forest Service. 1971. *Prescribed burning symposium proceedings.* Southeastern For. Exp. Sta., Asheville, N.C. 160 pp.

Vincent, A. B. 1965. Black spruce: a review of its silvics, ecology, and silviculture. *Can. Dept. For. Publ.* 1100. 79 pp.

Weaver, H. 1951. Observed effects of prescribed burning on perennial grasses in the ponderosa pine forests. *J. For.* **49**:267–271.

Wright, H. A., and A. W. Bailey. 1982. *Fire ecology, United States and southern Canada.* Wiley, New York. 501 pp.

Zahner, R. 1955. Soil water depletion by pine and hardwood stands during a dry season. *For. Sci.* **1**:258–264.

Zasada, J. C., *et al.* 1977. Forest biology and management in high-latitude North American forests. In: *Proc. Symp. North American Forest Lands North of 60 Degrees.* Univ. Alaska, Fairbanks. Pp. 137–195.

CHAPTER 9

Choice of Species and Genetic Improvement

FITTING SPECIES TO THE SITE

The ideal goal of silviculture is putting the right tree in the right place with just the right amount of growing space at each stage of development. Questions about what is right are matters involving much opinion and analytical thought about ecology and human demands (Champion and Brasnett, 1958). One extreme view is that the genetic material composing a stand should consist of those species and genotypes best adapted to survive and reproduce on the site as a result of many generations of natural selection. However, the attributes that provide for survival of the species are not necessarily those needed to meet human requirements. The natural composition is, furthermore, not static or well-defined; it is instead dynamic and subject to continual changes resulting from the developmental processes that are usually initiated by competition or regenerative disturbances. This means that even if one adheres to the natural composition, a choice must still be made about which developmental stage to imitate. Another problem with this approach is that human disruptions have often made it very difficult to know what natural compositions might have been.

At the other extreme is the view that silvicultural prowess makes human wishes about species composition come true, as is the case with agriculture. However, problems such as those of getting woody perennials to survive through dormant seasons cause foresters to stop short of full emulation of the agronomy of herbaceous annuals. One may move maize from Central America to Minnesota but not Honduran mahogany. In fact, one cannot safely move trembling aspen from a moist site to a dry one within a Minnesota farm woodlot or, for that matter, let a natural lightning fire cause the same thing to happen.

Since neither natural factors nor human wants can be ignored, the most logical courses lie between these extremes and also vary with the circumstances. The first step in this course-setting is analysis of limitations imposed by the environmental factors that collectively constitute the site. This restricts the number of species considered in the second step, which is the choice of species (plural or singular) that will most nearly meet the humanly ordained objectives of stand management. A third step is consideration of the degree of artificial control that will be exerted over the genetic constitution of the species that are selected. Such control may range from simply accepting the existing situation to highly artificial manipulation of the genetic constitution of plants to be propagated from tissue culture.

The basic objective is to use genetic material that will not only survive and thrive on the site but also yield the wood or other benefits at some optimum rate. However, it should be noted that suitable "genetic material" is seldom any single genotype even if only a single species is to be used. It may also be some combination of many species. The choices that are made determine not only the species but they also affect, and interact with, decisions about the programs of silvicultural treatment that will be devised.

Nature of Site or Habitat

Anything that is done in silviculture should be based on knowledge of the capacities and limitations of the site or habitat in which the trees are to be grown. While the term **site** is the traditional one denoting the total environment of a place, **habitat** more fully connotes the idea that the place is one in which trees *and other living organisms* subsist and interact.

As far as trees and other green plants are concerned, the site is controlled mainly by the totality of the physiologically available supply of solar energy, water, carbon dioxide, and various chemical nutrients. Sites can be usefully categorized in terms of the limitations imposed by shortages of one or more such factors. The supply of water is generally the most important factor differentiating between sites. Its physiological availability is limited by absolute shortages, immobility induced by low temperature, or inability of roots to take up water that does not have enough oxygen to allow root respiration.

Carbon dioxide is probably always deficient but its amount varies so little that it is not a factor that seems to cause differences between sites. The sunlight regime is subject to much geographical and topographical variation, which sets absolute limits on the number of hours per year that photosynthesis might proceed if water or other factors permit. The effect of soil nutrients is exceedingly variable; in general they affect growth rates more than species composition and tend to be less important than water.

The interaction of all of these physical and chemical factors clearly determines what kinds of organisms can survive on a site and also how well they can grow there. However, these organisms themselves become additional constituents of the local environment, especially in how they may affect the particular species that the forester may be trying to grow there. In this sense, browsing animals, mycorrhizal fungi, insects, or other plant species, to name only some of the biotic forces, become part of the site or habitat. Regardless of the semantics, they are part of the environment that must be carefully considered in any silvicultural decisions. Choices of species are commonly dictated by pests and other damaging agencies.

Methods of Identifying and Classifying Sites

The first consideration in site classification is the regional climate. The climate classification of Koeppen (Trewartha, 1968) is especially helpful because it is an attempt to categorize physical climatic data by using the vegetation as the most sensitive measuring device. If the silviculture relies entirely on genetic material already on the site there is little real necessity of assessing the climate because it can be presumed that the plants are well adapted to it. However, it becomes critical to know about the climate when artificial movements of species and strains thereof are involved. However, site classification also involves making much finer distinctions among the different components of the intricate patterns of variation that exist within climatic regions (Fig. 9-1).

FIGURE 9-1 a–e   A series of photographs showing old-growth forest vegetation character-istic of markedly different sites, each requiring correspondingly different silvicultural treat-ment, all in northern Idaho. Fig. 9-1a (above) shows the lowest and driest site with a pure stand of ponderosa pine. The other pictures in the series are on subsequent pages. (*Series of photographs by U. S. Forest Service.*)

FIGURE 9-1b An open stand of ponderosa pine on a south-facing, dry slope at middle elevations. A closed stand of the so-called western white pine type, such as shown in Fig. 9-1c, occupies the opposite north-facing slope.

There are various ways of assessing site (Spurr and Barnes, 1980). Most of them depend on using either the composition or the productivity of the vegetation as the basis of classification. This is not surprising because the practical objectives of classification in silviculture are usually to guide decisions about what species composition ought to be and to predict production rates. The analytical techniques are used so often to predict yields that their utility for guiding decisions about silvicultural treatment and species composition is often overlooked. Such classifications can provide diagnostic clues about those factors of the particular site that will control susceptibility and vulnerability of trees to damaging agencies, the nature of problems with competing vegetation, and response to various silvicultural treatments.

The most common purpose of site classification is to predict the timber yield of a given single species that one has already decided to consider growing or continuing to grow on the site. The best criterion of this is a recorded history of production on the specific tract itself, but this is available only where detailed records of careful management have been kept for many decades. In the absence of such records, it has become common to use rates of growth in height, of the dominant trees of a given species, as substitute indicators of stand productivity.

FIGURE 9-1c A mixed stand of the so-called western white pine type on a mesic north-facing slope. The nearest tree is a western larch; the one to its left, a western white pine; the one with vertically striped bark to the left of that is a western red-cedar; some of the understory saplings are white firs.

The most common parameter of this is **site index,** the average height of the tallest trees of an even-aged aggregation of trees at some index age. This age is 50 years unless otherwise stated; the logical index age is ordinarily somewhat less than that of a normal rotation. Curves of average dominant height over age, such as those in most yield tables, are used to extrapolate from heights observed at other ages to those at index age. There are enough uncertainties about such extrapolation and other aspects of "measuring" site that it is well to refrain from the spurious precision implicit in expressing site index to the nearest digit. One common antidote to this is the practice of assigning sites to not more than five or six rather broad categories of **site quality classes** (denoted by Roman numerals).

The site index concept is based on the observation that the rate of height growth of the leading trees is well correlated with the productive potential of a site but is not altered significantly by ordinary variations in stand density. In the simplest application it is necessary that the index trees have always been the leading trees and not been suppressed at any time. This poses problems with trees that were slowed in their initial height growth. One way of eluding this problem is to assess age at breast height and to take that point as the zero height level. One can also use the **height-intercept method** in which the index variable is the number of years required to grow from one stated height level to another. This method works best if the two levels are close to the ground and the trees produce one internode annually.

If the index species is not present on the site, procedures can be developed to predict the site index of one species from that of another (Doolittle, 1958; Foster, 1959); usually another tree species is involved but there is no fundamental reason why some other, preferably a deep-rooted perennial, could not be used.

Plant Indicators

Species composition can also be used to assess potential productivity, limitations set by environmental factors, and species suitability. The best-known methods, originally developed in Finland, involve use of the lesser vegetation that grows beneath stands. The principle behind this appears to be that some of the small plants are much more sensitive to variations in site factors than large trees. Some such plants have high indicator significance while others, presumably highly adaptable, have little. This mode of classification seems to be most successful where climatic conditions are restrictive, as in the boreal forests where it was developed.

FIGURE 9-1d A 225-year-old stratified mixture of the western white pine type on a mesic valley-bottom site below the stand shown in Fig. 9-1b. Among the other species present are western larch, western red-cedar, and western hemlock, with the two latter species in the lower strata.

FIGURE 9-1e A pure stand of white-bark pine characteristic of very cold sites at high elevations in the same locality as the other pictures of this series.

Especially in western North America (Daubenmire and Daubenmire, 1968; Steele *et al.*, 1981), there has been some success with what is termed **habitat classification** in which the late successional plant communities are used as the basis of what amounts to site classification. Understory shrubs and other lesser vegetation are of high diagnostic importance. There are, for example, large areas covered with Douglas-fir but much information can be obtained about the site, without measuring trees or digging soil pits, by noting whether the understory has sword ferns or rhododendrons. Tree species that are closely restricted to good or poor site conditions also make good indicators (Leak, 1980, 1982). This mode of site analysis works best where species composition is mostly the result of natural processes.

Soils and Topography

Site classification is most difficult where the forest is gone and where there is no forest or other vegetation to "assay" the site. Ways have been developed to predict site index by measuring the soil properties and other site factors such as aspect or degree of slope (Armson, 1977; Pritchett 1979). The predictive mechanisms take the form of multiple regression equations in which site index is determined from two or three independent variables that ordinarily define the depth of the rooting stratum and its capacity to store water.

This approach to site classification often requires deep digging to examine the soil structure and to collect samples for determination of physical, and sometimes chemical, properties (Soil Survey Staff, 1975). The detection of hardpans or other impediments that determine rooting depth is especially important. This technique depends in practice on assessing factors that can be measured on one visit to the site. Therefore, the annual regime of soil moisture must somehow be deduced from appropriately selected semipermanent, observable parameters of soil and site.

The shape of the terrain is often a key to site classification, usually because it tells so much about the water relations that are the chief ruling factor. The lower slopes of most hillsides are, for example, concavities that usually receive more water than falls on them from the sky. That from the convex hilltops seeps down to them leaving those areas robbed of water. The boundary between convexity and concavity sometimes defines differences in species composition and productivity. Unless they have sandy soils, very flat areas can be poor sites full of ponded, oxygen-deficient water. If the terrain is steep, the slopes that face the sun can be very dry while the opposite shaded ones are comparatively moist. This difference can be enough to induce grassy brushfields on sunny slopes and closed forest on shaded ones in climates with long dry seasons.

An analytical and predictive understanding of site variables can often be based on good knowledge of geology and especially geomorphology. The factors that have weathered rocks and moved the products of natural erosion around determine the composition and shape of the parent materials from which soils are formed. If it is known how they got where they are, one may know much about the extent of a particular kind of terrain form and its ability to support forest growth merely by viewing it from a distance or on a topographic map. Knowledge of geology helps not only in determining species composition and predicting yield but also in building roads, planning logging, and controlling erosion.

Combinations of Methods

In the classification of sites it is usually best to combine as many of these different methods as possible (Spurr and Barnes, 1980). No one of them is perfect but information from one technique can help greatly in strengthening that derived from another. The choice of one descriptor of site quality or habitat does not limit one to just that one single mode of analysis or parameter of site. For example, if the descriptor is site index of one species, it should not be presumed that this can be determined only by measuring heights and ages of that species. Indicator species, soil examination, history of wood production on the site, or topography can also be used and, for convenience, translated into site index.

SELECTION OF SPECIES AND PROVENANCES

The first consideration in determining which species to grow is their degree of adaptability to the site, especially to the pests and other damaging influences that are inevitably part of the site. This is the best means of reducing the losses that sporadically plague trees during their long lives. Trees must survive winter and summer while spread over large areas that can receive little artificial protection. Events such as the coldest day of the century, infrequent outbreaks of defoliating insects, rare severe fires, or bad windstorms must be anticipated. For the forester these should not be surprises but events for which one plans by fostering resistant species or other means. What happened once can happen again.

In a more positive sense, the next consideration is that the species be drawn from those that can grow well on the site, preferably better than others. This will also mean that they will tend to resist sources of damage that afflict trees of low vigor but not necessarily those that attack trees of high vigor.

While it is often ranked as the first consideration, it is usually best that the utility of the species be somewhat secondary to its adaptability. Difficulties commonly arise if one species is planted up hill and down dale because it brings the best price at the time of planting. Some of the worst problems ascribed to "monocultures" come from this. Choices made by non-forester executives in distant offices have a bad history of squandering the monies that these people are so proud of being able to manage. There is, however, reason for bias in favor of evergreens over deciduous species because they tend to be more efficient in production. Species that have straight stems will also be favored over the crooked.

Fortunately the technology of wood utilization becomes continually more versatile and less selective as to species. When paper was first made from wood on a large scale, only long-fibered spruce would do; now almost any species can be used, although long-fibered woods usually have to be significant ingredients. Users of solid-wood products are no longer as discriminating about species or even as able to identify them. It appears that wood utilization tends to change faster than foresters can change the forest composition. If this be the case, it is perhaps best to let the site determine which inherently useful species to grow.

At the other extreme, it is sometimes advocated that stand composition might as well be determined entirely by natural tendencies because any chosen species may fall from favor. The instances in which this has actually happened have been ones in which the utility of the wood was limited to specialized purposes. For example, some of the first silviculture in America was based on fast growth of insect-deformed, branchy eastern white pines on abandoned agricultural lands in New England. The wood was all right for making wooden boxes but not much else. Such material became technologically underemployed when it became cheaper to ship goods in corrugated boxes made of brown kraft pulp. However, the market for good white pine lumber remains excellent. It is perhaps logical now to consider whether hard pines should be grown in ways that make them suitable only for pulp for making cardboard boxes.

Yellow-poplar has recently declined in price mainly because particle board has often been substituted for it as the core-stock of veneered furniture. The improved adhesives that enabled development of particle board also made the excellent gluing properties of the honeycomb structure of yellow-poplar wood less necessary. It is unlikely that the high production rates and other fine qualities of yellow-poplar will go unused. Coincidentally the red oaks with which yellow-poplar is often associated in the forest have become much more fashionable for furniture and paneling. The prospect of such changes over time makes it easier to accept the diversity that results from matching species with sites. However, it is unwise to be content with growing trees that are, regardless of species, of such poor quality or small size that their use is limited to some very specialized purpose.

Frequently the choices to be made are not entirely of species themselves but of the stage of natural succession or of natural stand development to maintain. In general, the later the stage the easier it will be to maintain, but some earlier stage may have more productive or valuable species. The choice can also be viewed as that of an assemblage of species, both plant and animal, with pests and parasites included, representing some stage of vegetational development and not just a species for timber. This kind of choice is also one of the overall program of silvicultural investment and practice.

The safest and most conservative choices are species and strains thereof that are native to the site. These have the virtue of a long history of inherited adaptation to their environment. However, this kind of adaptation seems to be in favor of their reproducing their own kind and not necessarily for high production rates, straightness, or other humanly desirable attributes. Native species may even be too well accommodated to sources of damage that do not hamper reproduction.

Exotic Species

Plants that have been introduced where they are not native are called **exotics.** The success of introductions is governed exclusively by natural factors; political boundaries and similarities of human culture are of no significance. The red spruce of the North Carolina mountains is, for example, more foreign and less adapted to the coastal plain of that state than *Cryptomeria japonica.*

Generally an introduced species or variety should come from a locality with climate and soils closely similar to those of the home region. The climatic criteria are: (1) both mean and extreme temperatures, (2) length of growing season, (3) amount, distribution, and effectiveness of precipitation, and (4) latitude (which controls day length and is associated with some of the other factors).

The most successful introductions have generally involved moving species to the same latitude and position on the continent that they occupied in the native habitat because this is most likely to provide similarity of climate. For example, many conifers from western coasts of North America have been remarkably successful at the same latitudes in western Europe.

The forest economy of many countries in the Southern Hemisphere is heavily dependent on pines introduced from localities of comparable climate in the southern United States, California, and Mexico. It should be noted, however, that the famous Monterey pine has not done well everywhere in the Southern Hemisphere or even in Florida; it has thrived only in places with the same climatic pattern as the California coast.

Dramatic movements of species across major geographic barriers have been more successful than seemingly modest extensions of their natural ranges. If the new regions are climatically similar to the indigenous, the results are sometimes very favorable. The intercontinental movements mentioned in the previous paragraph have placed the species where they were adapted to grow but had been prevented from doing so only by the intervening oceans and inhospitable land surfaces. On the other hand, extensions of the natural ranges of some American conifers by distances of 100 miles or less have sometimes proven highly unsuccessful. This has been true of the southward extension of the range of red pine and the northward movement of loblolly and slash pine as well as of coast redwood.

Sometimes exotics grow better in the new environment because the climate is more favorable than the native one or because native pests have been left behind. The territory suitable for Monterey pine as an exotic is much larger and often with more rainfall than the small parts of California to which geologic events have confined it. The vast oceans of the Southern Hemisphere provide a mild climate that is very favorable to species from the more harshly variable climates of the Northern Hemisphere. The advantages of leaving the pests behind are important. Nevertheless they can be temporary as well as counterbalanced by collision with new pests.

All phytosanitary precautions must be observed in transporting plant materials for introduction; the movement of seedlings or soil is generally prohibited. Disinfected seeds are the most safely transported forms; however, this poses problems with attempts to move some tropical species that have seeds that can survive only a few hours.

Introduced species and strains should be used with caution until their safety and superiority have been tested in trials extending over all or most of a rotation. It cannot be presumed entirely on the basis of generalizations that some exotic species will or will not be successful. Only long-term tests will tell and even then there can be no certainty that the most appropriate genotypes were being tested.

There is one temptation to be avoided. Foresters should not carry their favorite species with them if they move from one region to another

Provenances

Most species are so adaptable and extend over such wide geographical ranges that they develop different forms suitable for different environments (Fig. 9-2). The geographical origin of seed is its **provenance.** As is

FIGURE 9-2 Foliage characteristics of two distinct races of ponderosa pine . *Left*: Open, plumelike arrangement of long, slender needles of North Plateau race from eastern Oregon. *Right*: Compact, brushlike arrangement of short, thick needles typical of race found east of the Continental Divide from Colorado northward. The needles shown on the right are more resistant to frost and winter injury than those on the left. (*Photographs by U. S. Forest Service.*)

the case with species, the safest provenances are the indigenous ones. If a native species is being used, there is usually temptation to bring in seed from a provenance of more favorable climate. This is because the characteristics, such as waxy foliage or early stomatal closure during times of water stress, that favor survival also reduce photosynthesis and growth. The wisdom of using the faster-growing plants may depend on whether thinning, fertilization, or some other treatment will reduce the risks being incurred.

Seemingly small moves can be dangerous if there are no natural barriers that would have prevented natural movement of the material. If the strain was thoroughly adapted to the site it may be presumed that it would be there already. There are exceptions to all generalizations. Provenances of loblolly pine from the western and northeastern parts of its range are, for example, more resistant to fusiform rust than those of the central portion. The disease is so serious that the distant provenances are used in developing strains for planting in the central area.

GENETIC IMPROVEMENT

Choices of species are choices of genetic material made on a grand scale. Those discussed in the previous section are rather passive because they involve finding and using the best populations already available. It is possible to do better than this with more active intervention in the processes by which trees pass their characteristics to their offspring. The techniques of forest genetics and tree improvement are described in detail by Wright (1976) and by Zobel and Talbert (1984).

Most tree improvement depends on finding good-looking trees and testing their progeny to determine whether the good characteristics are transmitted by sexual regeneration. Progeny do not always resemble their parents. In fact, there are so many possible combinations that, except for identical twins from single seeds, no two products of sexual regeneration are exactly the same genetically.

From the genetic standpoint, the best that can be said of an outstandingly good tree found in the wild population is that it is a good phenotype; the effects of its genetic constitution *and its environment* have produced a good outward appearance. If the sexually or vegetatively reproduced progeny of good phenotypes display characteristics superior to others when grown in uniform environmental conditions, this shows that the parents were desirable genotypes. The opportunities for genetic improvement depend on the degree of variability in the wild populations. If the variability is large, there are more genotypes from which to pick so that the possibility of genetic gain is large. However, the same high variability is also likely to mean that each individual is strongly heterozygous and may thus transmit mixtures of good and bad characteristics to its progeny. Trees have not been subjected to the centuries of human selection that have tended to produce strains of agricultural crop plants that are genetically very uniform or homozygous.

Tree breeding is quite different from that of agricultural crops, at least if sexual regeneration is involved. While it may be making a virtue of necessity, the same variability that impedes improvement by breeding also helps maintain populations likely to include individuals that can endure the vicissitudes of the long lives of trees. Furthermore, stands of woody perennials, unlike those of most crops of agricultural annuals, start with many more individuals than are present in the final crop. A small number of the fastest growing are likely to endure natural or artificial thinnings. In fact, if the trees of a stand are too similar, especially in their rate of height growth, most will survive and tend to stagnate in diameter growth. There are, in other words, important reasons to try to maintain some degree of genetic variability.

Procedures

The first step in most tree improvement programs is **mass selection** in which promising phenotypes are rigorously selected from the wild population. Seedlings or cuttings from the chosen trees are then tested in uniform environments in the field to assess the degree to which the good characteristics are hereditary. If the heritability of the desired traits proves to be high this usually means that even more intensive mass selection is promising and that the prospects of progress are good. If the degree of variability proves to be low for a particular trait this means either that the prospects for gain by selecting for that trait are small or that improvement should be sought from a different wild population.

If the results show a high degree of variability but that good parents have variable offspring, this usually means that it will be best to shift to **family** or **clonal selection,** which is based on breeding from material with good genotypic characteristics as demonstrated in the progeny tests. A good **family** would be the progeny from seeds of the same good female parent. A **clone** consists of vegetatively reproduced offspring of one tree; both parent and offspring are of the same genotype. Breeding of families by sexual reproduction is a kind of two-edged weapon. It provides opportunity for further improvement by combining the good traits of different parents. However, it can also allow the reintroduction of any poor traits that still lurk in the heredity of seemingly good parents. Asexual propagation of clones, on the other hand, locks in whatever progress has been made during whatever selection led to the chosen genotype or individual but it allows no further improvement.

The least costly kind of family selection involves choices of **half-sib families** ("half-sibling") in which only one parent, usually the female, is known but pollination is not otherwise closely controlled. Such selection might be practiced, for example, by planting a mixture of good phenotypes together and then thinning out those female parents that did not have offspring that showed up well in progeny tests. In such a case the pollen parents are not controlled except to the extent that only ones of good phenotypic characteristics were planted in the test area.

Selection among the progeny of **full-sib families** becomes possible if cross-pollination is done artificially so that both male and female parents are known. Since this requires collecting pollen and applying it to flowers that have to be covered to prevent wild pollination, it is costly. However, it offers better opportunities for genetic improvement than less intensive measures, especially if both male and female parents have already been developed from some sort of rigorous selection.

None of this is as simple as it may sound. Successful genetic improvement programs are highly specialized research projects. The desirability of making selections from large populations and the high cost of the work have made it expeditious for different workers and entities to share genetic material on an organized basis. Such sharing helps develop the combinations of diversity coupled with good hereditary characteristics needed in the planted progeny.

The improvement programs are usually continuing operations that switch from one approach to another and back or use different methods simultaneously. Regulated sexual regeneration aimed at seeking genetic gains may alternate with vegetative regeneration to preserve the gains that have been made. The use of rooted or grafted cuttings often speeds up progress at certain stages of the effort. Progress is slowed if cuttings will not root or prove to be incompatible with root-stocks on which they are grafted.

As testing progresses, many of the genotypes that were the best in the initial stages may be replaced either by new ones or by superior ones derived from their progeny. Usually swiftest improvement comes when selection is made for only one or two different characters.

Hybrids

There have been improvements from a few cases of interspecific hybridization. Some hybrids have developed spontaneously when some exotic member of a genus has been planted with another. The vigorous Dunkeld hybrid of Japanese and European larch is the best example. The only artificially cross-pollinated hybrid in mass production is the use of two American species in Korea. This is a cross aimed at combining the straighter form of loblolly pine with winter-hardiness of pitch pine. It is also being tested for use north of the natural range of loblolly pine.

These and some other hybrid combinations have been practical because the crosses can be made merely by planting the different species together or because the hybrids themselves produce plenty of fertile seeds. Except for certain pines, larches, and spruces, most tree species are sufficiently far apart in their heredity that they do not easily cross or, if they do, produce hybrids that are infertile. However, hybrids of cottonwood poplars and willows are easily perpetuated because their cuttings root very easily.

Multiplication and Tissue Culture

One of the impediments to application of genetic improvement is the cost of the important step of multiplying the desirable progeny. Seeds produced from tree improvement programs are very costly unless they come from species such as birches and eucalypts that produce vast quantities of very small seeds. This generally means that the seedlings must be reared in nurseries and planted simply to achieve high efficiency in use of expensive seed.

The development of techniques for propagating whole plants vegetatively from culturing small bits of tissue in nutrient media *in vitro* is especially promising for silviculture (Bonga and Durzan, 1982). Breeding programs that must rely on sexual regeneration of tree species, which are characteristically very heterozygous and slow to bear seeds, are maddeningly slow. It can be especially frustrating if some excellent genotype has attributes that seem to be the result of some rare combination of genes not likely to be produced again even by the same parents. If such a genotype can be multiplied by mass-production tissue culture many problems of breeding depending on cross-pollination can be bypassed. However, this kind of tissue culture is still under development and seedlings produced by it are not likely to be cheap. It is significant nevertheless that most varieties of woody plants used as ornamentals and for fruit production are vegetatively propagated, commonly from aberrant seedlings detected by observant growers.

Use of Genetic Improvement

If stands are to be established by planting it is folly not to use the best genetic material available. Conversely, the availability of improved material can be a reason to regenerate stands by planting.

Genetic improvement of forest trees has not produced either the spectacular gains or the worrisome over-reliance on intensive cultural practices associated with improved strains of agricultural crops. Most of the genetic manipulations of agriculture have been aimed at diverting more of a fixed amount of plant production into fruits, seeds, roots, or other plant parts that happen to be most useful (Silen, 1982). Where total production has been increased, often immensely, it has been mainly by improving the site through such measures as fertilization or irrigation or by controlling weeds and other pests. However, it has also been accomplished to some extent by breeding varieties that can not only take advantage of intensive cultural measures but are also heavily dependent upon them.

Perennial woody plants are different but not completely so. Most of the production of trees is tied up in the stem. This is the very organ that is usually viewed as wasteful in agriculture and is "robbed" to increase some other parts. In a certain sense this may mean that trees are hard to improve because nature has already done much of the work. It might help timber production if trees had more main stem, fewer branches, but somehow supported the same amount of sugar-producing foliage. It would also be good if seed production could be turned on and off at will by either genetic control or application of appropriate hormones.

Much effort in forest genetics has been aimed at increasing production per hectare. Because of the variable effects of site conditions on production it is very difficult to determine whether this objective has been or can be achieved. Increases in the sizes of individual trees definitely have been produced but these do not increase productivity per unit of land area unless trees of given size occupy less growing space because of using it more efficiently. It is worth noting that if the individuals grow faster they may attain optimum sizes on shortened rotations even if the production per unit area is not improved.

Since different species clearly differ in production per hectare, it seems likely that similar differences must exist within species (Cannell and Last, 1976). Much of the evidence suggests that the faster-growing material is generally that adapted to favorable sites or parts of the natural range. So long as the growing conditions happen to remain favorable or can be kept so by artificial measures, these faster-growing varieties can persist on sites where they might ordinarily suffer damage from drought, cold, or similar problems. In other words, this kind of increase in forest production may come with the same kind of price paid for such improvement in agriculture.

Tree improvement work has also been aimed at important factors other than productivity. Stem form and branching characteristics have frequently been improved. Resistance to pests and damage is always important. Among the other characteristics that can be improved are wood density, fiber length, color of Christmas trees, and resin production, to name but a few. There is evidence that success comes most rapidly with artificial selection for characteristics, such as stem quality or wood density, that have so little importance for survival that nature has not selected for them. In such cases, traits that are both good and bad from the human standpoint are very likely to have persisted in the natural populations.

The benefits of applied forest genetics do not depend entirely on whether improvements are made on nature. Many of them come simply from the fact that genetic resources are being managed knowledgeably and purposefully. Some of what counts for improvement may actually be either the remedy or prevention of degradation. There are grounds to suspect that many Scotch pines are crooked because much seed was once gathered from easily climbed, short, stunted trees in the early days of European forestry. Crossings within small populations or between parents and offspring can produce "inbreeding depression," the manifestation of undesirable traits carried by genes that remain recessive unless an organism inherits a double dose of them.

Genetic improvement is not a substitute for other kinds of silvicultural treatment. In fact, its benefits are usually not available without such treatment. It is not a substitute for natural selection either. Natural populations are still such a storehouse of hidden genetic resources that it is increasingly obvious that significant samples of truly natural forests must be maintained or allowed to redevelop. Only in this way can potentially important gene combinations be maintained. Such places are also where silviculturists can truly learn about the natural developmental processes that they claim to be able to direct.

BIBLIOGRAPHY

Armson, K. A. 1977. *Forest soils: properties and processes.* Univ. Toronto Press, Toronto. 390 pp.

Bonga, J. M., and D. J. Durzan. 1982. *Tissue culture in forestry.* Nijhoff/Junk, Hingham, Mass. 420 pp.

Cannell, M. G. R., and F. T. Last (Eds.) 1976. *Tree physiology and yield improvement.* Academic, New York. 567 pp.

Champion, H., and N. V. Brasnett. 1958. Choice of tree species. *FAO Forestry Development Paper* 13. 307 pp.

Daubenmire, R., and J. B. Daubenmire. 1968. Forest vegetation of eastern Washington and northern Idaho. *Wash. Agr. Exp. Sta. Tech. Bul.* 60. 104 pp.

Doolittle, W. L. 1958. Site index comparisons for several forest species in the southern Appalachians. *Soil Sci. Soc. Amer. Proc.* **22**:455–458.

Eyre, F. H. (Ed.) 1980. *Forest cover types of the United States and Canada.* Society of American Foresters, Washington, D.C. 148 pp.

Foster, R. W. 1959. Relation between site indexes of eastern white pine and red maple. *For. Sci.* **5**:279–291.

Holdridge, R. R., *et al.* 1971. *Forest environments in tropical life zones—a pilot study.* Pergamon, New York. 747 pp.

Leak, W. B. 1980. The influence of habitat on silvicultural prescriptions in New England. *J. For.* **78**:329–333.

Leak, W. B. 1982. Habitat mapping and interpretation in New England. *USFS Res. Paper* NE-496. 28 pp.

Pearson, G. A. 1951. A comparison of the climate in four ponderosa pine regions. *J. For.* **49**:256–258.

Pritchett, W. L. 1979. *Properties and management of forest soils.* Wiley, New York. 500 pp.

Rowe, J. S. 1959. Forest regions of Canada. *Can. For. Br. Bul.* 123. 71 pp.

Silen, R. R. 1982. Nitrogen, corn, and forest genetics. *USFS Gen. Tech. Rept.* PNW-137. 20 pp.

Soil Survey Staff. 1975. Soil taxonomy: a basic system of soil classification for making and interpreting soil surveys. *USDA, Agr. Hbk.* 436. 768 pp. (Reprinted by Wiley, New York)

Spurr, S. H., and B. V. Barnes. 1980. *Forest ecology.* 3rd ed. Wiley, New York. 687 pp.

Squire, R. O. 1983. Review of second rotation silviculture of *Pinus radiata* plantations in southern Australia: establishment practice and expectations. *Australian For.* **46**:83–70.

Steele, R. W., *et al.* 1981. Forest habitat types of central Idaho. *USFS Gen. Tech. Rept.* INT-114. 137 pp.

Trewartha, G. G. 1968. *An introduction to climate.* 4th ed. McGraw–Hill, New York. 282 pp.

Wright, J. W. 1976. *Introduction to forest genetics.* Wiley–Interscience, New York. 282 pp.

Zobel, B., and J. Talbert. 1984. *Applied forest tree improvement.* Wiley, New York. 505 pp.

C H A P T E R   1 0

Production of Planting Stock

The most expeditious and certain way of establishing trees is to rear them in some protected environment and then plant them where they are to grow. In this way they are not exposed to the rigors of the site to be restocked until they have successfully passed the most critical early stages. Many of the microenvironmental problems that were discussed in the previous chapter are simply bypassed. The planted trees can be **wildlings** of natural origin dug from elsewhere in the forest. It is usually more efficient, however, to raise planting stock in nurseries or greenhouses. The plants that are moved can be rooted or non-rooted cuttings as well as seedlings.

The habits associated with most of silviculture, in which crops of modest value are grown with procedures of low intensity, should not be carried over into growing seeds and seedlings for forest planting. They are crops more valuable, per hectare, than almost any of those of agriculture and require investments, treatments, and specialized management of the highest intensity. The very condensed discussion of these matters in this chapter is merely an introduction to them.

Most forest planting involves conifers because the prospects of successful establishment and high yield are ordinarily greater with them than with hardwoods, or at least with deciduous hardwoods. This chapter will deal with conifers and planting stock grown from seed, except in those cases where it is stated otherwise. One source of detailed information on continuing developments in artificial regeneration is the U. S. Forest Service periodical *Tree Planters' Notes*. There are also comprehensive accounts by Wakeley (1954), Goor and Barney (1968), Williams and Hanks (1976), Abbott and Fitch (1977), and Duryea and Landis (1984).

SEED COLLECTION, PRODUCTION, AND TREATMENT

Harvesting of Wild Seed

The investments and opportunities involved in planting are so great that it is important to start with seed of the best genetic and germinative qualities available. It is common that it must be gathered from ordinary forests under circumstances that do not make it easy to obtain the proper kinds and amounts. Purchases from unsupervised collectors and poorly known dealers often give bad results.

The most serious difficulties can be avoided by attention to the provenance of the seed. Unless better sources are known, seeds should be harvested from trees of native stock growing under conditions as similar as possible to those of the planting site. When seeds are purchased there should be precise specifications about provenance, date of collection, germination capacity, and any other useful information that can be logically required. Regardless of how the seeds are obtained, records should be kept of the origin of the plants of each plantation. Only in this way can future advantage be gained from experience with trees of different origins.

Most problems arise from the difficulties of gathering cones or fruits from tall standing trees. The work is hard and the period during which it can be done is usually short. Crops worth harvesting are likely to be produced only during sporadic good seed years.

The best seed producers are ordinarily dominant trees that have attained middle age and are healthy. The likelihood that their seeds will be of acceptable genetic quality is greatest if similarly good trees are close by to provide pollen. If good, younger, and shorter trees produce adequate amounts of fertile seed, it is all right to collect from them. Unfortunately, it is simpler to gather seed from short, deep-crowned trees of poor form than from tall, straight, well-pruned ones. Although poor form in the parent does not necessarily indicate that its progeny will be undesirable, there have been enough bad experiences to suggest that it can increase the possibility.

Seed Quality

The germinative capacity of seeds depends on many factors, but mostly on the extent of successful pollination between trees. Self-fertilization by pollen from the same tree often results in a high percentage of infertile seeds. There are many factors, such as non-synchronous development of male and female flowers, that tend to prevent self-pollination. Because of this and the desirability of having a broad genetic base for the population that is to be planted, it is desirable that there be a number of fertile trees of the same species in close proximity to induce ample cross-pollination.

Large seeds generally germinate better and yield larger seedlings than small. However, most evidence indicates that these differences are neither permanent nor indicative of other hereditary characteristics. The effect of initial differences in seed weight on the size of seedlings usually disappears after several years if the large seedlings do not compete with the small ones. Ordinarily both seeds and seedlings can be safely segregated on the basis of size to secure any cultural advantages associated with large size or uniformity without affecting the proportion of good genotypes.

It is well to examine the seeds of a tree in the field before effort is expended in gathering more of them. This can often be done by cutting them in cross section to see how many are hollow or "blind." Operations can be expedited by finding stands with large crops of sound seeds a few weeks before the seeds mature.

Harvesting of Seeds

Cones or fruits must be collected after the seeds have completed their development but before they have been dispersed. The time during which the work can be done varies greatly. If it is short, it is sometimes possible to collect somewhat prematurely, but only if there is guidance from research or previous experience. With some species, collections can begin when the stored substances in the seeds are no longer milky. Most cones and some fruits are mature when desiccation starts; if so, this is the most reliable indicator. The cones of most hard pines are, for example, best gathered when their specific gravity has decreased to about 0.85, a condition that can be assessed by determining whether they will float in oil, such as S.A.E. 20 lubricating oil, of about that density. Changes in color are also associated with maturity, but are more difficult to define and evaluate.

Cones or fruits can be harvested with long poles fitted with hooked blades; climbing is often necessary. The simplest method is to collect from trees felled just when the seeds have matured. Discrimination should be exercised in gathering seeds that have been cut down or stored by animals because they may start their work prematurely.

The best practices for the collection or any treatment of tree seeds vary widely with species. It is always desirable to consult such sources as the *Manual of the Seeds of Woody Plants* (U. S. Forest Service, 1974) for the details of treatment and quantitative data about the seed of various species.

Seed Production

Gathering large quantities of seed from the wild forest is usually as inefficient as it is difficult to control from the genetic standpoint. It is generally better to set aside or establish stands of good genotypes and manage them, often intensively and exclusively, for seed production (Faulkner, 1975).

The simplest kind are **seed production areas.** These are good stands, chosen by phenotypic characteristics, that are thinned or otherwise treated to stimulate seed production; sometimes poor phenotypes are removed.

The term **seed orchard,** is restricted to stands planted for seed production and composed of trees known to be of desirable genotypes from tests of their progeny (Fig. 10-1). They are usually established by vegetative propagation from the chosen genotypes and must be isolated from sources of contaminating pollen. To maintain a large and variable gene pool in the

FIGURE 10-1 Southern pine seed orchard.

progeny, it is desirable to start with enough trees to have at least a dozen genotypes represented in each orchard unit that ultimately becomes a single seed source. Trees of the same genotype should be separated so as to enhance cross-pollination. Progeny testing usually continues while an orchard is in production. Unsatisfactory genotypes can be detected and eliminated or "rogued" as the orchard develops.

Stands used exclusively for seed production can be treated like fruit orchards. The trees should be widely spaced and can sometimes be pruned in such a manner that they are easily climbed with the crowns remaining close to the ground. If they are of good genotypes, the fact that they may not resemble good forest-grown trees is of no consequence to the genetically or environmentally controlled characteristics of their progeny. The trees produce crops of extremely high value and should be cared for intensively. Carefully regulated chemical fertilization and timely applications of appropriate insecticides (Ebel *et al.*, 1975) and fungicides are generally mandatory. Irrigation and close regulation of ground-cover vegetation are usually necessary. All these treatments not only stimulate seed production but also greatly reduce the extent of the annual fluctuations encountered in wild stands.

It may be expedient to mechanize the harvesting. Lift-buckets can elevate the pickers up beside the tree tops. There are machines that can shake whole trees enough to cause cones or fruits to drop. If, as is the case with loblolly pine, these break the branches before the cones will drop, it may prove useful to wait until the seeds can be shaken from the opened cones onto rolls of plastic screening spread beneath the trees. Machinery exists to remove the seeds from the screening as it is rerolled.

Treatment of Seed

The methods of extracting seeds depend on the nature of the fruit. The cones of most softwoods and the dry fruits of some hardwoods will shed their seeds if dried in open air or in kilns. However, the seeds of the closed-cone pines can be extracted only in kilns. The seeds of trees with freshly or pulpy fruits may be removed by macerating or crushing them in special machines. Seeds borne in pods or husks can be extracted by threshing.

The cones of many softwoods can be opened merely by spreading them out in trays in open sheds or in direct sunlight, although this method is less satisfactory than heating them in kilns. The cones of many species must, however, be air-cured for about two weeks in this manner before they are placed in kilns. If they are too moist when kiln-dried they have a tendency to become "case-hardened" on the outside, which makes them very difficult to open by any means. The simplest type of kiln is operated entirely by convection. Such kilns consist of little more than rooms in which the cones are spread out in layered trays; they are heated from below, and the heated air rises up through the cones and goes out by way of ventilated slots in the top. The danger of uneven heating is so great that the process must be carried on slowly at relatively low temperatures.

Greater efficiency is obtained in forced-air kilns, which are ventilated by means of fans and operated at temperatures of 45° to 75° C. Definite schedules of drying are followed just as in drying lumber in kilns. Equipment has also been developed for extracting seeds by heating the cones directly with infrared lamps or solar heat collectors.

Regardless of how the cones are opened it is desirable to shake the seeds out as soon as possible. This is usually done in cone tumblers, which consist of rotating cylindrical screens.

After extraction the seeds of most species must be de-winged and then cleaned of waste material. This is usually done with special equipment designed to rub off the wings, screen out the coarse debris, and blow out the fine or light particles of waste. It is also possible to agitate the seeds on special tables designed to separate them according to specific gravity and eliminate those that are empty. These treatments reduce the volume of useless material that must be stored and processed; they also facilitate testing and sowing, ultimately making it possible to increase the uniformity of nursery seedbeds. However, all such treatments should be as gentle as possible because each seed is a living plant likely to be killed by rough handling.

Seed Storage

The infrequency of seed crops makes it advantageous to have some means of storing seeds for several years without loss of viability. Fortunately the seeds of most commercial species can be stored for periods of 3 to 10 years and sometimes longer if held at low temperature and low moisture content in sealed containers. The proper moisture content varies from 4 to 12 percent, depending on the species, and the temperatures should be below 5°C, preferably in the range from −17° to 0°C. It is important to dry the seed uniformly and to prevent fluctuations in moisture content during storage. Under these conditions respiration continues at the low level necessary to keep the embryos alive and only small amounts of the stored carbohydrates are converted into carbon dioxide in the process. Polyethylene bags make good containers because they are impermeable to water but not to oxygen and carbon dioxide. These attributes prevent changes in moisture content but allow slow exchange of the gases involved in respiration to continue through the container walls.

If seeds can be stored against times of shortage, there is less reason to resort to use of poor or unsuitable seeds. Unfortunately some species are difficult or impossible to store for long periods. Large nutlike seeds, such as those of oaks or walnuts, are difficult to store past the time when they would normally germinate in the first spring. They must, in any case, be kept in cold, moist media; if kept thus in refrigerated storage they may last a year or two. Some species, such as the true poplars and many species of moist tropical forests, are so thoroughly adapted for immediate germination after seedfall that their seeds are virtually impossible to store or even to transport. The seeds of most species of woody plants deteriorate if stored either moist or dry in heated rooms.

Treatment of Seeds to Hasten Germination

The seeds of a number of species often germinate slowly or fail to germinate at all after sowing. There are two main causes of this phenomenon, which is referred to as **dormancy**. The first is called **internal dor-**

mancy, a condition ordinarily resulting from incomplete digestion of the fats, proteins, and other complex insoluble substances stored in the seed. Before germination can occur these materials must be broken down into simpler organic substances, such as sugars and amino acids, that can be translocated to the embryo. The essential conditions are created by storing the seed in cool, moist substances just as it would be in the natural forest floor. Moist peat moss or sand at temperatures somewhat above freezing is the best medium for this treatment, which is called **stratification.** The enzymes that catalyze the breakdown of the complex stored substances are capable of functioning in a cool, moist environment, but those that cause the rapid respiration necessary to release energy for growth of the embryo are not. Consequently the stored substances eventually used by the embryo are mobilized without being converted into carbon dioxide by useless respiration. The seeds of many conifers develop internal dormancy to some extent.

The stratification of seed in mixtures with other materials is a time-consuming procedure and also has the disadvantage of exposing the seeds to mold. With some species it is fully as effective and much simpler to soak the seeds at 5°C for 1 or 2 weeks. Good results have also been obtained with some species by soaking the seeds briefly in water and then refrigerating them in polyethylene bags.

Seed disinfectants may have to be used to prevent molding during stratification. The most convenient way to avoid difficulty is to sow in moist soil during the dormant season after danger of premature germination of any nondormant seed is past. Stratification becomes necessary only with dormant seed when sowing is going to be delayed until the normal germination time or done at some abnormal time in a greenhouse.

The second cause of imperfect germination is **seed-coat dormancy.** This occurs in seeds that have protective coverings so impervious that either oxygen or moisture is excluded from the embryo. This kind of dormancy can be broken by mechanical abrasion or chemical softening or etching of the seed coat. The first kind of treatment is referred to as **scarification** and may be accomplished by grinding the seeds in mixture with coarse, sharp sand or in machines equipped with disks or cylinders of abrasive paper. Chemical treatment is usually done by putting the seeds in concentrated sulfuric acid. Black locust is one species that can be treated effectively by either method. Some species, notably basswood, have both kinds of dormancy and require stratification after treatment of the seed coat. It is probable that seed-coat dormancy is broken in nature by the grinding of seeds in the gizzards of birds or by the gradual decay of seed coats by fungi. It is most common in hardwoods with hard seed coats and rare in conifers.

The natural function of both forms of dormancy is the prevention of germination at times unsuitable for seedling survival. Seeds can often be induced to germinate if one analyzes and imitates the conditions of normal storage in nature.

Seed Testing

Before the amount of seed to be sown on a given area can be calculated, the **germinative energy** of each lot must be known. This value is the percentage of seeds in a well-mixed sample that will germinate under optimum conditions during the period of most active germination (Fig. 10-2). This period extends through the first 7 to 35 days following the start of the test. Little advantage is gained by determining the total percentage of seeds potentially capable of germinating, that is, the **germinative capacity.** Only those seeds that germinate quickly are likely to produce seedlings vigorous enough to survive competition in the open.

Germination tests can be made in a variety of ways. The most convenient method is to determine how many seeds will actually germinate during a given period in the greenhouse or laboratory. The germination medium may be acidic sand or peat moss in flats, or absorbent paper in special germinators. In either case the seeds are kept moist (but not submerged in water) at temperatures that are either maintained at 30° continuously or allowed to drop to 20°C at night. Results may sometimes be obtained with less delay through the treating of the seeds with hydrogen peroxide. The cutting test by which sound seeds may be distinguished from blind seeds

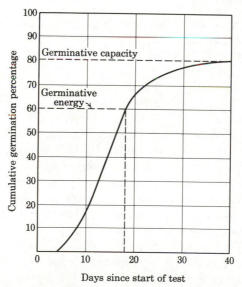

FIGURE 10-2 Cumulative germination curve showing relationship between germinative energy and germinative capacity. In this example, only those seeds which had sprouted by the time, 18 days after the start of the test, when the rate of germination started to decline should be regarded as capable of germination in the nursery.

gives results that are far too high; it is valuable only for making quick estimates when seed is being collected.

It is desirable to centralize the testing of seed at large, specialized laboratories. These usually have automated means of weighing, drying, counting, sorting, and germinating seeds. Sometimes they are equipped to make seed tests by X-ray examination or the application of stains to excised embryos that color only the living ones.

The germination percentages derived from laboratory tests are usually 50–100% higher than the percentage of sound seed that will survive in the nursery and result in plantable seedlings. The discrepancy between germinative energy and nursery survival tends to be smaller the higher the quality of the seed and, of course, the more carefully it is cultured. The actual proportion of germinating seeds that produce plantable seedlings in the nursery is the **tree percentage.** It is perhaps the most desirable criterion of all. Efforts to record and predict it for each lot of seeds and seedlings can greatly enhance the efficiency of nursery management.

It may also be desirable to determine the purity of each lot of seed. This is expressed as the percentage, by weight, of clean, whole seeds, true to species, in an unbiased sample consisting of both seeds and impurities.

NURSERY OPERATIONS

The management of forest-tree nurseries is highly specialized and more akin to agronomy or horticulture than to silviculture. Every nursery has its own special problems that must be worked out by experimentation. Efficient, economical production can be achieved only in large nurseries that grow millions of trees per year and represent large investments in land and equipment. High labor costs have caused need for much mechanization. Small nurseries with simple equipment (Fig. 10-3) can have the advantage of being close to the planting sites but usually have high production costs. If only a small amount of planting is to be done it is ordinarily better to buy planting stock than to start a nursery.

Seedlings that are grown outdoors are usually grown from seed and are **bare-rooted,** which means that their roots are separated from the soil when they are to be carried to the ultimate planting site. It is comparatively easy to transport large numbers of such seedlings to the field. However, the reestablishment of contact between roots and soil is such a crucial and fundamental problem that the planting can be done only during those short seasons during which there is rapid root growth. The alternative techniques of growing planting stock from vegetative cuttings or in containers will be considered in later sections.

Location and Establishment of Nurseries

The most important consideration in growing forest planting stock is the selection of a suitable location for the nursery. More problems have

FIGURE 10-3 The equipment used at a very small nursery within a forest in Bavaria.

been created in nursery management by poor judgment in this matter than by any subsequent errors. Nurseries should be located on level or gently sloping ground suitable for agriculture. The best soils are loamy sands or sandy loams free of stones and moderately well drained. They should be inherently fertile and also acidic, with a pH between 5.0 and 5.5, particularly if conifers are to be grown. The site should also have reasonably good air drainage so as to reduce the risk of injury to the stock from extreme temperatures. The soils and vegetation should be carefully investigated to determine whether problems with insects, fungi, nematodes, or weeds are likely to be serious; recently cultivated land is often undesirable from this standpoint.

It is very important that the location of the nursery be synchronized with that of the forest areas it is to serve. This means that, if root elongation is a seasonal phenomenon, the roots of seedlings in the nursery must be growing or ready to start growth at the same time that they would be at the planting site. This will not be true if the climatic conditions of the two places are greatly different, as they may be in mountainous regions. If the soils freeze or can be too muddy for machinery, it will help if the nursery soils thaw or dry slightly earlier than those of the planting sites, so that the seedlings can be lifted as soon as needed.

Nurseries should be readily accessible to main highways, electric power, and sources of seasonal labor. It is essential that there be an abundant supply of clear water of as low alkalinity as possible. The area should not be cramped and it is well to have enough both for expansion and to allow some land to remain fallow for a year or more. It is better to spend time and money to get a good site than cope with the deficiencies of a poor one chosen on the basis of expediencies.

Fertilization and Management of Nursery Soils

The problems of managing nursery soils are more critical than those of most agricultural soils because of the very large investment and value represented by an acre of tree seedlings. The delicately balanced nutrient cycle of the forest does not exist in forest nurseries. Large quantities of inorganic nutrients are removed in the tops and roots of each crop of seedlings; adhering soil colloids are also lost. In fact, tree seedlings exhaust the soil more rapidly than many agricultural crops. Therefore, fertility can be maintained only by proper application of both organic matter and inorganic fertilizers. It can seldom be maintained solely by using composts and other organic materials because of the expense and practical difficulties encountered in obtaining and handling the large volumes of material that would be necessary.

Attention must also be given to preventing deterioration of the physical properties of the soil. The main purpose of applying organic matter and plowing in cover crops is to protect the structure and water-holding capacity of the soil. It is also desirable to protect the soil surface as continuously as possible with seedlings, cover crops, or mulches. If the nursery is on sloping ground subject to erosion, it is necessary to construct low terraces.

The problems of treating nursery soils must be worked out by careful and skillful study at each nursery. The pitfalls are many and usually invisible to the uninitiated. A fuller appreciation of this topic may be gained from such references as those of Armson and Sadreika (1979) and Pritchett (1979).

Preparation and Sowing of Seedbeds

The seedbeds should be as long as possible so the machinery that passes over them can be used with a minimum of turning. For purposes of record keeping and of controlling treatments, it may be useful to mark subdivisions with standard lengths, such as 12 feet. The standard width of seedbeds has been 4 feet to facilitate hand weeding, a practice not completely obviated by herbicides. Any alternative width might be set by the space between the wheels of suitable and standardized nursery tractors. The space between beds should be wide enough for workers to walk and the wheels of machinery to roll.

Relatively deep plowing and thorough harrowing are necessary to offset the effects of soil compaction by machinery and to mix supplementary organic matter back into the soil. If drainage is impeded to any extent, the beds should be slightly elevated and convex in cross section; otherwise they should be flat. The form and even the spacing of the seedbeds can be created with rototillers or specially constructed scraping devices called **seedbed shapers** or **formers.** Seedbed surfaces should be lightly raked before sowing so that the seed will land in crevices where it will stay in place and may later be covered slightly.

There are two general methods of sowing. In **broadcast sowing** the seeds are distributed uniformly over the seedbed either by hand or by relatively simple machinery. In **drill sowing** more precise machinery is used to arrange the seeds in parallel lines that run the length of the beds. Broadcast seeding produces uniform stands of seedlings that usually close together soon and then tend to protect themselves from damaging agencies and weed competition. The main disadvantage of this arrangement is that it prevents such operations such as weeding with mechanical cultivators, root-pruning of rows of individual trees, and spraying pesticides beneath the tops of the seedlings. Drill sowing is usually done with equipment that sows the seeds at any desired rate, covers it to the proper depth, and rolls the beds, all in one operation. The drills are usually 10–20 cm apart, although much wider spacings are used for certain hardwoods.

The amount of seed to be sown should be calculated for each lot of seed from the data on germinative energy and other characteristics. The general formula employed is:

$$W = \frac{A \times S}{C \times P \times G \times L}$$

in which W = pounds of seed, A = square feet of area to be broadcast with seeds *or* linear feet of drill, S = number of seedlings desired per square *or* linear foot, C = number of seeds per pound of absolutely clean seeds at the moisture content of sowing, P = purity of sowing lot expressed as a decimal, G = germinative energy expressed as a decimal, and L = the estimated survival percentage expressed as a decimal. Allowances should be made for reductions in production from irregularly distributed catastrophic losses; however, this should be done by sowing a larger area at the optimum rate, rather than by increasing the amount of seed sown on a fixed area.

The desirable density of seedbeds varies from 10 to 100 seedlings per square foot, depending on the size of seedlings of the species involved. For most conifers the optimum density varies from 25 to 60 per square foot. As many as 200 per square foot may be grown if the seedlings are to be transplanted in the nursery before being sent into the field. The costs of treatments of seedbeds are logically viewed as costs per square foot of bed. Therefore, the aim should be to obtain the level of density that will yield the greatest number, per square foot, of seedlings larger than whatever size is set as a minimum standard. If the seedlings come up too densely after broadcast sowing the seedbeds can be thinned by means of devices fitted with whirling hammer knives or minute jets of burning gas designed to kill seedlings along very narrow, parallel lines.

After the seed is sown it is pressed into the soil with rollers. The depth of sowing should be one to four times the diameter of the seed; deeper sowing is necessary in sandy soils than in those with a high clay content. With broadcast sowing, it is often necessary to add a covering of clean sand after rolling.

After sowing is completed the seedbeds may be covered with straw or long strips of burlap until germination commences. If the seed is sown in the fall and there is danger of frost heaving, a covering of pine needles or straw is necessary. Seedlings must be mulched in similar fashion during subsequent winters if there is danger of loss from frost heaving.

The seedbeds should be protected from birds and rodents during the germination period. Wire screens are effective but expensive. It is most effective to poison the rodents if the open conditions of the nursery do not discourage them sufficiently. It may be necessary to shoot or scare away the birds, although seed coatings of arasan and similar chemicals can protect against them and damping-off fungi.

Control of Weeds

The barren soil of nurseries is an open invitation to the invasion of annual weed plants, which can easily overwhelm small tree seedlings. It is important to manage the entire nursery area, including its roads and edges, so that weed seeds are not imported and the weeds that inevitably appear are discouraged from multiplication. It is desirable to locate nurseries on soils that are as free as possible of the stored seeds of agricultural croplands. After that there should be suspicion about anything brought into the area.

There will nevertheless be weedy annuals. One of the best ways to reduce them, initially and subsequently, is with the use of other annuals, such as rye and other tall grassy plants, to suppress the weeds after they germinate but before they bear seed. The cover crops themselves must be plowed under or otherwise killed before they too produce seeds. The use of cover crops is at least as important for this purpose as it is for recycling nutrients, controlling erosion, and rebuilding soil organic matter.

It must still be anticipated that weeds will germinate in the seedbeds. The best time and way to control them is with periodic applications of preemergence herbicides that act at the soil surface to kill weed seedlings just as they are germinating. If this is successful, the weeds are killed when they are almost invisible. This effect can be produced with sprays of light oils containing 14–25 percent of aromatic hydrocarbons. Better results can generally be obtained with some of the more complex herbicides used in agriculture. Some species of tree seedlings are resistant to particular herbicides and some are not. If they lack resistance it is usually necessary to grow the seedlings in drills and apply sprays with nozzles that pass between the rows and below the seedling crowns. As is the case with any chemicals used in nursery management, the dosages must be carefully regulated and the compounds chosen on the basis of what amounts to a continuing research effort at each nursery.

Mechanical cultivation is sometimes used where seedlings are grown in drills. It may also help in paths between seedbeds, although heavy organic mulches applied there not only control weeds but also help maintain soil organic matter.

In spite of such efforts it is usually necessary to eliminate some resistant weed species by hand weeding. Complete reliance on hand weeding is possible in labor-intensive circumstances and may be necessary with seedlings grown in individual containers. Weeding done by hand or mechanically can cause seedling damage. Although it is easy to fall into the trap of having excessive seedling density, problems with weeds can be reduced if the seedlings close together soon and start to suppress weeds themselves.

In the common situation in which not only weeds but other biotic pests develop excessively high populations it is necessary to resort to soil sterilization with fumigants. At many nurseries this must be done annually before the seeds are sown, even though subsequent weed control remains necessary. Before herbicides were developed it was often regarded as prudent to move nurseries to new locations when weed problems got out of control.

Watering and Shading of Seedbeds

It is generally necessary to irrigate seedbeds, especially during germination and while the seedlings are succulent. Agricultural sprinkling systems with light, movable pipes are best unless fixed overhead systems already exist. Watering to maintain seedling growth is done with comparatively large, infrequent applications before there is any threat of deficiency. Excessive irrigation wastes water and leaches nutrients. Although water should not be withheld to the extent of preventing normal growth it is well to reduce or cease irrigation near the end of the growing season to help induce dormancy.

Irrigation to maintain soil moisture is best done at cool times of day when evaporational losses are low. Evaporational cooling from light watering can, however, protect young, succulent seedlings from heat injury during sunny weather if soil surface temperatures approach 50°C. Light watering can also protect them from the kind of frost damage caused by radiational cooling on calm, clear nights. The liquid water contains much heat energy and a very large amount of it must be lost before water at the freezing point is converted to ice at the same temperature. Ice-coated seedlings are not likely to freeze because bulky ice cools slowly and the freezing point of cell sap is less than 0°C. Light watering of seedlings under conditions of extreme surface temperatures do not damage them significantly.

Seedlings of species that are tolerant or moderately tolerant of shade, such as some northern conifers, must often be grown under partial shade for several months or even longer. This protection is desirable not only to protect these small, slow-growing seedlings from temperature extremes but probably also to prevent excessive respiration caused by high temperatures. Various kinds of screens, preferably ones that can be rolled up to facilitate storage and weed control, can be placed about a foot above the ground. This kind of shading is such a costly nuisance that it is best removed as soon as possible. In some circumstances in the tropics it may help to use the partial cover of special structures (Fig. 10-4) or trees for shade.

Control of Insects and Management of Fungi

Protection of nurseries from injurious organisms demands constant vigilance, intensive technology, and good knowledge of the pests. The majority of pests can be controlled by applying well-chosen pesticides at certain critical stages. The nursery manager must develop specialized knowledge of the pests characteristic of the locality so as to be able to

FIGURE 10-4   Mahogany and eucalyptus seedlings growing in polyethylene bags beneath a shade house at a nursery of the Puerto Rico Forest Service.

recognize and combat them either routinely or on short notice (Peterson and Smith, 1975). It may be necessary to apply some sort of pest management, such as eradication of the alternate hosts of stem rusts of conifers, in buffer zones around nurseries. Damping-off of young, succulent seedlings is one disease of almost universal occurrence. It is caused by a wide variety of weakly parasitic fungi that ordinarily infest organic debris. The means of control vary with the different groups of fungi. However, they tend to be most serious in soils with a pH higher than 5.5. Therefore, it is desirable to avoid the use of lime, fertilizers with basic reaction, or even alkaline sands and peats. Covering seed with clean, relatively sterile sand tends to reduce damping-off. The fungi ordinarily develop most rapidly in moist, shaded conditions and very dense seedbeds. Difficulties are reduced by any procedure, such as fall sowing, that favors prompt germination and produces relatively sturdy seedlings. Captan or other fungicides applied to the seeds can be effective.

Problems with various soil fungi, including those that cause damping-off and some leading to root rot in plantations, can be great enough to require sterilization of the soil before sowing. This is usually done with fumigants. These compounds are injected at spacings and depths in the soil of about 25 cm and the soil is covered with plastic sheets to retain the vapors. Fumigation is standard annual practice at many nurseries, especially in the southeastern United States. The purposes of control of weeds and insects are usually secondary to suppressing soil fungi. Soil sterilization does not prevent problems from pests that invade after treatment. It is, in fact, often viewed as a necessary evil that may increase such problems by eliminating the advantageous effects that preexisting organisms may have in excluding harmful ones.

Soil sterilization eliminates the soil fungi that live symbiotically with tree roots in the mutually beneficial mycorrhizal relationship. In the open and in forested regions where spores of appropriate species of fungi are in the atmosphere, these relationships are usually soon reestablished. If they are not, it is generally necessary to inoculate the soil artificially. Sometimes it is possible to introduce fungus strains or species superior to the natural ones. *Pisolithus tinctorius*, for example, can improve the survival and growth of planted seedlings of several pines (Marx *et al.*, 1984). The greatest problems with lack of mycorrhizae arise when tree species are introduced to regions where they do not occur naturally and the appropriate fungi are lacking (Harley and Smith, 1983). Under such circumstances, survival after planting may be very poor before inoculation and remarkably good thereafter.

Lifting, Transplanting, and Pruning of Bare-rooted Seedlings

The uprooting or lifting of bare-rooted seedlings from the nursery is a crucial step that must be timed to fit the phenology of root development. Sometimes the lifting season or "window" is only 2-3 weeks long.

Although seedlings may be dug out of seedbeds with spading forks, tractor-drawn machines do a cheaper and better job. The simplest devices are large, slightly tilted, blades that are drawn horizontally beneath the seedlings so that they are slightly lifted and the bottoms of their root systems are severed. After the soil is loosened with these or with hand tools, the seedlings are pulled out by hand. Even if the work is done carefully, unduly large proportions of the roots are broken off. This problem can be corrected by use of more sophisticated machines, sometimes modified potato diggers, which lift the whole seedbed up and shake the soil out of it. These usually separate the seedlings from the soil without breaking off many roots. However, some species, especially those adapted to very dry soils, have such extensive root systems that it may be almost impossible to lift the seedlings without leaving most of the roots behind.

In any step in handling bare-rooted seedlings it is vital that the roots always remain visibly moist. They should not be uncovered for more than 2–3 minutes at any time whether it is just after lifting, in the packing shed, or when it is finally being planted. Even briefer exposure is preferable. No lifting should be done when the air temperature is below 0°C because the effect of freezing the roots is to desiccate them. Tree roots are so easily killed that it is remarkable indeed that many millions of bare-rooted seedlings survive planting. Ideally nursery seedlings should be grown so that they have tops and roots of just the right sizes for field planting when they are first lifted. Regulation of seedbed density is not always sufficient to achieve this effect; some sort of corrective treatment is necessary if the seedlings or just their tops or roots are either too small or too large.

If the roots are too small for the tops or are too sparse and sprawling, the seedlings are lifted and replanted in widely spaced rows elsewhere in the nursery. This step is called **transplanting** in silviculture and the word "planting" is reserved for final placement. Trees produced by this method are called **transplants** to distinguish them from **seedlings** grown in the original seedbeds. Sometimes seedlings grown in containers are also transplanted. Transplanting has the effect of making the root systems more compact with a larger number of fine lateral roots than found in seedlings. If the seedbeds were crowded, the tops and, more important, the roots become larger. The rearrangement takes place if the plants are left to grow in the transplant beds for a year or, less commonly, two years.

If transplanting is done by hand it can be expedited by inserting the seedlings in notches appropriately spaced along the edges of portable **transplant boards** in such a way that the roots hang down and can be placed against the vertical walls of trenches before the soil is packed against them. However, transplanting is such an expensive operation that it is usually done, if done at all, with modified agricultural transplanting machines. With such machines the seedlings are usually strung in spaced pockets on revolving disks that set the plants down into trenches where they are tamped into place by packing wheels. Transplant beds usually do not need watering but the rows should be uniformly spaced to facilitate mechanical weeding.

Transplanting was much more commonly done when labor was cheap and chemical weeding had not been developed. Spindly seedlings were grown in seedbeds that were deliberately crowded to minimize hand-weeding costs. Transplanting is not only expensive but it must also be done right at the time of lifting when the nursery workload is most crushing. The only good reason for growing transplants is to produce planting stock that must be unusually tall.

Root pruning or **wrenching** is a good and inexpensive substitute for transplanting. If the object is to make the root systems shorter but more compact this can be done by drawing thin, horizontal blades, mounted behind tractors, under the beds at appropriate depths. The root systems can be trimmed off at the sides as well with U-shaped blades if the seedlings are in drills. The seedlings are then usually left to grow for another year.

The purpose of doing these treatments, or not doing them, is to produce seedlings with truly viable root systems small enough to be planted conveniently, deep enough to contact a stable moisture supply, and ready to start elongating. One basic problem with bare-rooted stock is that it inevitably leaves the nursery with the roots damaged and the ability of the tops to respire and transpire scarcely diminished.

Sometimes it helps to bring the tops into better balance with the roots by mowing back the tops of the terminal shoots. If this is done when the tops are succulent and actively growing, most conifers will form new buds that create new tops that are nearly normal. This treatment can also be resorted to if the plants must be left in the nursery for another year and might otherwise grow too large. If the roots are longer than the anticipated depth of the planting holes it is desirable to prune them back enough to facilitate planting. Since the ends of the roots are often killed by the lifting operation, root pruning may not actually sacrifice any useful tissue.

Tropical angiosperms, such as teak, often grow very large before they become dormant at the end of the first growing season. If they are capable of sprouting, they can be trimmed down to plantable size by severe pruning of both top and roots. The stubby **stump plants** thus created typically have 5-cm tops and roots 15 cm long.

Grading of Seedlings and Transplants

The primary classification of nursery stock depends on its age and treatment. These two characteristics are designated by two figures, the first showing the number of years through which the plants grew as seedlings, and the second the number of years they were grown as transplants. For example, "1-0" indicates a 1-year-old seedling that has not been transplanted; "2-1" designates a 3-year-old transplant grown 2 years as a seedling and 1 year as a transplant.

Before nursery stock is shipped into the field it should be culled to eliminate trees that will not survive after planting. It is always essential to remove plants infested with virulent insects or fungi, those that have been badly damaged in handling, and those with distinctly poor roots. Further segregation is generally an uncertain proposition because the external appearance of seedlings is not a very dependable indicator of the internal physiological qualities that affect survival (Duryea and Brown, 1984). The viability of seedlings can be assessed by various techniques of determining their moisture content.

If seedlings are to be machine-planted it may help to sort them into batches of uniform size. This problem is better anticipated by separating the seeds into different size classes and sowing them in separate beds.

Ordinarily plantable conifer seedlings should be at least 3 mm in "caliper" or basal diameter, between 10 and 35 cm in above ground height, and with a ratio, by weight, of top to root not more than 4 to 1.

The culling, grading, and counting of seedlings is usually done on conveyor belts under shelter as part of the packing operation. These operations should be done swiftly under cool, moist conditions with a minimum of handling of the seedlings. It is possible to set up rapid ways of counting seedlings or procedures in which the numbers of seedlings ready for packing are determined by weight on the basis of carefully drawn samples that are both counted and weighed.

Transportation and Storage of Nursery Stock

If bare-rooted stock is to remain viable, respiration and transpiration must be held to a minimum during transit. It is for this reason that success usually depends on doing both lifting and planting during periods when the trees are dormant or nearly so. The crucial test is the ability of the roots to renew rapid growth after planting. Bare-rooted stock quickly loses this capacity if water and carbohydrates are squandered in transit because the tops are growing too actively or because the plants become too warm or dry.

Planting stock is usually packed in ways that keep the roots moist and cool but with provision to allow oxygen to enter the package while carbon dioxide and other gases that might be noxious can escape. The traditional practice is to construct bales consisting of two tight bundles of trees arranged in opposite directions so that the tops are open to the air at each end and the roots of each bundle are in close contact with each other. The roots are then covered with sphagnum moss or similar water-holding material and the whole bales are wrapped with waterproof paper. Provided that they can be kept cool, the seedlings can simply be sealed in bags of paper impregnated with polyethylene; such bags are appropriately permeable to oxygen and carbon dioxide but impermeable enough to water that it is not necessary to include any moss.

The best way to store packages of nursery stock, whether for a few days or over winter, is under refrigeration at temperatures slightly above the freezing point. It is usually important to avoid freezing bare roots. While they may survive brief periods of freezing if thawed slowly, it is often best not to waste time planting seedlings that were accidentally frozen after lifting.

Leafless hardwoods endure storage better than conifers. Storage of this sort is most commonly done to hold the stock dormant for a few weeks until planting sites in localities colder than the nursery become ready for planting. Seedlings lifted in the fall are stored over winter to plant on sites that become ready for planting in the spring before stock can be lifted in the nursery. Short-term storage is also useful in evening out the flow of operations during the brief shipping season. If facilities for refrigeration are lacking, stock can be kept for a few days in cool sheds or cellars. However, it is best to reduce the time between lifting and planting as much as possible.

Nursery Records

Efficient management of nurseries requires the maintenance of complete maps, records, inventories, and accounts of costs. It should be possible to determine the history of each seedbed in detail so that present developments can be interpreted in terms of past treatments. Occasional inventories must be made of the amount of stock in various stages of growth by random sampling of the beds. Records of the history of each lot of stock, from seed to packing shed, are indispensable for improvements in practice; whenever possible these histories should be extended to the plantations in the field. The annual schedule of operations should be carefully planned with particular reference to the unavoidable seasonal peaks of activity.

PRODUCTION OF CONTAINERIZED STOCK

Under most circumstances, bare-root planting is cheapest and most expeditious. However, it has the fundamental disadvantage that contact between root and soil is broken. This contact can be dependably reestablished only if the bare-rooted seedling is planted during or just before short periods of active root growth. This constraint and other problems can be mitigated by growing the seedlings in containers of soil or soil-like medium and planting them without ever breaking contact between root and soil (Tinus and McDonald, 1979; Scarratt, Glerum, and Plexman, 1981; Guldin and Barnett, 1982). Containerized planting has often been necessary in arid regions (Goor and Barney, 1968). It is commonly used in the humid tropics, where many species become too large to move several weeks after germination and must be planted when actively growing and succulent. This problem has traditionally been overcome by planting both container and seedling. Pots made of quickly disintegrating bamboo or paper are often used, although polyethylene bags (Fig. 10-4) can be used if they are removed or thoroughly slit before the planting.

Trees more than a meter tall usually have to be dug out of the ground with large amounts of attached soil, which is then wrapped or "balled" in burlap for moving and replanting. The root systems of such trees are so much wider than the tops that many roots are left behind. The chief advantage of this very costly kind of containerized planting is that it gives instant beauty. It is a long time before the trees are able to resume active growth and their initial survival may depend on frequent irrigation.

The old technique of planting container-grown trees has now been systematized and mechanized for a variety of reasons. The most important is that seedlings individually grown in containers in greenhouses can be cared for much more effectively and grown faster than in the open. Among other things, it takes much less seed, an important consideration given the high cost of seeds produced in genetic improvement programs. It is easier to keep small batches separate and thus provide seedlings more closely fitted to specific site requirements. Even though containerized seedlings are costly to produce they can be much cheaper to plant if labor costs are high. Much of the cost of hand planting is that of making a hole in the ground. It takes less labor to punch or bore a hole to install a plant unit of uniform size, sometimes with a special planting tool or machine, than to dig holes for stock of irregular sizes.

Since the containerized seedlings can be grown faster and planted when smaller than bare-rooted seedlings, it is possible to meet requirements for emergency reforestation and fluctuating demands more effectively and quickly. While the capital cost of a given production unit may be high, efficient units can be smaller than is the case with large outdoor nurseries and they do not require as much land, machinery, or packing facilities.

With intensive production under cover it is possible to control closely the environment of seedling growth. In cold regions, like Finland, it has been possible to produce plantable seedlings faster and more dependably if the seedlings are germinated early under plastic-covered greenhouses. The same effect is, incidentally, also produced by similar early covering of ordinary seedbeds. In regions with longer growing seasons, such as the southern United States, it may be possible to produce two or more crops of seedlings annually under cover. Some of the acceleration of growth comes from heavy fertilization. Close control of pests not only improves growth but is also mandatory in the greenhouse environment. Another important advantage is that containerized seedlings can be planted at times when the roots are not growing rapidly enough to permit bare-root planting.

Kinds of Containers and Their Characteristics

Containerized seedlings can be grown (1) as **tubelings** in plantable pots or tubes, (2) as cohesive **plugs** pulled out of the containers before planting, and (3) in **blocks** of pressed peat or pulp that are both container and rooting medium. Somewhat similar are "mud-pack" seedlings, which are bare-root stock individually wrapped with soil around the roots in the nursery shed.

The ideal plantable pot would be one that held together through the act of planting and either disintegrated promptly thereafter or did not impede the subsequent growth of roots through the walls. Porous, biodegradable materials such as paper and pressed peat have generally been more suitable than plastic in this respect. When they are planted, however, the container tops should be covered with soil so that they will not act as evaporational wicks desiccating the soil. The volume of the containers should be uniform and is often about 40 cc.

Plastic containers are usually best for producing plugs. These are often grown in cylindrical cavities closely spaced and inset vertically in reusable molded blocks of Styrofoam. Another kind, sometimes called a book planter, is made of two hinged sheets of thin molded plastic that form proper kinds of cavities when folded together. When it is time to remove the plugs, or replace an empty one with one that has a developing seedling, the "book" is opened. Sometimes these containers are just individual tubes that can be set into, and taken out of, fixed racks.

There has been much concern about the tendency of elongating roots to be deflected into spiral arrangements when they impinge upon the walls of any kind of container. They can even be deformed if they grow against a minute flange on the inside of a plastic casting. Since root systems thus deformed tend to remain so and produce "pot-bound" seedlings, ways have been developed to counteract the tendency. One of these is to cast vertical grooves or ridges into the container walls so that the roots are directed downward. Another, called **air-pruning**, involves containers with holes in the bottoms supported such that there is air beneath them. When the roots reach the air, they simply stop growing but retain the capacity to resume when planted in contact with soil. The same effect is sometimes produced by painting the container walls with toxic copper salts in latex.

Theoretically one could avoid growing pot-bound seedlings by getting them out of the containers before the roots touch the walls, but this calls for impossibly good timing. Such problems are reduced by growing seedlings in blocks of peat or pulp, but not if the blocks are confined inside containers.

It should be noted that most methods of planting bare-rooted stock also lead to deformities of root systems that are merely less obvious than those caused by containers. Unless they have been confined within long-enduring plastic tubes, containerized seedlings probably ultimately develop more symmetrical root systems than bare-rooted seedlings that are usually planted with their roots in a single vertical plane like a crack in a rock.

Growing Containerized Seedlings

In mild, humid climates it is possible to grow containerized seedlings in the open with some provision for partial shade and watering. However, during recent years it has been common to use greenhouses, usually of transparent plastic, to control the environment more intensively. Sometimes seeds are germinated in highly controlled facilities and then the containers are moved to less costly ones, being replaced by new batches for germination. There are some installations in which the environment is so fully controlled that even the light is artificial. These use so much capital and energy that they are usually regarded as too costly.

The filling of the containers is usually either mechanized or systematized. Because of its light weight and high capacity for holding water, finely divided peat is the most common medium. It is used either pure or mixed with such ingredients as vermiculite and ordinary soil. It may be necessary to use mineral soil, in spite of its weight, for containerized stock to be planted in clay soils. The roots may not grow out of the containers if there is too much incompatibility of physical characteristics controlling water movement between the container medium, any container wall, and the soil of the planting site.

There are problems in arranging to have one seedling in each container regardless of whether the seeds are sown by hand or with machines. If two or three seeds are sown in each, subsequent hand thinning is necessary. With one seed each there will be blanks, although the numbers of these can be reduced by using good seeds and cleaning out the empty ones. It helps to have container systems in which empty individual containers can be replaced with ones with seedlings. Small seedlings can be transplanted by hand or with tweezers by poking their roots into holes made with wires of the right sizes. This kind of transplantation may be a good substitute for sowing seeds if, as with eucalypts, the seeds are too small to handle individually or if plantlets produced by vegetative reproduction are used.

Irrigation must be frequent and the containers are usually carefully fertilized to promote rapid growth. If the conditions are biologically sterile enough, inoculation with mycorrhizal fungi is desirable. The containers are usually open at the bottoms and grown on some sort of surface porous enough that they will be well drained and there will be less tendency for roots to grow out of the container bottoms. Pesticides must be used to forestall attacks by all the fungi, aphids, and other pests characteristic of greenhouses. Although the artificial environment is conducive to rapid growth and close control it must be vigilantly tended because the whole crop can be swiftly killed if something goes wrong.

The most artificial arrangements for control of light and temperature are used to grow seedlings at unnatural times of the year. It may be necessary to regulate the photoperiod or length of artificial day and night to accomplish this. Control of photoperiod, temperature, or both in combination may be necessary to induce or break seedling dormancy at times appropriate for either planting or storing the seedlings. For example, as is the case with some seeds, dormant seedlings of certain species require a chilling period before they will break dormancy. If they were caused to develop dormancy in a warm greenhouse and were kept there and then planted, they might grow abnormally or not at all the first year. More commonly, however, the problem with container seedlings is getting them to harden; succulent seedlings are too fragile to withstand much handling.

Handling of Containerized Seedlings

If necessary, dormant containerized seedlings can be stored under whatever conditions of low temperature or dryness they are conditioned to endure in nature. Since the roots are not exposed, there is no need to worry about freezing them if they are dormant and of species adapted to frozen soils. If they are about to be planted, however, they should be nondormant or capable of surviving from the time of planting until the normal, natural ending of dormancy.

Containerized stock is shipped to the field whenever suitable times have come for planting. There are no problems with the timing or mechanization of lifting as with bare-rooted seedlings. Sorting and culling have ordinarily been done previously during the growing process. If the containers are shipped out with the seedlings, packing consists only of putting the material in crates or on racks designed to prevent crushing the seedlings. If the containers are for growing plug seedlings and are to be reused there is the problem of returning them.

If plug seedlings are withdrawn from the containers before shipping it is usually necessary that they still be dormant. Ordinarily they are packed in polyethylene bags. They can be put in cold storage if desirable. Since such plugs lose their cylindrical shape during all this handling, they cannot be planted in standardized holes. The chief advantage is that the containers do not have to make a round trip to the planting site. Whether the containers make the trip or not, the main shortcoming of containerized planting is the simple but arduous and costly task of moving so much material to the characteristically dispersed spots where trees are planted.

VEGETATIVE PROPAGATION OF PLANTING STOCK

Planted trees do not always have to be grown from seed and there can be important advantages in bypassing sexual reproduction and germination problems. In the first place, asexual regeneration by vegetative propagation is the only sure way of perpetuating trees with exactly the same genetic qualities as the parent. Second, if cuttings form roots readily, it may be much easier and give quicker results to plant them rather than to grow seedlings. Even if it is difficult to make pieces of old plants form new ones, it can still be advantageous to do so in order to multiply a few individuals with highly desirable genetic characteristics.

With some species, such as the cottonwood poplars, one can merely plant long pieces of thin shoots and depend on most of them to form roots and become established (McKnight, 1970). This usually happens only on moist soils and with some species of those sites. Sometimes, as with syca-more, dense stands can be created by burying branches in shallow trenches and having each branch produce many shoots.

Most tree species are not propagated from cuttings anywhere near as easily. In fact, for most it is so difficult that it is done mainly for multiplying valuable ornamentals, fruit trees, and trees for seed orchards. Such work is usually done in greenhouses with the use of rooting hormones such as indoleacetic acid.

Even more sophisticated techniques are being developed to produce whole new plants from culturing small fragments of tissue in media that supply the complex biochemical substances needed for plant development. The techniques are basically similar to those used to culture the mycelia of fungi from spores in laboratories or to growing yeast in factories, but more complicated (Bonga and Durzan, 1982). Plantlets produced by such intensive methods are normally transplanted to containers for further growth before planting.

The potential advantages of creating forests from asexually regenerated genetic material are very great. Forest tree species have not been subjected to millenia of genetic selection by people and are therefore very heterozygous. This genetic heterogeneity can make it difficult to replicate individuals with many fine genetic traits through controlled pollination schemes. With vegetative propagation it is not possible to improve upon nature but the best individuals found there can be multiplied. However, the very nature of the development of stands of woody plants and the vicissitudes that long-lived perennials must endure are such that it would be highly undesirable to create stands of single genotypes. Genetic resources must be managed both to multiply good genotypes and to preserve diversity as well.

The production of trees from the results of artificial gene-splicing is a development of tissue culture that lies in the future.

BIBLIOGRAPHY

The bibliography for Chapters 10 and 11 is at the end of Chapter 11.

CHAPTER 11

Tree Planting

Planting is one of the most crucial investments made in silviculture. The success of a whole rotation can depend on decisions made at planting time. The establishment of a plantation commits the forester, and usually the forester's successor, to many subsequent actions and investments. Errors made at the time of planting usually cause future problems that are not always soluble. As is the case with all powerful techniques, planting affords opportunity not only for great success (Fig. 11-1) but also for the biggest blunders in silviculture.

Most of the blunders come from overzealous site preparation or putting species on sites to which they are not adapted. The only insuperable drawback of planting is the almost inevitable deformation of the root system. This can reduce windfirmness and may possibly increase root rots. Other difficulties are generally correctable. One of the greatest of these is the necessity of organizing people and facilities for a major effort during each short planting season.

Selection of Planting Stock

After the all-important choices about species have been made (see Chapter 9), the next have to do with the kind of planting stock. There may be a choice between bare-rooted seedlings and containerized ones or between different sizes and ages of seedlings of either category. Sometimes the choice is difficult because real alternatives exist. If, however, as is often the case with loblolly pine, dormant 1-0 bare-rooted seedlings are of the right size to plant and large nurseries exist to grow them, only special circumstances would cause one to consider anything else.

As indicated in Chapter 10, there can be a variety of reasons for using containerized stock or even of being limited to its use (Tinus, Stein, and Balmer, 1974). If, as in the humid tropics, seedlings remain succulent while growing very large, there may be no other alternative. Truncated stump-

FIGURE 11-1 A crew planting loblolly pine by the bar-slit method on eroded land in southwestern Tennessee (top) with a view of the same spot 6 years later (bottom). (*Photographs by U. S. Forest Service.*)

plants may enable evasion of the problem, as is the case with teak. At the other extreme, the problem of securing survival on very dry planting sites may dictate use of containerized stock and sometimes very large, *deep-rooted* plants grown in big cans.

Bare-root planting is generally most common if only because there is no need to move heavy, bulky material from nursery to field. The choices that are made among different kinds of bare-rooted stock have mainly to do with age and size; but older and bigger are not necessarily better. The object is to put a viable root system into the soil. As far as the top is concerned, all that is basically necessary is some shoot, bud, or apical meristem capable of growing into a tree. There must also be enough carbo-hydrate and other nutritional material stored in the seedling to enable it to resume growth.

Size of Plants

The first consideration is that the roots go deep enough and grow fast enough to maintain contact with the soil-moisture supply. The most important virtue of large or tall tops is the greater likelihood that they will not be submerged by competing vegetation. A tall seedling can also project above strata in which certain kinds of damage by animals or other agencies is common. A common example has to do with planting saplings for street trees (Grey and Deneke, 1978); it is desirable to have the tops taller than the level of people's eyes and the itchy fingers of children. Short seedlings are sometimes too easily buried by fallen leaves or eroding soil.

Plants often have to grow to a certain size before they have enough stored material or conductive tissue to resume development after the dam-age inevitably associated with planting. Deep-rooted seedlings tend to be less susceptible to frost-heaving and those with thick bark to heat injury. Otherwise, large planting stock can be more costly to grow and plant and less likely to thrive. This problem generally arises when too much root tissue has been left behind in the nursery but the top is intact. A planted seedling with a large top is readily visible and looks very impressive, especially if one paid a lot to get it planted. Even if it survives it may be so slow to resume active growth that a smaller planted seedling may actually overtake it in size. The only common case in which bigger is better is the one in which the planted tree must contend with tall competing vege-tation.

Many of the crucial characteristics of nursery stock have little or noth-ing to do with seedling size. Physiological health, capacity for root regener-ation, and similar attributes are controlled by the conditions of nursery culture, the care in lifting and handling, and the degree to which the timing of lifting is synchronized with the seasonal regime of plant devel-opment.

Season of Planting

Success in bare-root planting depends on doing it when periods of 2–3 weeks of active root elongation can be anticipated. The use of containerized stock gives more latitude in timing than is available with cuttings or bare-rooted material. Roots normally elongate at times of the year when supplies of physiologically available water can be captured, except when rapid shoot or leaf formation commands most of the supply of constructive materials (Jenkinson, 1980).

In climates with periods of cold or dryness that induce dormancy, the period of most rapid root growth occurs just after the breaking of that dormancy. This is the best time for planting not only because of the opportunity for quick reestablishment of contact between plant and soil but also because the whole growing season is available for further development. The ideal would be to have the plant break dormancy the instant it was planted because it is safer to move it while it is dormant.

Bare-root planting must, in other words, normally be done in a rather short period at the beginning of the growing season. If the period must be brief it means that the operations that start with lifting and end with planting are apt to be frantic, costly, and sloppy. If the lifting must be delayed by waiting for the nursery beds to thaw or stop being muddy, the season can be shortened even if the planting sites are ready. Given such problems it is well to consider ways of extending the planting season.

It may help to lift the stock much earlier and store it so that the lifting operation does not delay the start of planting. If soil moisture and temperature are adequate, there may be another period of active root growth in the autumn so that trees with tops that have hardened and formed buds can then be planted with fair success. However, not all of the species of the same locality have the same seasonal regime of root growth; thus it is best not to plant at any time without knowing from experience or direct observation that the roots can be depended upon to grow actively after planting. One of the major advantages of both containerized planting and direct seeding is that they can be done at times when bare-root planting cannot. It can be advantageous to have planting crews do that kind of work when bare-root planting cannot be done.

Bare-root planting should not be done when the air temperature is below 0° C or when hot or dry winds are blowing. If trees are planted in the fall in cold climates there can be serious problems with frost-heaving during the winter, especially if the trees are not well covered by an insulating blanket of snow. Another drawback of fall planting is that the seedlings do not get the benefit of a full growing season of carbohydrate production. It is well to remember that trees are generally storing carbohydrates at those times of year when water is available but formation of new tissues has ceased. They may continue to gain in weight late in the autumn even if their dimensions do not change. In some warm climates, such as that of the southeastern United States, spring and autumn are, in effect, merged so that planting is done during the mild winters. It may, however, have to be suspended during temporary periods of freezing weather.

Site Preparation

Successful planting often depends on measures such as reduction of competing vegetation, removal of physical obstacles to planting, and drainage of water toward or away from the planted trees. Most of these measures are dealt with in Chapters 5 and 8.

Control of competing vegetation is especially important because many planting sites are already crowded with preexisting vegetation. Unless a distinct vacancy is found or created in the growing space it will be difficult for a planted tree to survive or grow satisfactorily. Site preparation as a means of controlling competing vegetation is usually done as a separate operation. However, planting spots can also be prepared at the same time as the planting by spot applications of herbicides or **scalping,** which is the scraping of competing vegetation from the planting spot with hand tools. Scalping is moderately effective against thin grass or very low shrubs and can be done with the grub hoes sometimes used in planting.

Methods of Planting

An increasing number of kinds of hand tools and machines (American Society of Agricultural Engineers, 1981) can be used for making planting holes, inserting trees, and packing soil back around their roots.

Successful planting depends on the ability of the roots of the planted trees to regain contact with the soil so that the uptake of water and nutrients can resume. The first essential is that the roots be placed in soil in which water is immediately available; the moisture content must be well above the wilting coefficient and the soil temperature must be sufficiently far above the freezing point that water can move readily. Moist mineral soil is by far the most dependable water source; organic materials are not reliable unless they are peats and mucks situated so that they remain close to saturation without being deficient in oxygen. The soil must also be warm enough for roots to grow.

When bare-rooted stock is lifted from the nursery virtually all of the mycorrhizal mycelia and root hairs that constitute the main absorptive portion of the roots are destroyed. Until these unicellular strands resume proliferation the plant has no means of absorbing water that is not immediately adjacent to the existing roots. Fortunately suberized roots can absorb water directly (Kramer and Kozlowski, 1979), thus providing planted trees with an immediate, but very limited, supply of water. Presumably this is one of the reasons why it is so essential that the soil be solidly packed around the roots of planted trees.

Until the main absorbing network redevelops the planted tree has only slightly more ability to capture moisture than a cut flower in water. The development of new root hairs depends on the elongation of new adventitious roots and root tips, many of which are broken off during lifting. New root tips must form before any major portion of the permanent and readily visible part of the root system can extend itself in the soil. The root system of a planted tree elongates rapidly only during rather limited periods when the normal physiological rhythm of growth is directed toward this process. It is, furthermore, reasonable to suppose that such elongation of the roots will not become very active until photosynthesis becomes vigorous again. Since this is not likely to occur until resumption of active absorption of water and nutrients, any complete dependence on elongation of root tips and on absorption by root hairs would seem to jeopardize survival of planted trees.

Mycorrhizal mycelia are probably much more effective than root hairs in reestablishing early contact with the soil. The fungi that are involved need only a supply of reserve foods from the plant and the proper conditions of moisture and temperature to resume growth; consequently they do not depend on resumption of full activity by the whole plant. It is significant that most tree species normally enter into mycorrhizal relationships and may require them to survive or thrive after planting. There are many cases in which the planting of exotic species has become spectacularly successful, after earlier frustrations, as soon as nurseries were inoculated with the proper fungi.

The ability of the root system to redevelop is diminished more than appearances would suggest because so many bare roots die between lifting and planting. The roots closest to the root collar are more likely to survive than those farther from the stem.

Trees should usually be planted at the same depths they occupied in the nursery. Sometimes it is advantageous to plant them more deeply but never more shallowly. This is generally because the greater the depth the more dependable the water supply; buried stem tissues of a few species can form roots.

If the roots are more than 20–25 cm long they may be too long to plant properly and it may be desirable to prune the ends off before giving the seedlings to the planting crews. Every precaution should be taken to ensure that the roots remain *visibly moist* at every stage of handling. Seedling roots should be left covered until the moment comes to stick them into the planting hole and *not waved around in the air*. Moist soil should be packed around the roots and the planting holes should be filled completely, leaving no air spaces. Litter or other organic debris should not be used for filling the holes unless the soil itself is organic. The trees should be planted firmly enough to resist a gentle tug with thumb and forefinger.

Trees are ordinarily planted with stems erect and roots spread out in a vertical plane. Most species, but not all, will straighten up if the stems are not perfectly erect. Putting the roots in a vertical plane is distinctly unnatural and may be disadvantageous. It is least desirable with species such as the spruces, which have roots that characteristically grow in a horizontal plane; there are special techniques for planting trees with the roots so arranged.

Any mode of root placement resulting from planting is unnatural (Van Eerden and Kinghorn, 1978). The greatest problems arise from sloppy planting that leaves the roots curved upward like the letters J, L, or U. The first major difficulty that results from this is that the roots are not placed as deeply as they might be. The second is that the roots may not develop the symmetrical structural pattern necessary to confer windfirmness so that the trees become unusually vulnerable to uprooting after they grow tall. This problem is greatest with root systems that are wound up into balls when stuffed into shallow holes or left to grow too long in containers. Natural growth of new roots does correct some of the abnormalities of root systems of planted trees but seldom perfectly.

Regardless of whether the work is done by hand or with machines, there are two general methods of making holes and planting bare-rooted trees. In the **compression methods** holes are made for the plants by pushing the soil aside with sharp instruments driven into the ground; after the tree is inserted the soil is pressed back around its roots. In the **dug-hole methods** the soil is actually removed and set aside to be repacked around the roots after they are arranged in the hole. All methods of "hand" planting also involve the use of the heel to repack the soil around the tree roots.

Several types of heavy dibbles or planting bars, such as the two shown in Fig. 11-2, can be used for planting trees in sandy soils. The most common technique for using such tools, known as the **bar-slit method** and shown in Fig. 11-3, illustrates the basic principle of the compression methods. In this method of planting it is important to push the seedling down

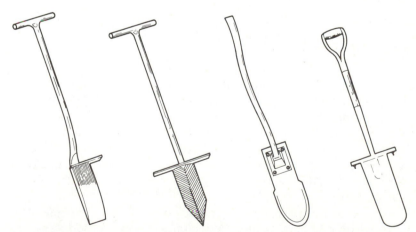

FIGURE 11-2   Four tree-planting tools (left to right): (1) planting bar, (2) a pointed planting bar useful in stony soils, (3) the Rindt grub-hoe (L-shaped) for making straight-sided planting holes, and (4) a tile spade planting shovel for digging deep holes for large planting stock.

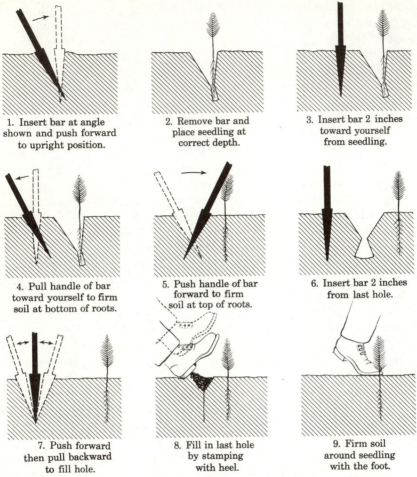

1. Insert bar at angle shown and push forward to upright position.

2. Remove bar and place seedling at correct depth.

3. Insert bar 2 inches toward yourself from seedling.

4. Pull handle of bar toward yourself to firm soil at bottom of roots.

5. Push handle of bar forward to firm soil at top of roots.

6. Insert bar 2 inches from last hole.

7. Push forward then pull backward to fill hole.

8. Fill in last hole by stamping with heel.

9. Firm soil around seedling with the foot.

FIGURE 11-3   Steps in the use of the bar-slit method of planting seedlings in sandy soil. (*Sketch by U. S. Forest Service.*)

deeply into the hole and then withdraw it so that the roots will go as deeply as possible but be straightened out and free of J- or U-bends. One continues to hold the seedling upright while pushing the soil back against its roots. Since one person does all the work in this method, it helps to carry the seedlings in a shoulder bag rather than in a container that is set on the ground; by using a planting bag one bends over only once for each tree. One can plant an average of 1500 trees daily with this fastest method of hand planting.

On soils that are stony or have so much clay that they become compact under true compression, a modified compression method can be employed. Grub hoes are used in this **grub-hoe-slit method.** The soil is partially lifted from the hole rather than pushed aside. One can plant about 700 trees daily with this method. One advantage of the grub hoe is it can be used for scalping whereas planting bars cannot.

The main drawback of the compression method is the difficulty of being certain that the holes are completely closed or that the roots of the seedlings are free of U-shaped bends. These problems can be partially avoided by refraining from working the blades of the tools back and forth after driving them into the soil. The holes created as a result of this error are shaped like hour glasses in vertical cross section; it is difficult to work the roots past the constriction and impossible to be certain that the lowest part of the hole is closed. Even the grub-hoe modification of the compression method may be difficult to apply in soils that are very stony or cohesive.

The dug-hole methods must be used where the compression methods fail to give good results. Grub hoes (Fig. 11-2) are usually employed in this method, sometimes with one person digging the holes and another doing the planting. The holes are made sufficiently wide and deep to accommodate the roots of the trees. The mineral soil is piled beside the hole separate from leaves, sod, and other debris to facilitate planting and to ensure the use of clean soil next to the roots. The trees are planted entirely by hand immediately after the holes are dug so that the exposed soil will not become desiccated. There are several different methods of shaping the holes and planting the trees.

The simplest dug-hole method is called the **side-hole method** (Fig. 11-4). One side of the whole is left smooth and vertical; the roots are spread

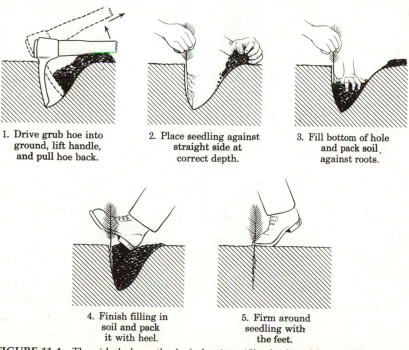

1. Drive grub hoe into ground, lift handle, and pull hoe back.

2. Place seedling against straight side at correct depth.

3. Fill bottom of hole and pack soil against roots.

4. Finish filling in soil and pack it with heel.

5. Firm around seedling with the feet.

FIGURE 11-4 The side-hole method of planting. (*Sketch adapted from one by U. S. Forest Service.*)

out against this wall and packed firmly in place by hand with a thin layer of fresh, loose soil. The rest of the mineral soil is then scraped into the hole and trodden down with the heel. Usually the area around the hole is scalped before the hole is dug so that the tree will not be immediately adjacent to competing vegetation. This method can be used under almost any conditions but is not rapid; the production rate is about 600 seedlings per person-day. It leaves the roots in a vertical or slightly curved plane.

The best way to get the roots placed in three dimensions is the **center-hole method.** This involves putting the tree in the middle of the hole and then sifting and packing the loose soil between the various strands of roots, without forcing them into either a horizontal or vertical plane. This is a slow method but it can be speeded by boring the holes with motorized augers (Fig. 11-5).

The **wedge method** is another way of improving root placement. It involves creating a hole with a ridge in the middle such that a vertical cross section would resemble the letter W (Fig. 11-6). Half of the roots of the tree are spread out on each sloping surface and tamped down in that position. Rudolf (1950) described how the holes could be made with several strokes of a grub hoe and stated that 500 trees per person-day could be planted by

FIGURE 11-5 Motorized augers being used for planting after timber cutting in an open forest of ponderosa pine and Douglas-fir in Montana. One auger can produce enough holes for a crew of 3 planters. (*Photograph by U. S. Forest Service.*)

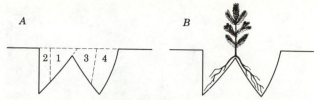

FIGURE 11-6 The wedge method of planting as described by Rudolf (1950). *A* shows the sequence in which soil is removed by four strokes of a grub hoe to create the hole. *B* shows the tree ready to be tamped in place.

this method. This procedure is well adapted to the planting of species like the spruces, which have distinctly horizontal root systems.

Perfectly horizontal placement of roots can be achieved by techniques such as making T-shaped incisions in turf with spades, turning up the resulting flaps, spreading out the seedling roots, and then tamping the flaps back over them.

Containerized seedlings (Barnett and McGilvary, 1981) can be planted with special devices of varying sophistication, provided that the units to be planted have retained sufficiently uniform size and shape. One of the advantages of using uniform units is the opportunity to poke standardized holes in the ground rather than doing laborious digging. This advantage is of greatest importance with stony soils where digging is difficult. The simplest devices, suitable only for small stock, are nothing more than rods stuck in the ground; these can also be used for planting cuttings of such easily rooting species as the cottonwood poplars. However, there are also devices that make holes by compression and either set or shoot the containerized material into the soil.

If the container units to be planted are not or have ceased to be of uniform size and shape they are put in the ground by the same hand or machine methods used for bare-rooted seedlings. This is common with plugs which were extracted from the containers before being shipped to the field.

Except where labor is cheap or where the terrain is unsuitable, it is both better and cheaper to plant trees with machines than it is to do the work by hand. A few well-trained people can do a better job than some large crew recruited for a brief flurry of work. It is also more nearly possible to compress bare-root planting into the phenologically appropriate period.

Most planting machines operate by the compression method. The typical machine (Fig. 11-7) cuts a narrow trench through the soil with a specially designed plowshare. The trees are placed in a slot just behind this trencher and the soil is pressed firmly back around the roots by two wheels that are mounted close together in slightly tilted positions at the rear of the machine. The machines are heavy enough that seedlings are usually planted more firmly than is possible with use of the human heel.

FIGURE 11-7 The essential parts of a tree-planting machine. The round cutting disk or coulter at right cuts through roots and sod ahead of the split plowshare or trencher. The trees are put down in the slit made by the trencher between the two wings or guides at its back end. They are then tamped in place by two tilted rubber-tired wheels mounted at the rear of the machine. (*Photograph by Louisiana Forestry Commission.*)

Planting machines work best on level sandy soils that are relatively free of stones, stumps, and tree roots. They can plant from 4 to 15 thousand seedlings per day, depending mainly on site conditions. One of the common purposes of preliminary site preparation is the removal of impediments to machine planting. However, various brush-clearing devices or plows can be fitted to the tractors that draw the planting machines so that site preparation and planting are done simultaneously.

There are other kinds of planting machines such as ones that do not make continuous slots in the soil. Holes suitable for side-hole hand planting can be scooped out with straight lugs projecting downward from the wheels or tracks of tractors. However, the general kind shown in Fig. 11-7 is most common and most large-scale forest planting is done with them using bare-rooted seedlings. The hand methods are usually used for small areas, for terrain that is steep or rough, and for soils that are full of stones or clay.

The use of containerized seedlings is usually associated with conditions that require hand planting but there is no basic reason why they cannot be planted with most kinds of planting machines. A large and highly automated machine has been developed in Finland to plant containers on ground prepared in the same pass of the apparatus. The containers are injected into the soil through flexible planting tubes. The planting head at the end of each tube has a sensing device that moves the head to a different spot if it is pushed into a stone or other impediment.

Placement of Seedlings and Reshaping of Soil Surface

Most techniques of treating vegetation and soil in preparation for planting were discussed in Chapter 8. Some involve making furrows or ridges. With these, the basic principle is that if the soil is too wet one plants on the tops of the little ridges, but if it is too dry one plants in or near the bottoms of the furrows. If, however, furrows are made to use folded-over sod or other turf to uproot and also cover competing vegetation, it becomes logical to plant trees along the lines where the ridges and troughs intersect. In this way the trees are kept as far as possible from competition and also have their roots partly in a double thickness of the soil strata in which nutrients tend to be concentrated.

In very dry situations, it is desirable or essential that trees be planted in basins or on terraces designed to concentrate runoff of water (Goor and Barney, 1968). Stones can be put around the seedling stems to shade them, redistribute water, and restrict direct evaporation from the soil. It is well to note that it is possible to store significant quantities of water in depth if infiltration can be concentrated in localized spots. Water sinks into the soil only to the depth that can be wetted to **field capacity** (the amount of water that can be held against gravity). If it is uniformly distributed over the surface, much of it is likely to be lost to direct evaporation or transpiration by shallow-rooted vegetation. However, if it can be concentrated around planted trees, deeper columns of water can be stored in such a way that trees may still have some available even when the soil close by has been depleted. The same general effect can be achieved where snow is caused to accumulate in drifts. The main point is that trees can be started in semiarid conditions if water can be concentrated in the planting spots.

Care of Planting Stock in the Field

If it becomes necessary to store planting stock after it is received from the nursery, it is better to store it in the original bundles in cool places, as previously described, than to store it in the field. As an emergency measure, it can also be heeled in on the planting site and taken out as needed. **Heeling-in** consists of storing planting stock in a trench dug in moist soil with a smooth, vertical or slanting side as deep as the roots of the trees are long. The plants are put in this trench in a relatively thin layer and covered with soil up to the root collars. Protection from sun and wind is essential. Ordinarily a day's supply of planting stock, if kept shaded and moist while on the planting site, will be damaged less in the original bundles than if heeled in.

The trees can be carried in pails, baskets, bags, or trays by the planting crews. It is essential that the roots of the planting stock remain moist until planted. Moss, a puddle of wet mud, or other material for keeping the roots moist is often placed in the container.

Protection of New Plantations

Plantations are fully as subject to damage from biotic and atmospheric agencies as natural stands in the same stage of development. It is an open question whether they are more subject to injury. Losses in artificially regenerated stands represent greater wastage of direct investment than comparable damage in stands that have been reproduced naturally. The large investment in plantations requires a correspondingly heavy outlay for protection that should be regarded as part of the cost of artificial regeneration.

Wild animals are an important cause of damage to plantations. Although there are no certain measures of control, it is generally best to attack the problem by indirect means. Rodents tend to prefer dense cover and can sometimes be discouraged by eliminating low vegetation before planting (Barnes, 1973). Temporary reductions of the rodent population may be achieved by distributing poisons on the planting area or applying them directly to the stems of the trees. Progress in the development of effective repellants for use against larger animals has been slow, but some compounds now on the market are occasionally satisfactory. Another solution is to avoid planting in places or during years of high population of the most damaging animals. The use of fences against wild animals is common in Europe but usually far too costly under American conditions.

Damage by domestic grazing animals may be controlled by proper herding or fencing, depending on the custom of the locality. Sometimes livestock can be lured away from planted areas by putting out salt elsewhere.

One of the most important causes of plantation failure is the competition of woody vegetation. All too often plantations are established and then left to fend for themselves. Even those made in open areas can be overwhelmed by fast-growing vegetation. The costs of releasing plantations can be reduced by avoiding the temptation to plant up brushy areas or spots adjacent to clumps of existing trees. The underplanting of trees beneath brush or existing stands of trees involves a definite commitment to carry out subsequent release cuttings.

The holes that develop in new plantations from scattered losses of planted trees can sometimes be corrected by subsequent planting or **refilling** ("beating up" in British terminology). If this is not done soon (and sometimes even if it is) the trees of the second planting get overgrown by the initial survivors. Since most losses occur during the first year, the ideal time for refilling comes a year after the first planting. Even then it is well if what is planted has some enduring advantage in its height or will grow faster than the original species.

Refilling is most effective in replacing large patches of dead trees, provided that the causes of the initial failure are identified and either corrected or evaded. If only scattered trees have survived, it is often best to eliminate them and start a whole new plantation. In the common situation in which losses are well scattered, refilling usually proves to be a costly waste of effort.

Density of Plantations

One of the advantages of planting is that it gives close control of the initial density and spatial arrangement of the new stand (Sjolte-Jorgensen, 1967). There is a tendency to think primarily of how many trees will be needed to achieve early occupancy of the growing space. However, it is really better to envision the kinds of trees wanted in the stand late in the planned rotation and work backwards from this goal. The question of whether or how the stand may be thinned is a crucial consideration in planning the initial arrangement (Lundgren, 1981).

If the trees are planted close together, their crowns and roots will soon close and full occupancy of the site will be achieved early. The high stand density will induce small branches, slow diameter growth, a low degree of stem taper, and rapid upward retreat of the bases of the live crowns. The trees may seem to be taller than those of less crowded stands but this is ordinarily an optical illusion, at least as far as the dominant and codominant trees are concerned. The height growth of the leading trees of a stand is not affected by stand density except that it can be reduced if stands are extremely crowded or open.

Total production wood is maximized at the highest stand density that does not cause diminution of height growth of the leading trees. If the stand density is any lower there will be some loss of potential total production during the delay in achievement of full occupancy. Although the rate of production is generally the same after attainment of full occupancy, there is no way of recovering any production loss resulting from such delay. However, there would be purpose in very dense planting to maximize total production only where wood was precious and the labor of gathering it so low that saplings could be harvested profitably.

If timber production is an objective, the purpose is to optimize the *yield* of economically utilizable wood. Since this always involves some specification of a minimum diameter of tree or piece and since tree value increases rapidly with diameter, total production is deliberately sacrificed to enhance diameter growth. The greater the target diameter, the lower the initial density. An owner seeking only sawlogs from trees more than 25 cm D.B.H. would plant at wider spacing than one who could also use pulpwood from trees as small as 10 cm. The ideal number to plant is precisely that which can be grown to the smallest size that is profitably utilizable. However, unless small trees can be utilized, it is commonly necessary to plant more trees than this in order to restrict branch size and prevent the trees from being too tapering. Sometimes it is even desirable not only to plant these extra "trainer" trees but also to kill some of them in later precommercial thinnings.

The question of whether thinnings will be done has bearing on the initial density. If thinnings cannot be made, it is best not to have the stands close any earlier than is essential for the development of acceptable stem quality. Conversely, if early thinnings are profitable it can be desirable to plant more trees just to enhance the yield from thinning.

The initial density of stands can also be varied with differences in site quality. In general, the better the site the greater is the number of trees that can grow at an acceptable rate. The crucial variable ordinarily seems to be the amount of foliage that can be supported by the soil moisture supply. Those parts of alluvial flood plains that have continuously moist but well aerated soils are good examples of the sites where trees can be close together but still grow well. At the other extreme are dry sites where it may be most logical to plant trees so far apart that their crowns never close even though the roots presumably do. In such cases, artificial pruning is sometimes substituted for the natural pruning associated with closed canopies.

Planting density should also vary with species but so many kinds of factors are involved that it is unsafe to generalize about them. Theoretically stands of shade-tolerant species can be denser than those of intolerants because they can carry more foliage. However, considerations such as product objectives, crown forms, or cost of planting stock control the choices. If there is small genetic variability in height growth, as is the case with red pine, there is risk that all of the trees will grow slowly and uniformly in diameter if there are too many of them. For such species one should plant fewer trees than with one in which there was enough variability to cause early differentiation into crown and diameter classes.

Spatial Arrangement of Plantations

Consideration should be given not only to the number of trees to plant but also to their geometric arrangement in relation to the future management and development of the stand. The arrangement that normally comes first to mind is the very common square spacing. The chief advantage of this is that it facilitates control of the work of hand-planting crews. It is the only arrangement compatible with driving such things as grass-mowing machines in straight lines both between and across the rows. Such treatments may be needed in plantations of Christmas trees.

Square spacing does provide for early development of uniformly closed stands. However, two other patterns provide even more uniform closure while others offer alternative advantages coupled with adequate development of uniformity. The ideal arrangement is the **equilateral** spacing (Fig. 4-14), in which the tree crowns fit together in the horizontal dimension like hexagons and stand at the corners of equilateral triangles. As the crowns expand, competition is joined uniformly around them and not unevenly as with square spacing. This kind of arrangement requires costly measurements and is seldom used. The **staggered** arrangement is virtually the same and much easier to establish. This involves planting the trees in parallel rows with each tree opposite the middle of the gaps between trees in adjacent rows.

The important thing to note about the hexagons and uniformity of crown coverage is that they are much more crucial considerations in the later stages of stand development than at the beginning. When the stand is planted the important final-crop trees are mixed with and generally indistinguishable from the more numerous ones that will be thinned out as the stand develops. Any total production lost because some spots are tardily closed over by trees is likely to be in the form of small-diameter wood of debatable economic utility. Only as the stand gets older does it become essential to secure uniform coverage and to strive for the ideal equilateral spacing in thinning. The numbers of trees for various spacings and kinds of arrangement are shown in Table 11-1.

There is almost always some early mortality so it is necessary to consider how this might affect the number planted. If it can be anticipated that the losses will be randomly and uniformly distributed, it is then logical to plant enough extra trees to equal the expectable losses. However, if, as is usually true, the losses tend to occur in patches, it is best not to deviate from the optimum spacing. If one planted more tree to allow for patchy losses the remaining parts of the stand would be too dense and the vacant patches would still be empty. In such cases, it is better to consider refilling the gaps by supplementary planting.

Table 11-1 Numbers of trees per unit of area for different approximately equivalent spacings

Spacing		Number of trees	
Feet	Meters	Per acre	Per hectare
2 × 2	0.6 × 0.6	10,890	27,778
3 × 3	0.9 × 0.9	4,840	12,346
4 × 4	1.2 × 1.2	2,722	6,944
5 × 5	0.5 × 1.5	1,742	4,444
5 × 10	1.5 × 3.0	871	2,222
6 × 6	1.8 × 1.8	1,210	3,086
6.6 × 6.6	2.0 × 2.0	1,000	2,500
7 × 8	2.1 × 2.4	778	1,984
7 × 10	2.1 × 3.0	622	1,587
8 × 8	2.4 × 2.4	681	1,736
8 × 10	2.4 × 3.0	544	1,389
9 × 9	2.7 × 2.7	538	1,372
9 × 10	2.7 × 3.0	484	1,235
10 × 10	3.0 × 3.0	437	1,111

In English and S.I. Systems, with arithmetic perfectly consistent only within each system.

Regularity of spacing should not be pursued to the exclusion of other considerations. Deviations from the spacing pattern are fully justified if patches of poor soil or competing vegetation can be avoided or if trees can be placed on the shady sides of obstacles like stumps or logs. The effect of natural obstacles and mortality among the planted trees always reduces the initial stocking below the figure theoretically indicated by a given interval of spacing and should be taken into account in planning operations.

Uneven spacing of trees does cause their crowns to become asymmetrical. This can occasionally be detrimental. Any associated asymmetry in the stem appears to be most pronounced in the region of the basal butt swell, while the upper part of the stem remains more nearly circular in cross section.

If trees are planted in furrows such as those created by most planting machines it is cheaper to have them close together within the rows but with the rows far apart. This reduces the number of times the machine needs to be turned around as well as the power used to plant a given number of trees. Furthermore, if the rows are far enough apart, it may be much easier to enter the stand with logging machinery at the time of the first thinning. In determining the number of trees required for this kind of row planting, one can think of it as a **rectangular** spacing even though the placement within the rows is not likely to produce a perfectly rectangular arrangement.

When trees are planted in rows it may be desirable to leave wide spaces between some rows to provide avenues for later thinning operations or similar activity.

One of the advantages of close initial spacing is that of inducing better stem form in trees destined for the final crop. However, if there are only going to be 200 of these per hectare it seems rather wasteful to plant trees 1 m apart (10,000 per hectare) to secure such training effects. If there were 200 dense clusters per hectare, centering about 7 m apart, this would seem sufficient. When such arrangements are created, trees are planted at much wider spacing between the clusters than within them with the plan that they will grow well in diameter but be removed in the thinnings. The clusters may have to be precommercially thinned when the training effects have been secured. This kind of **cluster** arrangement has the drawback of complexity. When it is used the clusters are often planted first and the more widely spaced trees afterwards.

Mixed Plantations

Although most plantations are pure "monocultures," there is no fundamental reason why they cannot be mixtures of species. It is more complicated to create and manage mixed plantations but the benefits can outweigh the complexities.

As is set forth in more detail in Chapter 17, the management of intimate mixtures of species usually depends on recognizing that no two species grow in height at the same rate. The best and simplest way to deal with this disparity is to take advantage of it by mixing quick-growing intolerant species with tolerant ones that start off more slowly. The fast growers form an upper stratum and, if they are sufficiently far from each other, grow as rapidly in diameter as if they had been heavily thinned. The slower starting ones stay in the lower stratum; there they fill the remaining growing space and usually retain the capacity to grow rapidly once the overstory species have ceased rapid growth and have been removed. Ordinarily the overstory species should be distinctly less numerous than the understory one.

There are parts of Europe where it is customary to develop mixed plantations in which the different species are kept in the same canopy stratum. This requires careful attention to removing the trees that race ahead and encouraging those that would lag behind in nature. It can also be done by planting different species in different years.

Another way of creating mixed plantations is by planting small, pure clumps so that interactions between species are reduced in extent. When different species are planted in alternate rows, they usually interact as they would in even more intimate mixtures. Paradoxically the most successful mixtures are generally those in which the species are of very different ecological status. The more nearly they are the same, the more likely it is that the one with a slight advantage in height growth will suppress the other. Some of the least compatible mixtures are of species from the same genus.

The Role of Planting

The establishment of new forests, or the regeneration of old, by planting is one of the most costly steps in silviculture. These costs, however, vary widely depending on factors such as cost of labor, capital, nurseries, planting stock, equipment, site preparation, and treatment after planting. The benefits can be very high and there are plenty of cases in which the ratio of benefit to cost is higher with planting than any other technique of stand establishment. The least costly kind of planting is that done where labor costs are low and some devastating event such as fire or agricultural use has made good, level soil free of competitive vegetation. At the other extreme are cases requiring control of aggressively competitive vegetation and hand planting of containerized stock with costly labor.

The financial case for planting can seem very dismal when the investment, necessarily made at the very beginning of a rotation, is carried at compound interest to the end. Society, foresters, and landowners exhibit some ambivalence in dealing with this matter. Sometimes the planting of trees seems so soul-satisfying that society evades the problem with various direct or indirect subsidies.

If regeneration after timber harvest is required by law or ownership policy, the cost is often counted simply as a cost of the harvest rather than as an investment in the future. Sometimes this is a case of evading the issue. It is, however, also common that analysis of the situation has shown that a combination of expeditious logging with seemingly expensive planting is the least costly way of harvesting the old stand *and* starting a new one.

Sometimes, as with reforestation where forests have been eliminated or in afforestation of sites never forested, the only choice is whether to undertake the effort. Direct seeding is an option but often not feasible. Planting can similarly be virtually the only way of starting stands of newly introduced exotics or plants produced in genetic improvement programs. This latter reason is, in fact, one of the most common reasons for using planting rather than natural regeneration to replace good stands.

Distinction should be drawn between planting for timber production and that for the forestation of devastated areas to prevent erosion and similar sources of injury to adjacent lands and waters. Where timber is the goal, priority is given to those areas where planting is most rewarding. Reforesting vacant areas commands priority over replacement of degraded stands and that, in turn, usually gains more than planting to regenerate good stands that already have desirable species. If the objective is protection of soil and water, the first areas treated should be those that are the source of greatest damage. Planting for wildlife or aesthetic improvement, just as that for timber, is done first where the prospective benefits are greatest.

Part of the appeal of planting springs from a popular belief that trees will not grow unless they are planted. Difficulties can arise when foresters also become addicted to this idea but are unable to get owners to provide sufficient funds to plant all areas in need of regeneration. Under such circumstances it is an abdication of professional responsibility to give up in frustration and fail to explore other means of regeneration.

Planting is the most common means of growing Christmas trees and trees that are grown for wood, fruit, or other purposes, in combination with grass or agricultural crops.

Most planting, especially in the temperate zone, is done with conifers. This is partly because coniferous wood is usually in high demand and short supply. It is also because conifers are often easier to plant successfully than angiosperms, especially where there is grass competition. Sometimes the best way to establish hardwoods on old grassy fields is to grow a rotation of planted conifers first and then depend on adjacent natural seed sources of hardwoods to provide advance growth for the second rotation. However, except for problems with grass, there are no basic natural barriers against the planting of angiosperms. In the warmer parts of the world such important species as the eucalypts (Jacobs and Métro, 1981), teak, mahoganies, fast-growing pioneers such as *Gmelina arboraea*, and various multipurpose leguminous trees are commonly planted (Evans, 1982; Nwoboshi, 1982).

BIBLIOGRAPHY

Abbott, H. G., and S. D. Fitch. 1977. Forest nursery practice in the United States. *J. For.* **75**:141-145.

Abrahamson, L. P., and D. H. Bickelhaupt (Eds.) *North American Forest Tree Nursery Soils Workshop Proceedings.* State Univ. of N. Y., Coll. of Env. Sci. & For., Syracuse. 333 pp.

American Society of Agricultural Engineers. 1981. *Forest regeneration, proceedings.* Amer. Soc. Agr. Engineers, St. Joseph, Mo. 376 pp.

Armson, K. A., and V. Sadreika. 1979. *Forest tree nursery soil management.* Ont. Min. Nat. Resources, Toronto. 179 pp.

Barnes, V. G., Jr. 1973. Pocket gophers and reforestation in the Pacific Northwest: a problem analysis. *U.S. Dept. Int., Fish and Wildlife Serv., Spl. Sci. Rept.—Wildlife No.* 155. Denver. 18 pp.

Barnett, J. P., and J. M. McGilvary. 1981. Container planting systems for the South. *USFS Res. Paper* SO-167. 18 pp.

Benzie, J. W. 1977. Manager's handbook for red pine in the North Central States. *USFS Gen. Tech. Rept.* NC-33. 22 pp.

Bonga, J. M., and D. J. Durzan. 1982. *Tissue culture in forestry.* Nijhoff/Junk, Hingham, Mass. 420 pp.

Bonner, F. T. 1984. Glossary of seed germination terms for tree seed workers. *USFS Gen. Tech. Rept.* SO-49. 4 pp.

Cleary, B. D., R. D. Greaves, and R. K. Hermann. 1978. *Regenerating Oregon's forests.* Oreg. State Univ., Ext. Serv., Corvallis. 286 pp.

Duryea, M. L., and G. N. Brown. (Eds.) 1984. *Seedling physiology and reforestation success.* Nijhoff/Junk, Hingham, Mass. 322 pp.

Duryea, M. L., and T. D. Landis (Eds.) 1984. *Forestry nursery manual.* Nijhoff/Junk, Hingham, Mass. 368 pp.

Ebel, B. H., *et al.* 1975. Seed and cone insects of southern pines. *USFS Gen. Tech. Rept.* SE-8. 41 pp.

Evans, J. 1982. *Plantation forestry in the tropics.* Clarendon, Oxford. 472 pp.

Faulkner, R. (Ed.) 1975. Seed orchards. *U.K. For. Comn. Bul.* 54. 149 pp.

Goor, A. Y., and C. W. Barney. 1968. *Forest tree planting in arid zones.* Ronald, New York. 504 pp.

Grey, G. W., and F. J. Deneke. 1978. *Urban forestry.* Wiley, New York. 279 pp.

Guldin, R. W., and J. P. Barnett (Eds.) 1982. Proceedings, Southern Containerized Forest Tree Seedling Conference. *USFS Gen. Tech. Rept.* SO-37. 156 pp.

Harley, J. L., and S. E. Smith. 1983. *Mycorrhizal symbiosis.* Academic, Orlando, Fla. 496 pp.

Heidman, L. J. 1976. Frost heaving of tree seedlings: a review of causes and possible control. *USFS Gen. Tech. Rept.* RM-21. 10 pp.

Heidman, L. J., F. R. Larson, and W. J. Rietveld. 1977. Evaluation of ponderosa pine reforestation techniques in central Arizona. *USFS Res. Paper* RM-190. 10 pp.

International Poplar Commission. 1979. Poplars and willows in wood production and land use. *FAO Forestry Series* 10. 328 pp.

Jacobs, M. R., and A. Métro. 1981. Eucalypts for planting. *FAO Forestry Series* 11. 677 pp.

Jenkinson, J. L. 1980. Improving plantation establishment by optimizing growth capacity and planting time of western yellow pines. *USFS Res. Paper* PSW-154. 22 pp.

Kozlowski, T. T. (Ed.) 1972. *Seed biology.* 3 vols. Academic, New York. 1274 pp.

Kramer, P. J., and T. T. Kozlowski. 1979. *Physiology of woody plants.* Academic, New York. 811 pp.

Laurie, M. V. 1974. Tree planting practices in African savannas. *FAO Forestry Development Paper* 19. 185 pp.

Lundgren, A. L. 1981. The effect of initial number of trees per acre and thinning densities on timber yields from red pine plantations in the Lake States. *USFS Res. Paper* NC-193. 25 pp.

Marx, D. H., *et al.* 1984. Commercial vegetative inoculum of *Pisolithus tinctorius* and inoculation techniques for development of ectomycorrhizae on bare-root tree seedlings. *For. Sci. Mongr.* 25. 101 pp.

McKnight, J. S. 1970. Planting cottonwood cuttings for timber production in the South. *USFS Res. Paper* SO-60. 17 pp.

Mroz, G. D., and J. F. Berner (Eds.) 1982. *Artificial regeneration of conifers in the Upper Lakes Region, proceedings.* Mich. Tech. Univ., Houghton. 453 pp.

North Central Forest Experiment Station. 1982. Black walnut for the future. *USFS Gen. Tech. Rept.* NC-74. 151 pp.

Nwoboshi, L. C. 1982. *Tropical silviculture, principles and techniques.* Ibadan Univ. Press, Nigeria. 333 pp.

Peterson, G. W., and R. S. Smith, Jr. 1975. Forest nursery diseases in the United States. *USDA, Agr. Hbk.* 470. 125 pp.

Pritchett, W. L. 1979. *Properties and management of forest soils.* Wiley, New York. 500 pp.

Rudolf, P. O. 1950. Forest plantations in the Lake States. *USDA Tech. Bull.* 1010. 171 pp.

Scarratt, J. B., G. Glerum, and C. A. Plexman (Eds.) 1981. Canadian Containerized Tree Seedling Symposium. *Can. For. Service COJFRC Symp. Proc.* O-P-10. 460 pp.

Sjolte-Jorgensen, J. 1967. Influence of spacing on coniferous plantations. *Intl. Rev. For. Res.* **2**:43–94.

Tinus, R. W., W. I. Stein, and W. E. Balmer. 1974. North American Containerized Forest Tree Seedling Symposium. *Great Plains Agr. Council Publ.* 68. Denver. 468 pp.

Tinus, R. W., and S. E. McDonald. 1979. How to grow seedlings in containers in greenhouses. *USFS Gen. Tech. Rept.* RM-60. 256 pp.

U. S. Forest Service. 1974. Seeds of woody plants in the United States. *USDA, Agr. Hbk.* 450. 883 pp.

Van Eerden, E., and J. M. Kinghorn (Eds.) 1978. Proceedings, Root Form of Planted Trees Symposium. *Brit. Col. Min. For. and Can. For. Service, Joint Rept.* 8. 357 pp.

Wahlenberg, W. G. (Ed.) 1965. A guide to loblolly and slash pine plantation management in southeastern USA. *Ga. For. Res. Counc. Rept.* 14. 360 pp.

Wakeley, P. C. 1954. Planting the southern pines. *USDA, Agr. Monog.* 18. 233 pp.

Williams, R. D., and S. H. Hanks. 1976. Hardwood nurseryman's guide. *USDA, Agr. Hbk.* 473. 78 pp.

Williston, H. L., and W. E. Balmer (Eds.) 1979. Proceedings: symposium for the management of pine of the Interior South. *USFS Tech. Publ.* SA-TP-2. 221 pp.

PART THREE

Silvicultural Systems

C H A P T E R   1 2

Development of
Silvicultural Systems and
Methods of Reproduction

A **reproduction method** is a procedure by which a stand is established or renewed; the process is accomplished during the regeneration period by artificial or natural reproduction. The various methods consist of the removal of the old stand, the establishment of a new one, and any supplementary treatments of vegetation, slash, or soil that are applied to create and maintain conditions favorable to the start and early growth of reproduction. Any procedure, intentional or otherwise, that leads to the development of a new stand of trees is identifiable as a method of reproduction.

The term **silvicultural system** is more comprehensive and designates a planned program of silvicultural treatment during the whole life of a stand; it includes not only the reproduction cuttings but also any tending operations or intermediate cuttings. The reproduction methods employed have such a decisive influence on the form and treatment of the stand that the name of the method is commonly applied to the silvicultural system; the shelterwood system, for example, leads to reproduction by means of the shelterwood method of cutting.

A refined and intensive silvicultural system consists of a number of steps conducted in logical sequence. In an application of the shelterwood system there might, for example, be some early cleanings in established reproduction, followed by pruning and a sequence of free, crown, and low thinnings, all leading to the series of partial cuttings that are designed to establish natural reproduction under the old stand and are characteristic of the shelterwood method. A less intensive application of the same system might involve nothing more than final harvest cutting in two stages.

In the consideration of silvicultural systems and reproduction methods, heavy and perhaps undue emphasis is placed on the effect of spatial patterns of cutting on the establishment of regeneration. As was indicated in earlier chapters, this effect is important but not necessarily decisive in regeneration. The greatest significance of the pattern of cutting lies in its influence on a whole complex of biological, physical, and economic considerations, including logging problems, administration, manipulating of growing stock, and protection of the stand. The development of a silvicultural system is controlled fully as much by these considerations as by the vital objective of establishing new crops.

Classification of Reproduction Methods

Many different methods of reproduction have been developed, but they can all be reduced to a few broad categories. The details of applying the same reproduction method vary widely as they are altered for each species, region, and objective of management. Some variants are simpler than others, but no single, detailed procedure is the standard for each general method. As with many classifications, there is a wide range of variation in each category and some borderline cases may not clearly fit any pigeonhole.

The following list of six general reproduction methods has been simplified and is the classification that has been most widely accepted in North America. Each method might be further subdivided, and each will be more fully defined and discussed in subsequent chapters.

High-forest Methods—producing stands originating from seed.
  Even-aged stands:
    Clearcutting method—removal of the entire stand in one cutting with reproduction obtained artificially or by natural seeding from adjacent stands or from trees cut in the clearing operation (Chapter 13).
    Seed-tree method—removal of the old stand in one cutting, except for a small number of seed trees left singly or in small groups (Chapter 14).
    Shelterwood method—removal of the old stand in a series of cuttings, which extend over a relatively short portion of the rotation, by means of which the establishment of essentially even-aged reproduction under the partial shelter of seed trees is encouraged (Chapter 14).
  Uneven-aged stands:
    Selection method—removal of the mature timber, usually the oldest or largest trees, either as single scattered individuals or in small groups at relatively short intervals, repeated indefinitely, by means of which the continuous establishment of reproduction is encouraged and an uneven-aged stand is maintained (Chapter 15).

Coppice-Forest Methods—producing stands originating primarily from vegetative regeneration (Chapter 16).
    Coppice method—any type of cutting in which dependence is placed mainly on vegetative reproduction.
    Coppice-with-standards method—production of coppice and high for-

est on the same area with the trees of seedling origin being carried through much longer rotations than those of vegetative origin.

The foregoing classification is simplified mainly because almost any equally inclusive alternative would be intricate and complicated. The more detailed classifications that exist do not differ in basic principle from that just presented. Some of the most sophisticated ones are in the German literature (Spurr, 1956; Dengler, 1972). Troup (1952) described and classified silvicultural systems as applied in many parts of the world. Silviculture is a flexible art and scores of different methods and systems of treatment are continually devised to fit different conditions. The use of a simple classification of regeneration methods definitely does not mean that there are only six such routines; actually there are hundreds.

The names of silvicultural systems and reproduction methods are merely devices of communication for expediting the description of programs of treatment. Especially when the classification is simple, each term really does not tell much about the procedure and only starts the descriptive process. Each term covers many variants.

No one method or system has any single ideal routine of application akin to the precise "school figures" of figure skating. There is, for example, no one ideal mode of shelterwood system compared with which all others are necessarily inept, imperfect, or "less intensive" imitations. If they have these attributes it is not because of departure from some preordained ideal treatment program. None of these methods is a schedule or routine that needs only to be copied to produce success. Furthermore, it cannot be presumed that any of the listed methods can be safely applied to any kind of forest just because it is a "recognized method."

What really has happened, or should, is that treatment procedures are first devised to fit the circumstances and the naming of them is done afterwards. In other words, the terminology describes treatment, but does not dictate it.

The Basis of Distinction between Reproduction Methods

Classification of reproduction methods is usually based on (1) mode of origin of regeneration as well as on arrangement of cutting areas in (2) time and (3) space. The first distinction is between high-forest methods involving regeneration from seed and the coppice-forest methods, which rely mostly on vegetative regeneration from sprouts, root-suckers, or layered branches.

Coppice-forest methods are sometimes called "low-forest" methods because they were once thought of as being limited to growing short trees on short rotations for fuelwood. Since tall, old trees of many species, including oaks and redwoods, can be grown from sprouts, the old term is rather archaic. High- and coppice-forest methods are actually differentiated on the basis of the *predominant* sources of regeneration. The coppice-forest methods may include some regeneration from seed. Regeneration by planting goes with high-forest methods as does that from the release of advance growth and the resprouting of small advance growth.

As far as arrangement of cuttings in time is concerned, the most important distinction is between (1) methods in which reproduction cuttings extend throughout the rotation leading to the creation of uneven-aged stands and (2) methods for the maintenance of even-aged stands in which regeneration cuttings are concentrated at the end of each rotation. The methods for even-aged stands, however, also differ according to the number of cuttings that are required to replace one stand with another.

The complexity of many classifications results from recognition of the almost infinite variations that can be created in the horizontal, geometric pattern of cutting areas. Each general method can be applied so that openings and uncut timber are left either in uniform distribution or in concentrated strips, groups, wedges, and the like. In the general classification used here, simplicity has been achieved largely by restricting the amount of attention given to modifications involving differences in spatial arrangement.

The ultimate distinction between reproduction methods is the form of forest produced. The size, shape, and position of the areas cut over, as well as the proportion of the timber removed, determine the arrangement of age classes, which is in turn the chief factor governing forest form. Each reproduction method, when systematically applied, produces a characteristic stand structure.

Certain methods result, for example, in the maintenance of even-aged stands whereas others produce stands with varying numbers and proportions of identifiable age classes. If regeneration is established or released by cutting at infrequent intervals, the stand will have as many age classes as there are cuttings during the rotation.

Where the regeneration period is reduced to one year, as it may be in the coppice method or where a clearcut area is planted, the stand is absolutely even-aged. Most even-aged stands, however, contain trees of more than one age because the reproduction period can rarely be restricted to a single year. A stand may be considered even-aged for purposes of management if the difference in age between the oldest and youngest trees does not exceed 20 percent of the length of the rotation. The age of trees that grew from small "advance reproduction" developing under larger trees is generally best dated from time of release, although age at breast height sometimes gives a convenient approximation. Small trees of some species can respond rapidly to release after decades of suppression and the difficulty of ascribing an age to them does not diminish their utility. Their age is best reckoned from the time of release rather than that of germination.

Even-aged stands or aggregations of trees usually have a canopy top at a uniform height, especially if there is only one species. In mixed, even-aged stands some scattered trees of species with rapid or long-sustained height growth may project above the main canopy height. In uneven-aged stands the height of the main canopy top is usually variable, that is, low for the young components and tall for the older ones.

There are, however, special cases in which truly uneven-aged stands can be flat on top. One is that of very old stands in which all the age classes have attained culmination of height growth. Wind effects on exposed sites may produce such an effect by stunting height growth. In mixed stands, differences between species in the regimes of height growth could conceivably cause the leading trees to be of the same height but different age, if there were no truly young age classes.

As previously stated, diameter is an astonishingly poor criterion of tree or stand age. Perhaps the most that can be said is that a pure stand with a nearly bell-shaped curve of diameter distribution is almost sure to be even-aged in terms of effective age; all other such distributions need diagnostic interpretation by criteria not limited to diameter.

If one can see for some distance horizontally beneath a closed tree canopy, it may be suspected (but not proven) that the stand is even-aged. If such a view is blocked by occasional patches of seedlings or saplings that have blue sky above them and are surrounded by taller trees, the stand may well be uneven-aged. There is no substitute for knowing how old the trees really are and even the definition of "age" sometimes requires some combination of interpretation and arbitrary decision. It helps to bear in mind that most trees start, after regenerative disturbances, in even-aged aggregations that can be of very small area but are nonetheless even-aged throughout their lives.

The method of reproduction being employed in a given stand may not be evident to the casual observer. The identification and definition of a method of cutting depend fully as much on the results actually obtained, the intent of the treatment, and the nature of subsequent operations as they do on the pattern according to which the trees are removed. A reproduction cutting is a reproduction cutting only to the extent that it leads to the establishment of a new crop. If something intended as a thinning results in the establishment of vigorous reproduction it is best regarded as having been a shelterwood cutting. However, if the residual stand is then allowed to suppress the reproduction to the point of elimination, the initial cutting was indeed a thinning. A partial cutting aimed at starting the development of an uneven-aged stand is selection cutting only if subsequent operations are sufficiently consistent with the first to result in the ultimate creation of such a stand.

Identification of the silvicultural system is also complicated because it may be changed during the course of a rotation or because different parts of the same stand are reproduced by different methods. If the reproduction method originally contemplated is one that involves a sequence of partial cuttings, there is often latitude for changing to methods that are found to be more satisfactory during the sequence. Variation in procedure within a stand is most likely to be dictated by differences of site, accessibility, or the condition of the stand itself. Such changes of method in time and within stands are likely to be common when a forest is, like most American forests, being brought under management for the first time. In fact it is by means of such changes that the procedures are perfected and the pattern of stands molded into arrangements that are rational from the standpoint of site variations and economic management.

A cutting does not have to be planned by a forester or involve any thought whatever about the future to be recognizable as a kind of intermediate or reproduction cutting. It matters not how indiscriminately the trees are removed or how haphazardly the new growth develops. Of course, it is not necessarily easy to categorize cuttings that involve no future plans and the act of doing so may be purely an intellectual exercise; the point is that the use of standard cutting techniques as defined in silvicultural terminology is not merely a fancy refinement limited to intensive forestry.

One of the most important reasons why methods of reproduction and silvicultural systems are classified and recognized is that the act of doing so indirectly forces planning for the future care, development, and replacement of stands. The deliberate decision to employ a particular method of reproduction is in itself an act of planning without which the management of stands easily becomes a kind of rudderless drifting, governed more by current market demands than by any intentions about the future.

FORMULATING SILVICULTURAL SYSTEMS

A good silvicultural system is a long-term program of treatment designed to fit a specific set of circumstances. It is not likely to be something that has already been invented and can simply be selected ready-made from classifications or schematic descriptions of silvicultural systems. This book, for example and in spite of certain superficial resemblances, is definitely not a catalog from which such choices can be made or a cookbook for their application. It may help in constructing them.

A silvicultural system also evolves over time as circumstances change and knowledge of them improves. A classic history of the management of red pine in Minnesota, by Eyre and Zehngraff (1948), exemplifies this evolutionary process. It describes the emergence of a refined application of the shelterwood system from initial misadventures with the seed-tree method.

Formulation of a silvicultural system should start with analysis of the natural and socioeconomic factors of the situation. A solution is then devised to go as far as possible in capitalizing on the opportunities and conquering the difficulties found to exist.

When the important act of inventing the solution has proceeded far enough the less important step of attaching a name to it can be taken. The standard terminology should be used to the extent of its limited capacity for providing information in terms meaningful to all foresters and then supplemented with the additional, detailed information that is usually necessary.

Sloppy use of the terminology can corrupt it to the point where the words mean little. "Selective cutting" is a potentially useful term that slovenly usage has almost destroyed as a technical term. It seldom helps to try to redefine existing technical or common terms.

"Clearcutting" is really a word with many meanings as well as an ugly connotation; as a term of technical silviculture, it is an unhappy attempt at redefinition of a logging term and might better be replaced. Some years ago the term "unit area control" was applied to a silvicultural system that had been carefully devised for some mixed conifer forests of the Sierra Nevada of California (Hallin, 1959; Davis, 1959). It was mainly a variant of the group-selection system and perhaps might better have been named the "Dunning group-selection system" after its chief originator. The term "unit area control" probably caused more confusion than clarification but it did cause foresters to analyze what they were doing.

If silvicultural systems are not chosen ready-made from a manual, it is logical to examine the various considerations that enter into their construction and evolutionary development. In the first place, a rational silvicultural system for a particular stand should fit logically into the overall management plan for the forest of which the stand is a part. Second, it should represent the best possible amalgam of attempts to satisfy all of the following major objectives, each of which will be discussed in this and subsequent chapters. These basic objectives are as follow:

1. Harmony with goals and characteristics of ownership
2. Provision for regeneration
3. Efficient use of growing space and site productivity
4. Control of damaging agencies
5. Provision for sustained yield
6. Optimum use of capital and growing stock
7. Concentration and efficient arrangement of operations

These objectives are fully as likely to conflict as to harmonize with one another. It is for this reason that the procedures followed in applying systems with the same name vary widely depending on the relative importance attached to different objectives. Some of the contradictions are partially resolved in the development of management plans for the whole forest. The purpose of this chapter is to consider the ways in which silvicultural systems are affected by these conflicting objectives.

The situation can be clarified by categorizing the various objectives of forestry and becoming aware that they represent forces pulling the forester in different directions. Analysis will show that single-minded concentration on any one of these objectives can ultimately lead to ridiculous results.

The best solution obviously lies in finding the most appropriate blend of partial fulfillments of all significant objectives.

Harmony with Goals and Characteristics of Ownership

Choice among all the alternatives of silvicultural treatment is greatly simplified by clarification of the objectives of ownership. This logical first step automatically eliminates many of the possible alternatives. It also forces recognition of the fact that it would not necessarily be in the interest of two different owners to manage the same stand in the same way. On this continent laws seldom prescribe silvicultural practice in detail, so silviculture on different ownerships would be the same only to the extent that natural and economic conditions were similar. Except to the extent that law requires it, there is no justification for a forester to embark in arrogant wisdom on any "standard" procedure for the growing of a particular kind of stand regardless of whether the technique fits the owner's purposes.

It is commonly necessary to help owners select their own objectives before thinking much about formulating a silvicultural program. However, choices of objectives considered by owners must be limited to those that are reasonably attainable.

The objectives of ownership clearly dictate the relative amounts of attention paid to management for timber, wildlife (Schemnitz, 1980; Bailey, 1984; Shaw, 1985), forage (Paulsen, 1975), water (Hewlett, 1982), recreation, scenery (Bacon and Twombly, 1980), or other benefits that forests may provide. On land managed for multiple use (Myers, 1974) the silviculture logically differs from that on similar land that might be owned by a lumber company for growing saw timber; this would in turn differ from a paper company's silviculture for pulpwood production. If an owner is most interested in wildlife or in preserving some old-growth timber purely for aesthetic purposes, the forester should modify the silviculture accordingly. In fact, the forester's occupational bias in favor of efficient timber production can be more of a liability than an asset.

Analysis of the objectives of ownership will normally define the kind of vegetation to be maintained, the kind of trees that are to be grown, and the amount of time, money, and care that can be devoted to the process. The intensity of practice is determined by the amount of money that the owner is willing and able to place in long-term silvicultural investments as well as by the interest return required on such investments. If long-term investments cannot be made, the silvicultural treatment must be limited to those things that can be done in the process of harvesting merchantable timber. Where the future of the enterprise seems limited to the life expectancy of the owner, attention may well be restricted to securing maximum benefits during, the prospective lifetime of the owner; manipulating the existing growing stock would probably take precedence over securing regeneration.

Provision for Regeneration

Continuity of any forestry enterprise, including the maintenance of stands with museum-piece trees thousands of years old, ultimately and absolutely depends on replacing old trees with new. The objective of regenerating stands is seldom achieved without temporary, but resolute, sacrifice of some other objectives. It is necessary to reduce the competition from the old vegetation enough to provide sufficient growing space for the new. This often requires cutting some trees before they are mature or reserving others beyond the time of maturity as sources of seed or shelter for the new crop.

The efficiency of harvesting operations is almost always reduced by measures needed to ensure regeneration. Trees or groups of trees that might otherwise be cut may have to be left to provide seed or shelter; special measures are also necessary to protect such trees. Even with clearcutting and planting it may be necessary to harvest some trees at a cost in order to vacate growing space for regeneration.

The period during which reproduction is established can also be one in which full occupancy of the site is lost and the stand exposed to sources of injury that could be avoided if the stand did not have to be reproduced. Loss of full occupancy of the growing space can result from delays in the establishment of natural reproduction. Even if a stand is replanted the day after a clearcuttting some years elapse before the new trees expand enough to reoccupy the site fully. The environmental conditions necessary for reproduction may call forth a whole suite of damaging insects and other injurious agencies not found in older forests. Among other things, the risk and hazard of fire are usually greatest during the reproduction period.

Even if advance reproduction of desirable species appears without effort, it must often be released in a manner and at a time not ideal from other standpoints. The whole situation is a good argument for undertaking to reproduce stands no more often than necessary and for going about it with sufficiently single-minded purpose that it is accomplished in a short, well-defined period.

It is seldom a good idea to expect natural regeneration as the unearned or unplanned by-product of treatments done for other purposes at vaguely defined times during the rotation. The result of such casual procedure is usually either poor regeneration or costly and cumbersome efforts to release desirable trees from more shade-tolerant competitors.

It should be noted that there is no need to have seedlings present at all stages in the life of a stand. They are needed only during the period of establishment and may be no more than a symptom of wasted growing space at other times. It is only with uneven-aged management that one might want seedlings present at all times but even then it would be only in those parts of an uneven-aged stand that were actually being regenerated.

The various high-forest methods can be viewed as representing a complete gradational series of different degrees of exposure of the forest floor to solar radiation, precipitation, and air movement. They also differ with regard to the degree of root competition and supply of seed from trees left on the cutting area. The manner in which critical factors are altered by some methods of reproduction cutting are depicted in generalized fashion in Fig. 12-1, which is actually even less precise than the deliberately vague graduation of axes implies.

The clearcutting method, at one extreme, increases exposure of the site to the maximum, decreases root competition to the minimum, and eliminates any source of seed not already on the site or provided by adjacent stands. In the seed-tree method a supply of seed is produced on the area, although the remaining trees do not appreciably modify the environment. The outstanding feature of the shelterwood method, as the name implies, is the shelter provided by the residual stand, particularly against direct solar radiation. The vigor of the remaining trees, that is, the "shelterwood," may be improved sufficiently to increase the supply of seed to the maximum. Once reproduction has become established, the shelterwood is removed, leaving an essentially even-aged stand freed from competition with the residual stand; this essential characteristic of the shelterwood method is not indicated in Fig. 12-1.

The information presented in Fig. 12-1 does not fully depict all of the different effects on the regeneration environment created by various kinds

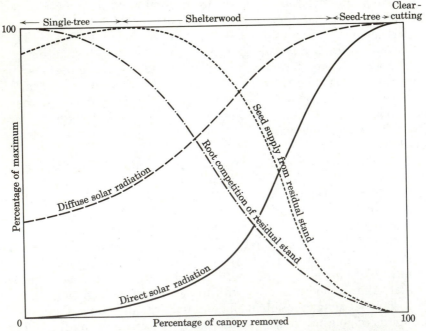

FIGURE 12-1 A simplified representation of the effects of the initial cuttings of various methods of reproduction, when applied uniformly over an area in a humid, temperate climate, on several factors usually critical in the establishment of natural reproduction.

of uneven-aged management under the selection system, the low forest systems, or treatments of complicated mixtures of species. Many of these will be considered in subsequent chapters. The various selection methods employed in uneven-aged systems are more nearly modes of forest administration than they are of control of regeneration environment. Although the so-called single-tree selection method, as indicated in Fig. 12-1, is the lightest of all kinds of regeneration cutting, it is entirely possible to provide the environment of a large clearcut area within the administrative framework of the group-selection method.

The relationships depicted in Fig. 12-1 are for forests in humid, mid-latitude climates. The curves for radiation would be different at high or low latitude and that for root competition would be higher on the graph if the water supply were more limiting.

Figure 12-1 shows some of the ecological effects of the choice of reproduction method but it should not be interpreted as meaning that this choice is necessarily a powerful determinant of the species composition of new stands. The choice of method is influential but not decisive. The microenvironmental factors controlling forest regeneration (Chapter 7) are affected by the condition and treatment of the surface soil and the lesser vegetation fully as much as they are by the patterns of removal of old trees. Furthermore, planted trees, vegetative sprouts, and preestablished advance growth do not respond to the environmental changes induced by cutting in the same ways as trees arising from new germination.

Efficient Use of Growing Space and Site Productivity

Forest vegetation usually seems to fill all of the available growing space rather quickly but not all of the plants that do this filling are equally efficient or desirable. The question of what plants are desirable depends on objectives of ownership; sometimes grass or shrubs may be fully as desirable as trees. Among tree species not all are equally productive or adaptable to a given site even within their natural ranges. A silvicultural system is, among other things, a program for the allocation of use of growing space throughout whole rotations.

One important goal, especially when forest lands are first being put under management, is to see that all available growing space is filled with useful plants. Areas not so occupied are an economic burden and may also invite invasion of undesirable competitive vegetation. It is better to have a unit of ground occupied by a poor but useful plant than by a useless or undesirable one. However, the objective of full occupancy should not be pursued so far that stands become overcrowded with desirable trees. What may initially appear to be understocking may correct itself as the trees increase in size; overstocking seldom corrects itself automatically. Damaging agencies may thin out overcrowded stands but only in erratic fashion; precommercial thinning is a better solution but it can be a costly one.

Overstocking can be just as inefficient as understocking. The best way to guard against either extreme is to develop and follow logical programs for regulation of stand density at all stages of development from regeneration to maturity. The regeneration techniques that produce the most seedlings are not always the best.

Even a stand that has no obvious gaps does not maintain full occupancy of the growing space or use it with equal efficiency throughout its life. The times when it does not are mostly likely to come when it is young and when it is old. As was described in Chapter 2, the biological productivity of a stand is highest at some period intermediate between youth and middle age. Optimum economic productivity is achieved somewhat later because of the effect of tree size on the utility of wood. However, in almost any terms, the annual productivity increases rapidly with age during youth and declines gradually with the onset of old age.

The decline in production in the later stages is caused mostly by various kinds of decreasing biological efficiency as well as the attrition of damaging agencies. Crown friction induced by increasing wind-sway reduces the amount of stand foliage even while the amount of meristematic tissue to be nourished increases. The margin between gross production and losses due to respiration and damage shrinks and a given age class of trees slowly loses its command of the growing space. The inefficiencies of old age are forestalled by determining logical lengths and simply replacing mature stands at the right time.

The inefficiencies of youth are more nearly inevitable and difficult to mitigate. A young stand does not arrive at the peak of biological efficiency until the root systems and live crowns of the trees have expanded both horizontally and vertically to occupy all the space they are capable of claiming. The presence and vigor of shrubs and other accessory vegetation can indicate the extent to which such stands fall short of full occupancy. The most serious loss of potential production occurs during any interval of time that elapses between the removal of the previous crop and the establishment of a new one.

The ideal way to reduce such loss of production and to retain some command of the growing space is to start the new stand under the old one as with shelterwood cutting. Such action is especially advantageous with species that grow slowly in youth. The next best way for most species, and the only way for some decidedly intolerant ones, is securing the promptest possible regeneration after removal of the old stand. With sprouting species this can be done by coppice cutting, a procedure in which the root system of the stand remains alive and in rather complete occupancy of the soil. True clearcutting and planting can also provide very prompt regeneration.

However, some loss of production is almost inevitable during the regeneration stage. Maddening delays in both artificial and natural regeneration are common enough to be arguments for not replacing stands any more often than is truly necessary.

Calculations of economic rotation lengths that are based on the assumption that regeneration is prompt and easy should be regarded with suspicion. One way of allowing for the effect of regeneration delays and production losses is to determine rotation length from calculations of present net worth for periods that extend over two or three rotations rather than just one. Such calculations can take into account the predicted effects not only of delays but also of the gain from telescoping rotations that results when new stands are started under old ones.

It may be noted that, at the time of regeneration, perfect efficiency would be achieved if the new stand arrived at full occupancy with just that number of trees that could ultimately be used profitably without any growing space having been wasted during the transition from the old stand to the new.

During the rotation increased efficiency in the use of growing space can be sought by application of intermediate cuttings, including thinning and techniques of adjusting stand composition. It is by such measures that silvicultural treatment usually makes its principal contribution to increasing the yield over that naturally available without treatment.

Soil and site conditions often vary so much that optimum efficiency in the use of growing space tends to dictate some variation in species composition from one part of a stand to another. This is most likely to be true of sites where many species can grow and least true of those with limiting moisture conditions that restrict species composition. Even if a single species is best for all parts of a large tract of land, there may be enough site variation to call for differences in rotation length in different places. For these reasons, maximum efficiency in use of growing space is more likely to be achieved with mixed or uneven-aged than with pure or even-aged stands. The extent to which one acts on these considerations depends on judgment about the relative cost and value of the simplicity of uniformity and the fine-tuning possible with irregularity. The fitting of species to sites is not merely a matter of choice according to gross productivity but is also strongly affected by ability to survive and withstand damage.

Many silvicultural systems, especially those that involve cutting a number of times during a rotation, contain important provisions for some reduction of the enormous loss and waste that hobbles forestry enterprises. There are not only the obvious losses from damaging agencies and natural suppression but also subtle losses of growth from excessive competition of poor trees with better ones. It is logical to attempt to reduce this loss by measures such as frequent cutting, thinning, salvage cutting, and integration of harvesting operations for multiple-product utilization.

Nevertheless, like all worthy motives, this objective can become a fetish if pursued too far. The cost of reducing waste and loss as well as that of capturing more complete occupancy of the site must always be weighed against the cost of the additional effort involved. The point of diminishing returns is reached long before the staff has hastened hither and yon through the forest planting up every vacant gap, salvaging every dying tree, thinning every stand, and utilizing every crooked top log.

Control of Damaging Agencies

Any successful silvicultural system is modified by the objective of creating stands with adequate resistance to insects, fungi, fire, or other injurious biotic and physical agencies. The management of some forest types is indeed governed by this consideration. The modifications are mostly specific steps taken against specific damaging agencies and cannot be safely based on generalities intended to apply to all forests. Appropriate procedures cannot be developed without detailed knowledge of the behavior of the damaging agencies.

Most of the generalizations about the damaging agencies of the forest are more nearly true than false, but they cannot be accepted as a basis for silvicultural procedure without being scrutinized for applicability in each instance. Among these are the view that vigorous, fast-growing trees are more resistant than less thrifty, slow-growing ones; that mixed stands are safer than pure; that uneven-aged stands are more resistant than even-aged; and that close duplication of natural conditions will safeguard against many difficulties. Exceptions are numerous.

The stem rusts of conifers and the white pine weevil are, for example, more serious pests of vigorous trees than of the less thrifty. There are some insects and fungi that alternate between different host species and are thus most dangerous in mixed stands containing the appropriate hosts. Pure stands of spruce are much less susceptible to the spruce budworm than are mixtures of spruce and the highly susceptible balsam fir. In uneven-aged stands there is excellent opportunity for infection of young age classes from older trees by pathogens such as dwarf-mistletoe that attack trees of all ages, thus enabling an infestation to remain established in a stand indefinitely. Highly artificial stands of well-tested exotics from faraway continents are sometimes less subject to damage than in the native habitat.

Consideration of biotic enemies best starts from recognition of the fact that the trees of any kind of forest represent a source of food to a wide variety of organisms. Owing to the availability of food, organisms ranging from minute viruses to large herbivorous mammals have evolved that are adapted to feed upon the forest. These organisms are so dependent upon the vegetation that changes in the forest cause changes in the populations of dependent organisms. Changing the forest does not get rid of parasites; it merely exchanges one group of parasites for a new set that may be more or less harmful and difficult to handle.

Fortunately, only a very few of the dependent organisms are harmful. In a sense, the parasite that kills its host and thus its supply of food is a poorly adapted one. This is why introduced parasites often cause much more economic loss than the native ones. The well-adapted parasites sometimes cause so little damage that they go almost unnoticed. However, there are also some of these that can cause substantial economic damage without threatening the life of the tree. For example, the heart-rots, which attack the nonliving wood inside a tree, can ruin the utilizable wood without significantly harming the vital processes of the tree. Need for modifying silvicultural systems to reduce losses to biotic enemies can normally be confined to those few that can cause serious economic loss; the vast majority that merely feed on the trees without economic damage can, from this standpoint, be ignored.

The most important silvicultural approach to reducing losses to damaging agencies is merely the evasive action of avoiding the conditions that are conducive to damage (Boyce, 1961). It is remarkable how much damage from pests or nonbiotic agencies can be traced to encouraging species or strains thereof that are not adapted to the sites.

If there is adequate knowledge of the situation it may be possible, by silvicultural measures, to create forests that are resistant to damage. In dealing with this or any other matter relating to damaging agencies, it sometimes clarifies thought to make a careful distinction between (1) **susceptibility to attack** and (2) **vulnerability to damage.**

For example, the best places for the overwintering and early feeding of the eastern spruce budworm are the male cone buds of large, vigorous balsam firs. However, because of their health, they are the last firs in the stand to die. They are thus susceptible to attack but low in vulnerability to the damage called mortality. If this were not known, one might overlook the fact that these are the very trees most likely to support a buildup of this serious defoliating insect. The first individuals attacked are, in this case, likely to be the last to die, which is the opposite of what intuition would suggest.

Another important distinction that is more than semantic is that between silvicultural and biological control of damaging agencies. **Silvicultural control** involves creation of forests and forest environments that resist either damaging agencies or the effects of damage by them. **Biological control** involves the use of biotic agencies that combat the damaging biotic agencies, that is, the introduction or encouragement of organisms such as fungi antagonistic to damaging ones, parasites of insects, or predators of herbivores. The distinction is not necessarily perfect; silvicultural measures can be used to encourage biotic enemies of pests.

Examples of control of damage by modification of silvicultural systems are many and varied. Among these are thinnings and other removals of old, damaged, or otherwise susceptible trees in southern and western pine stands to reduce damage by *Dendroctonus* bark beetles (Sartwell and Stevens, 1975; Thatcher *et al.*, 1981). With species, such as aspen and balsam fir, that are seriously threatened by heart-rots, it is logical to terminate rotations before the trees become old and highly susceptible. In many parts of the Southeastern Coastal Plain stands of hard pines are so continually threatened by fire that prescribed burning is essential for fuel reduction; this usually necessitates some form of even-aged management.

Damage by deer and other mammals, long important in parts of Europe, Australia, and New Zealand, is of increasing concern in North America. If other ways are not devised to control these animal populations it may be necessary to resort to fencing regeneration areas as is often done in Europe. Part of the solution is likely to involve favoring spruces or other tree species not palatable to the animals as well as regulating the forest that is the food base and shelter of the animals.

Another part of the solution is to recognize that large herbivores are as much pests of the forests as fires or introduced defoliating insects. Protecting deer can be like protecting gypsy moths. The animal population of the forest must be managed just like that of the plants; in fact, one cannot be managed without also managing the other. Hunting is an important tool of wildlife management for the same reasons that cutting is in silviculture, but, in the case of herbivores, it does not deal directly with the food base that supports the animals.

The pattern of any sort of cutting is sometimes adjusted to avoid wind damage to residual trees; gaps and weak places in stand borders are, for example, often the places where serious blow-downs start. The shelter-wood system is often used for eastern white pine partly because white pine weevils prefer to lay their eggs on pine leaders exposed to direct sunlight; the partial shade of a shelterwood tends to protect pine saplings from the serious deformities caused by this insect.

There are so many different kinds of modifications aimed at reduction of specific kinds of damage that it is hard to detect any general or consistent way in which they conflict with other objectives. However, these measures usually complicate harvesting and other operations. The indirect silvicultural measures of control are often the slowest to take effect but are the most enduring and automatic (Knight and Heikkenen, 1980). If they work they can usually be depended upon to do so even if no one remembers to do something at the right time.

Provision for Sustained Yield

Perhaps the most noble and ambitious goal of forestry is that of making each forest the source of indefinitely sustained and uniform flow of wood and other benefits. The difficulty of attaining this goal is greater the longer the production cycle of the benefit or the age of stand on which it is dependent. Water is a short-cycle benefit dependent mainly on having the kind of vegetation that will keep the forest soil porous; a low shrub cover is often better for this than a tall forest. Wild and domestic animals that feed on grass or browse thrive on young plants and have production cycles of only several years. It is not necessarily easy to manage forests for these kinds of "short-cycle" benefits but the measures do not require decades of juggling silvicultural regeneration schedules.

Sustained yield is most difficult to achieve with timber and other benefits available only from stands that have ages measured in decades or centuries. It is sobering to recognize that, for any given tract, virtually all of the trees that can be harvested for timber during the next quarter-century or more are standing now and there is no practical way of suddenly creating more. Some wildlife species depend on very old trees, a continuing supply of large dying trees, or the temporary conditions associated with some intermediate stand age. The stands of old-growth natural areas for scientific study or public edification do not live forever so there should be a continuing supply of younger age classes.

The ensuing discussion deals mainly with sustained-yield management for timber because that goal is usually within grasp and crucial in silviculture. The ideas are transferable to sustained yield of other benefits. The fundamental principle is simple. Each annual age class, from that of Year 1 to that of rotation age, which is being harvested and replaced this year, must be equally represented in whatever forest unit is the source of the sustained yield.

Even though this goal is simply stated its perfect achievement is practically impossible and not absolutely desirable. Both the technical details and the semantics become complicated. It is very important to recognize that foresters almost always manage forests that have maldistributions of age classes. This means that the goal of sustained yield is approached but not achieved. It also means that certain neat assumptions that may be made about the growth and yield of ideally arranged forests seldom apply.

Viewpoints about sustained yield vary from unquestioning worship to derision. Close adherence tends to prevail in places such as Western Europe, where society has not forgotten times of timber famine. In localities with much old-growth timber the importance of the idea has traditionally seemed remote. It is often attacked on the ground that only short-term economic considerations count and that technology can always find substitutes for depleted resources. Some of this doctrine appears to emerge from the observation that people are short-sighted so it is then an "economic law" that they should be short-sighted. Recognition of the fact that most of the substitutes for wood would come from oil or other nonrenewable resources has diminished the assertiveness with which that argument is made.

It is, however, important to note that rigid adherence to annual sustained yield would force timber onto the market when demand was low and withhold it when demand was high. The purpose here is only to show how silvicultural systems are modified to provide for sustained yield and also how these modifications can be applied in ways that either restrict or liberate silvicultural practice.

Over the centuries there has been nothing that has come closer to putting silvicultural practice in a straightjacket than attempts to develop sustained yield by simple, arbitrary procedures. Modern techniques for this purpose are much less restrictive and increasingly effective; they are also far easier to accommodate to other objectives than are the traditions lingering from the days when silviculture could not be so flexible. The idea of the silvicultural system itself emerged mainly as a device to ensure that the cutting practices followed in stands would be orderly enough to fit into the old, simple plans for sustained-yield management for whole forests. In fact, the reputation of the silvicultural system as a highly rigid procedure comes more from the requirements of this ancient kind of management than from any that have directly to do with growing stands of trees.

The first planned forestry in the western world appears to have developed in Western Europe late in the Middle Ages (Knuchel, 1953). The main objective was to guarantee the perpetual supply of indispensable fuel from sprouting hardwoods. Transportation was so primitive that the forest area tributary to a community was rigidly circumscribed. Sustained yield was achieved by what has come to be called the **area method of regulation** of the cut. This consists basically of dividing the total forest area into as many equally productive units as there are years in the planned rotation and harvesting one unit each year (Fig. 12-2). Such management under the coppice system worked very well, because there is no kind of reproduction more certain or prompt than that which comes from stump sprouts. It was almost equally successful in growing conifers by clearcutting and planting, a system that imitated agriculture and represented the next step toward more efficient forestry. With such a scheme reproduction was also reasonably sure and could be obtained without delay.

If sustained yield were the only objective or if the coppice system and that of clearcutting and planting were the only silvicultural systems, there would be no need of any other methods of regulating the cut to secure sustained yield. However, this regimented kind of forestry with its fixed rotations and annual cutting areas does not provide adequate latitude for dealing with the kinds of nonuniform stands that are either encountered or else created by partial cutting for the attainment of other management

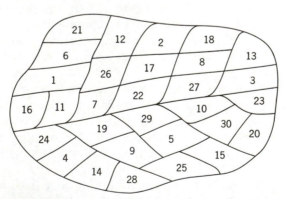

FIGURE 12-2 Schematic diagram of a forest divided into thirty equally productive stands so that the cut can be regulated by area to provide a sustained yield under purely even-aged management on a 30-year rotation. The numbers indicate the sequence of cutting according to an arrangement designed to avoid having contiguous areas under regeneration. Note that stand boundaries conform to the terrain of a typical stream drainage rather than any arbitrary pattern.

objectives. The method of regulating the cut by area can be applied without great difficulty if the stands are essentially even-aged and partial cutting is limited to measures such as thinning and most forms of uniform shelter-wood cutting. However, this method becomes very difficult to apply when the individual age-class units become too small for practical area measurement and some kind of uneven-aged condition prevails or is being created.

It may be necessary to work with the uneven-aged condition simply because the particular stands involved happen to be uneven-aged when placed under management. In other instances, the ecological requirements for natural regeneration, wildlife and watershed management, or protection of the forest necessitate diversity of age classes within stands. Efficient use of growing stock requires recognition of the fact that not all the trees of a stand are of the same age or even of the same size when the best time comes to cut them; this point tends to argue for some departure from the even-aged condition.

The area method is not applicable to those unhappy situations in which severe economic limitations dictate "high-grading," the kind of partial cutting in which only the biggest and best trees can be cut. This kind of cutting may have to be applied to even-aged as well as uneven-aged stands. While it is hardly an efficient kind of silviculture, it is one in which sustained-yield management is possible but both difficult and crucial. Careful scrutiny is necessary to ensure that the rate of removal of large trees is kept in balance with the ability of the smaller trees of the residual stands to supply a flow of large trees of adequate quality in the future.

European foresters became aware of the desirability of occasional departures in the direction of uneven-aged management more than a century ago. By that time, the concept of achieving sustained yield through regulation of the cut by area was firmly entrenched not only in the tradition of forestry but also in the laws regulating forestry. Consequently, the idea of introducing more latitude into silviculture could not then be adopted without substantial concessions to the objective of sustained yield and the traditional technique of ensuring it. The method of regulating the cut by volume and probably the concept of the silvicultural system were fruits of these concessions.

In the **volume method of regulation** of the cut the basic procedure is to determine the allowable annual or periodic cut in terms of volume of wood with due regard for the rate of growth, current and potential, and for the volume of growing stock, existing and desired (Davis, 1966). Ideally this means that if one could ever determine and create the appropriate volume of growing stock made up of the proper distribution of sizes and kinds of trees one could depend upon having it yield annually a certain well-defined volume of growth available for cutting. If this could be done, the length of the rotation and the amount of area exposed by cutting each year would not have to be known.

The volume method of regulation is really a sophisticated and indirect way of applying the area method. In fact, it depends so often on regulating diameter distributions that it might better be called "**regulating the cut by diameter distribution.**" Instead of working directly with the idea of cutting equal areas each year under a definite rotation, one determines the distribution of number of trees with respect to D.B.H. that would prevail in the forest at hand if it were composed of a series of even-aged stands with each age class, from age 1 to that of rotation age minus 1, represented by an equal area. The resulting diameter distribution defines the growing stock that should be left after each year's cutting if sustained yield is to be assured; the allowable annual cut is whatever volume this growing stock will yield in annual growth. The process of determining the appropriate growing stock and the allowable annual cut will be discussed in more detail and more critically in Chapter 15 on the selection method.

It is easy to get the impression that the volume method replaces the tyranny of the area method with the hard labor of creating and maintaining balanced uneven-aged stands by the selection system. It is true that this technique can be used to make a single stand into a self-contained unit of sustained yield. However, there is seldom any point in going to this extreme because the stand itself rarely needs to be a sustained-yield unit.

The main virtue of the volume method is that it frees the forester of the rigid restraints of either extreme. It enables him to make a sustained-yield unit out of a forest composed of almost any combination of patterns of even-aged, two-aged, balanced uneven-aged, and irregularly uneven-aged stands that might be desirable to satisfy other objectives. If it is most feasible to keep certain kinds of stands in the even-aged condition, it is still possible to have others elsewhere in the forest in an irregularly uneven-aged status without disrupting a sustained-yield program. The chief requirement is that all the stands of the whole sustained-yield unit in combination conform to the appropriate distribution of diameter classes. Although it calls for skillful forest management, this approach is better than trying to fit the forest into a highly arbitrary pattern merely to simplify the bookkeeping. Usually it also enables conversion of an existing growing stock into a properly distributed one much faster than would be the case if each stand had to be made perfectly even-aged or uneven-aged and balanced.

Many of the rigid ideas about regulation of the cut have been handed down from bygone times when transportation was poor and sustained yield was more of a crucial necessity than now. The techniques of both area and volume regulation were so crudely developed that the stands had to be uniform if they were to work at all. The age distribution and basic program of management had to be sufficiently simple and sharply defined to be clearly comprehensible to anyone concerned with regulating the cut. The silvicultural system was the vehicle of terminology by which this information was conveyed.

In a sense, the silvicultural system, once established, was looked upon as a binding contract between silviculture and forest management. This viewpoint still finds expression in the idea that there is some sort of requirement that a whole forest should be handled according to a single silvicultural system. The same tradition likewise makes it easy to conclude that stands must be reproduced at the end of the planned rotation regardless of market conditions or any other consideration. It also leads to the idea that it is mandatory that stands should be either distinctly even-aged or definitely balanced and uneven-aged and cannot be anything in between. There are plenty of good reasons for imposing some degree of uniformity and homogeneity on forest stands, but these are more likely to involve the objectives of increasing the efficiency of logging and other operations in the stands than just that of assuring sustained yield.

Both of these methods of forest regulation are simple in principle but maddeningly difficult to apply, even when ownership has embarked on sustained yield as a policy. Each method really works well only when the proper distribution of age classes exists. This kind of distribution can only be an artificial creation. It could come about in nature only if regenerative disturbances occurred annually and affected equal areas. Even as an artificial creation the balanced distribution is a goal approached only over one or more rotations; it would probably not be maintainable even if it was attainable. The forest manager is normally concerned mainly with some planned alteration of the existing distribution of age classes. This requires knowing what the distribution is and forecasting the consequences of various silvicultural alterations of the distribution.

The method of area regulation remains the most dependable technique of developing and guiding sustained yield. Aerial photographs have made it easier than it once was to map small areas of different age classes; associated techniques of digital analysis expedite area determinations. Computer simulation can also make the area method a much more effective way of analyzing and planning the future than when all calculations were so time-consuming that only simple ones were done. It is now possible to take an existing array of age classes and determine what would happen with different schedules of stand replacement. Optimum programs can be selected merely from trial and error or by use of some optimization routines of operations analysis such as linear programming.

Electronic computers also make the calculations of forest mensuration swifter and more sophisticated; this improves use of the method of volume regulation by volume or diameter distribution. It is still debatable whether these techniques are accurate enough for developing long-term programs of sustained yield. However, they have greatly improved the ongoing analysis of changes induced by growth, mortality, and harvest in the composition and structure of the growing stock of forests. In fact, management schemes that combine use of the area method for sustained-yield regulation and the techniques of volume regulation primarily for analysis of growing stock are normally better than reliance on either method alone.

Discussion of the volume method of regulation will resume in Chapter 15 on the selection system in which this method is very important. What remains for consideration here are those modifications of silvicultural treat-

ment that might seem to make no sense except for turning forests into
sustained-yield units.

Such cases are common in America because real forest management
has only recently started and with forests that rarely have any semblance of
the distribution of age classes necessary for sustained yield. Forests that
have not been subject to much cutting or severe natural catastrophe have a
plethora of overmature stands and a paucity of any age classes younger
than the ultimate rotation age. If heavy cutting has been concentrated in a
brief period during the past, almost all the stands will be of one age class
dating from the time of the cutting. Even where the inheritance consists of
uneven-aged stands of natural or accidental origin it is usually found that
some age classes are deficient and others present in surplus.

Much of the management of old-growth forests in the West involves
circumstances of this kind. Management might, for example, start with a
large virgin forest consisting almost exclusively of stands that originated
after a single massive fire that occurred 375 years ago. If one wished to
convert this to a sustained-yield forest on a 100-year rotation the simplest
way would be to clear-cut 1 percent of the area annually for the next
century. One could accelerate this process, among other ways, by cutting
over the old growth more rapidly and by planning to cut some of the first
new stands before they reached 100 years. However, it is still necessary to
leave substantial acreages of old timber standing for decades without any
real increase in value. If one considered only the silviculture of the individual stand, it would seem witless not to replace all the old timber with
young vigorous stands immediately. In this situation it is often useful to
conduct salvage cuttings to mitigate losses in overmature stands that are
being stored until the time of their scheduled replacement.

In other localities, such as much of the East and South, it is common to
find whole forest regions full of middle-aged stands that arose after heavy
cuttings some decades ago. Sometimes the growing stock has neither timber of mature size nor stands of seedlings and saplings. A period of time
may elapse during which the forester can do little but eke out meager
income from intermediate cuttings that do nothing to correct the distribution of age classes. The first stands that reach the condition where they can
be harvested and replaced may be removed long before they are mature,
just to get some new age classes started. Later some of the stands with
which one started may be held well beyond the logical rotation age in order
to have something to cut when the missing younger age classes would
have been coming to maturity. Almost any sort of adjustment of age-class
distribution is likely to involve holding some stands beyond rotation age
and, at other times, cutting some before they become mature.

The same kind of manipulations may be made in more detail with
particular diameter classes. Consider, for example, the case of a forest
being managed with the intention of growing the crop trees to 20 to 24
inches D.B.H. There is an overall surplus of trees in the range from 17 to 20
inches compensated by deficiencies in all diameter classes below 15 inches.

If trees smaller than 10 inches D.B.H. are ummerchantable, adjustments must be made mainly by the cutting of trees that are larger than 10 inches. The surplus diameter classes should be picked over very carefully with several definite objectives in mind. In general, the *best* trees should be reserved in sufficient quantity not only to replace the trees larger than 20 inches that are harvested but also to provide an adequate volume to compensate for part of the present deficiency of trees now less than 15 inches. The removals from the surplus diameter classes should, to the greatest extent possible, come from the poorest members and provide for the early establishment of new reproduction in order to accelerate recruitment of new age classes.

While the surplus diameter classes are winnowed quite rigorously the deficient diameter classes receive an indulgent treatment that might seem inconsistent. In these classes it even becomes desirable to try to make good trees out of poor ones. While one would be unlikely to release a 19-inch codominant in these circumstances, it would be logical to cut an 18-inch dominant to promote two adjacent 12-inch codominants to the dominant position. It might be justifiable to release and prune a 10-inch tree with a 25 percent live crown ratio if one could be sure that it would ultimately recover and develop into a good tree. This would not be the most logical way to grow good trees if one could start from the beginning with seedlings, but there is the necessity of creating a good 12-inch tree in 15 years rather than in several decades.

The adjustments are by no means always made with partial cuttings. In the example at hand, one might carry out the seemingly reckless and premature harvest of a fairly good stand with trees mostly in the 17- to 20-inch range even while carefully nursing a rather poor adjacent stand in which most of the trees were 8 to 15 inches.

One swift remedy for deficiencies of age or diameter classes is the acquisition of stands and forests that have them. It may, for example, be wiser to purchase a heavily cutover area than to clearcut an immature stand to balance the age distribution of the forest. Management for sustained yield is not an objective to be pursued unquestioningly but neither can it be ignored. The imbalances that exist must be identified and their consequences must be faced. Decisions have to be made about whether to rejuggle age-class distributions and, if so, how this will alter silvicultural treatment. Perfect solutions do not exist. Ugly surprises or hard times lie in wait for the successors of forest managers who ignore, aggravate, misunderstand, or neglect the problem.

Preoccupation with sustained yield for timber should not obscure the fact that the appropriate distribution of age classes also provides for a continuing flow of other benefits. It also brings about high diversity in those ecological factors governed by stand age. The benefits to the environment that go with using forests to sequester carbon and chemical nutrients are at their greatest if there is an equal representation of all age classes up to maturity. When annual losses of biomass to death and decay exceed annual production the stand begins to release carbon and nutrients back into the rest of the system. Some of the benefits associated with multiple use of forests require some stands older than optimum rotation age for timber. Some require an allocation of area to young stands that is too great for optimum sustained yield of timber. However, in general it can be said that age-class distribution that is best for sustained yield of timber is at least good for the sustained flow of all other important benefits of the forest.

Optimum Use of Capital and Growing Stock

It takes so long for trees to grow that silviculture, especially that for timber production, is powerfully affected by policies about long-term investment of capital. Two quite different kinds of capital investments are made and it is important to distinguish between them clearly and analyze them separately.

Treatment of Monetary Investments

The most obvious of the two kinds is the money actually invested out-of-pocket in the costs of growing trees and holding the land beneath them. A common way of analyzing these kinds of investments is by using compound interest calculations to compare the present net worths of different courses of action as was described briefly in Chapter 1. Such matters are treated in much more detail in various works on forest economics (Gunter and Haney, 1984; Leuschner, 1984).

One of the most important uses of this particular analytical technique is the financial comparison of different programs or systems of silvicultural treatments. It is a technique used by foresters and other planners but not in conventional accounting, which does not recognize unrealized future returns. Comparisons of present net worths help with choices about regeneration techniques, rotation lengths, thinning programs, logging methods, and many other decisions made in building a silvicultural system. Such analysis always militates in favor of securing early returns and postponing costs, with these effects being more pronounced the higher the chosen interest rate.

While the basic concept is thought of as an enemy of long-term forestry, the real culprit is often the sloppy use of it. Society and forest owners seldom use such analysis to determine whether to engage in long-term silviculture. That decision is usually based on some more intuitive kind of concern for and faith in the future. Once the decision has been made to invest in long-term forestry, analysis of present net worth becomes a forester's tool for comparing different ways of executing the policy and of detecting financially optimum solutions.

Before it has been decided to invest in silviculture, much is made of the difficulty of securing high rates of interest on such charges as regeneration costs. After regeneration has become a matter of policy or a legal requirement, the cost of it is quietly charged against the value of the timber of the previous crop. This is sound enough theoretically because the question of whether to invest in regenerating the stand has ceased to be debatable. However, the analyses done by calculating present net worths become ways of determining which mode of stand regeneration is financially best given that the stand is to be replaced. This is the purpose for which these methods were first developed; their use as a deterrent of forestry is an unfortunate by-product.

Investments in Growing Stock

The second kind of capital invested in timber management silviculture is the value tied up in merchantable trees that are standing and growing. This investment is in income postponed rather than in money actually invested, unless one had recently purchased the trees of the growing stock. A good growing stock has a high stumpage value and the money that could be realized from its liquidation represents a substantial investment. The value of the forest growing stock necessary to supply a given processing entity on a sustained-yield basis is often worth much more than the mills and equipment. This has not always been true; there are still places where stumpage is cheap and the objective is likely to be maximization of the return on investment in the processing equipment.

In either situation it is desirable to try to manipulate the growing stock so that its increase in value represents an acceptable rate of compound-interest return on its own value. The assessment of this situation is called **financial maturity analysis** (Duerr, Fedkiw, and Guttenberg, 1956). The liquidation value of a tree or a stand of trees is treated as the investment and the increase in value, through change in quantity and quality, during a decade or some other specified period, is taken as the return on that investment.

For example, a tree increases from 18 to 20 inches D.B.H., and from 53 to 63 cubic feet in volume. Quality and other factors associated with change in size increase stumpage value per cubic foot from 66 to 83 cents, so tree value increases from $35 to $52.20. If the tree has been growing 10 rings per radial inch and so continues, it takes 10 years for this increase to take place and the compound interest earned on the $35 would be 4%. However, if the tree can be made to increase in radial growth to a rate denoted by 6 rings per inch, the interest earned during the period shortened to 6 years would be 7%. If the goal rate was 6%, it would be logical to harvest the tree at the beginning of the period in the first case but to leave it if the growth acceleration of the second case seemed likely. In the first case the tree is economically mature, given the choice of interest rate; in the second case, it is not.

It should be carefully noted that the values of the two trees were *not* the costs of growing the trees. There is not necessarily any direct relationship between this kind of value and the growing costs. The analysis of actual monetary investments by calculations of present net worth, described in the previous section, represent the most common method of analyzing such costs. There is no complete or perfect method for long-term financial analysis of timber management. Neither of these methods is, but analysis of present net worth comes closer to completeness than financial maturity analysis.

Analysis of Investments

The two kinds of tests are separate ones. It is often logical to apply both simultaneously, that is, to require that there be a satisfactory return on the money actually spent and, by a different way of viewing the matter, on the value of the growing stock. These two tests are not, incidentally, additive. One cannot require a 13% return, arrived at by adding 8% demanded on actual costs to 5% asked on value of growing stock; that would be logical only if one could somehow harvest the trees and invest their value elsewhere yet leave them to grow.

As indicated in Chapter 4, financial maturity analysis tells nothing about the status of trees that are so small or poor that they have no positive value. However, on the first day that a tree is just large enough to acquire positive value, the compound interest that it earns on its initial value of zero is positive and infinitely high. Thereafter the rate of compound interest that the tree earns decreases with remarkable rapidity. In fact, if it is the goal to grow trees substantially larger than those of the minimum merchantable size it is often necessary to modify the silviculture in ways that will accelerate increase in value, reduce the size of the investments in growing stock, or do both at once.

The main silivicultural approach to the problem is typified by, but not limited to, the effect of thinning. The removal of trees of positive stumpage value in thinning definitely reduces the investment tied up in standing trees. Properly conducted thinnings have the simultaneous effect of increasing the rate of value production of the stand over what it would have been without thinning. Among all of the techniques of forestry, there is probably no more powerful contribution that can be made to the crucial problem of increasing the rate of interest on forest capital. Part of the increase comes from making good trees grow in size and value at a greater rate; part comes simply from the more passive effect of eliminating trees that are poor investments in favor of those that will yield good returns.

This same general approach is used to detect those stands that cease to return enough interest to remain as parts of the growing stock and should be replaced by new stands that will be better investments. Silvicultural techniques are modified most, however, when partial cutting is indicated.

The idea of regarding the individual tree as the basic unit of production is appropriate because the trees in a given stand are not exactly the same and do not grow at the same rate. They might do so in stands with the spacing of orchards but not in forest stands. Even in the finest stands most of the capacity to produce value is concentrated in a limited number of trees, usually the biggest and best. As a result, the trees of even the most uniform stands do not all reach the optimum time for cutting simultaneously.

As is the case with all useful ideas, this one can be pursued to absurdity and sometimes is. If one harvested every tree just as soon as it had positive value, the rate of interest earned on growing stock would be kept at an impressively high level. However, the rate of return on the money spent, especially on the costs of holding and administering the land, would ordinarily be dismally low or negative.

Choice of Interest Rates

The rate of compound interest demanded on the investments will not be the same for all owners; it may also be higher for actual monetary investments than for return on growing stock. A logical rate is one realistically available to the owner from some alternative investment that might actually be made, often in some other part of the forestry enterprise. It is possible to arrive at a choice by analytically subdividing an interest rate into its logical components.

The first component is a risk-free, pure rental rate for money that is usually regarded as 3%. Allowance must also be made for certain risks. One is the risk of unsalvageable loss of values from damaging agencies. Sometimes there are data about percentage rates of annual losses that can be used to help with estimating the size of this component.

Another kind of risk can be called the "market risk," the risk that the demand for the trees being grown will dwindle or vanish. People no longer want elm wood for wagon-wheel hubs and their use of Douglas-fir framing timbers might wane. This kind of allowance must be made but the percentage assigned to it depends on imaginative forecasting. One offsetting component that can be kept separate from or included in the market risk is allowance for any appreciation in value. This is, in general and on the average, more real and less elusive than the risk that some kinds of trees will become technologically underemployed. The stumpage values of saw-timber have increased at an average rate of 1–2% compound interest in terms of constant dollars for many decades (Williams, 1981). The inclusion of a prediction of price appreciation would require a subtraction from the overall interest rate.

There is one component of interest rates that normally does not belong in the rates used for analysis of investments in forest growing stock and silvicultural treatment. That is the allowance for inflation or other change in the value of money. If a tree is worth a pair of shoes today and it neither grows nor loses its utility during the next 10 years it will still be worth a pair of shoes. Allowances for inflation have to be made for investments in bonds, savings accounts, or other paper promises of future payment of money. Trees are inflation-proof equity. In times and places of rapid inflation they have long been a refuge for capital, provided that they are not taxed severely. In analysis of returns on investments made in growing trees or in value of growing stock, most of the effects of inflation are canceled by making the calculations with present costs and values for all investments and returns regardless of when they occur. The possibility that costs might increase faster than values is a factor included in the allowance for market risk.

Application of Financial Maturity Analysis

This approach can be used to choose between two trees that are competing for the same growing space; in such instances, the tree that will return the highest rate of interest is the one to leave. The concept is not limited to use in partial cutting; if a stand is to be reproduced by clearcutting it is logical to terminate the rotation when the rate of return on the investment represented by the whole stand falls below the acceptable rate.

This kind of analysis usually reveals in financial terms that fast-growing trees of high quality are more promising growing stock than those that are slower in growth or poorer in quality. The feeble 25-year-old intermediate that has just become merchantable for pulp-wood may have the same degree of financial maturity as the 30-inch veteran that has finally started to decrease in growth after 120 years. By means of such analysis it is often possible to restrain the subconscious tendency to carry most trees of a stand too long and to cut the best ones too soon.

This concept places so much emphasis on the variation between individual trees that it is altogether too easy to jump to the conclusion that it can be applied only by means of the single-tree selection system. If the stands are uneven-aged or irregular a pure application of the idea would tend to perpetuate and accentuate the degree of variation in age. Old, mature trees would almost automatically be replaced by new ones but not all of the trees of one given age class would grow at the same rate and be replaced simultaneously. If stands are even-aged, a series of thinnings leading to shelterwood cutting, all under even-aged management, would represent almost perfect application of the idea (Fedkiw and Yoho, 1960). In most overmature stands this approach would dictate a prompt clearcutting. If a perfectly balanced, uneven-aged stand of good quality already existed and was handled solely on the basis of financial maturity, the cuttings might automatically maintain the distribution of diameter classes appropriate to sustained-yield management of the stand. However, the distributions of age and diameter developed by such cutting are essentially an image of those that existed initially, although this image would gradually be blurred in random fashion by the effects of the differences in vigor and quality among the trees cut and left in successive harvests.

Cuttings based on financial maturity may thus have some effect in broadening the distribution of age classes but are distinctly different from the systematic manipulations necessary to create and maintain sustained-yield units. The objectives of securing sustained yield and optimum use of growing stock are distinctly separate. Unless there is the theoretically perfect distribution of age classes these two objectives even conflict with one another.

The concept of financial maturity is such a powerful analytical tool that it is prudent to call attention to its inherent shortcomings and to the errors that are often made in its application. In the first place, the concept does not take into account any investments other than those represented by the as yet unrealized value of standing timber. There is no accounting for the value of or carrying charges on land, roads, manufacturing facilities, or any of the other investments that make up the whole enterprise (Worrell, 1953). Some of the values that have been built into the tree by silvicultural treatment or by making it more accessible are brought into consideration, but that is all. This is why separate analysis of the funds actually invested out-of-pocket is desirable.

Financial maturity analysis is also difficult to use in any consideration of whole rotations. It tells nothing much about the status of any tree or stand that has not attained positive value. It is also a rather nearsighted mode of analysis that is much more useful for looking one or two decades ahead rather than at programs for whole rotations.

Preoccupation with manipulation of growing stock can result in failure to secure adequate, timely regeneration. This often comes about because the desire to fill the growing space with trees of good earning power does not leave vacant space for regeneration. Even when the concept of financial maturity is a good guide to partial cutting throughout the rest of the rotation it is often best to depart from it during the relatively brief interval in which the establishment of reproduction must become the ruling objective. In using the concept it is very difficult, although not impossible, to compare the prospective returns from continued growth of existing trees with those from the new trees that might replace them. To the extent that cuttings guided by financial maturity lead to the creation of irregular stands they may complicate and reduce the efficiency of logging and other operations involving the stands.

The application of the concept can be no better than the predictions of future growth and economic conditions on which it is based. The calculations are complicated enough that there is a tendency to engage in oversimplifications that lead to poor results. It should not be presumed that every unit of cubic volume is worth the same as all others or even that trees of equal size are equally valuable if their quality differs.

The necessary predictions of the growth of individual trees can be guided by observations of increment borings and bark characteristics (Fig. 6-3). However, where partial cutting is involved, one should not simply extrapolate from these indications of the growth of the recent past. As in all silviculture practice, the treatment must be guided by prediction of its effects, even if the predictions must be highly intuitive.

Pursuit of the objective of optimum use of existing growing stock easily becomes a means of rationalizing evasion of other crucial problems. This is especially true in stands in which only a fraction of the trees are of merchantable size or quality. Such trees are then gradually cut in the hope, often desperate and sometimes vain, that smaller or poorer residual trees will develop into good ones for subsequent harvests. If the likelihood of getting desirable reproduction is in doubt, the issue is sometimes evaded by prolonging the rotation and trying to bring all of the existing trees to good size. If such evasions are necessary they should be honestly recognized and corrections for their consequences made in management plans and cutting budgets; meanwhile basic solutions should be sought for the causes of the situation. It is too much to expect that the best course of silvicultural action will always be found to follow the path of least resistance.

The rather aimless procedure just described often occurs during the period when forests or stands are being placed under management. At such times it is often difficult to determine what kind of forest to create or, even if a goal has been decided upon, it may still take time to invent means of getting there. In complicated situations this period may extend over several decades, during which time it may be very difficult for even the professional onlooker to determine what is going on. The greatest danger in such procedures is that they may be complacently accepted as permanent solutions.

In summary, it is well to reiterate that forestry enterprises rely heavily on getting adequate returns from growth on growing stock. This depends ultimately on obtaining good growth in both size and quality from trees comprising a growing stock that is not allowed to build up to the point where it becomes a large amount of capital yielding low returns. The concept of financial maturity is a useful guide in modifying silvicultural practice to make best use of the growing stock, even though it is subject to errors in application. It usually constitutes a powerful argument *for* partial cutting and *against* those radical or deliberate changes in the age distribution of existing forests that would necessitate cutting financially immature trees.

Concentration and Efficient Arrangement of Operations

In previous pages attention has been devoted to a number of considerations that tend to argue more against than in favor of developing uniform stands. It has, for example, been pointed out that the objective of sustained yield is hardly cause by itself for inflicting the uniformity of the pure, even-aged stand on silviculture. In general, highly flexible procedures of cutting and arrangement of stands enable the closest approach to optimum use of both growing stock and growing space, and in lesser degree to the successful regeneration of the forest and its protection against damaging agencies. These objectives are, however, not always best attained by creating or maintaining irregular stands and there are other important reasons for resisting the temptation to do so. Most of these involve the point that all sorts of operations can become complicated and expensive in a stand or forest composed of a mosaic of age classes or species.

The harvesting of timber crops is usually the most expensive operation conducted in the forest; it is thus important to arrange stands so that costs, per unit of volume harvested, will be kept at the lowest level consistent with other objectives. Transportation is the component of logging costs most affected by the arrangement of stands. If the merchantable age classes or species are scattered rather than concentrated in a contiguous unit, the gross area that must be covered to harvest a given volume of timber in a single operation is correspondingly increased. This is especially true if terrain is difficult, if roads must be built or improved for each operation, or if the cost of shifting heavy equipment from one operation to another is high. If the heterogeneity of the stands dictates handling a broader range of sizes, qualities, and species of trees than is possible with a single set of machinery or a single procedure, there is the additional cost of having a wider variety of equipment or of trying to handle material with equipment not suited to the purpose. The cost of supervision also tends to increase the more scattered and complicated the operation.

The ideal stand from this viewpoint would be the largest even-aged, pure stand that could be harvested during the period before the harvesting equipment had to be moved for some reason not related to characteristics of the stand. One point of departure for thinking about this matter is the size and shape of the cutting area tributary to a single log-loading point. If the costs of roads to the loading points are high, there will be a tendency to make the cutting units, which often become new stands, even larger.

In considering the relationship between silviculture and logging costs, it is crucial to distinguish between area-related transportation costs and those of handling and processing, which are most strongly affected by tree size. For example, the costs of felling, bucking, loading, and manufacturing material from forty 20-inch trees are not significantly different if they come from 1 acre or from 10 acres of gross stand area. The larger the trees, the lower are these handling costs per unit of volume, at least up to the point where the trees are too large to be handled by the chosen kind of equipment.

If the only objective were to minimize transportation costs that depend on the volume of cut per gross acre, the solution would be to cut everything worth the cost of harvesting and processing. If, on the other hand, it were to minimize handling costs that depend on tree size, only the larger trees would be cut.

The question at hand, however, also involves planning for the production of material for future harvests. Not only does this mean that a host of other objectives must be considered but also that the plan covers a series of harvests rather than just one. In principle, it is no more difficult to accelerate the rate at which trees grow to good size in uniform stands than it is in heterogeneous ones. In other words, problems involving tree size can be solved in either kind of stand but those involving transportation can best be solved with uniform stands. It is, therefore, ordinarily better from the standpoint of a long-term program of harvests to work toward increasing the uniformity of stands.

Another cogent argument for uniform stands is the point that the forest industries are hamstrung at all stages of harvesting and manufacture by the tremendous variability of raw wood. This necessitates repeated sorting and complicates all handling operations; some of this extra work is avoided in the processing of competitive materials that are more homogeneous. To cite merely one example, the variability in size of trees dictates that logging equipment be made versatile at the expense of being completely efficient. Any silvicultural steps that can be taken to reduce variability are advantageous, although there are inherent biological limits to the uniformity that can be achieved. Variation in diameter as well as in other characteristics is bound to appear in any stand that fully occupies the site for any length of time and in stands where competitive effects are used to develop good stem form. However, it is certainly possible to counteract the heterogeneity often found in nature.

Many methods of partial cutting, especially thinning, can be applied to enable the systematic removal of various relatively uniform categories of trees at successive intervals, with a high proportion of the volume being harvested from favorable ranges of diameter. The main objective should be to obtain the widest possible margin between the costs of logging and the value of the material harvested during a series of harvests. This goal is not likely to be achieved if the sole consideration is minimizing the cost of each separate harvest without regard for those that will follow. Silvicultural measures to create trees of good size and quality, preferably in stands of high volume per acre, contribute fully as much.

Most other operations are facilitated by systematic arrangement of uniform stands. Silvicultural operations such as pruning, release cutting, and site preparation can become very complicated and expensive in heterogeneous stands. Such blanket operations as prescribed burning and application of chemicals from the air are expensive or impossible unless large areas can be treated as solid units. Timber marking is very time-consuming and expensive in heterogeneous stands, a situation aggravated by the fact that the more complicated the stand the more expensive the talent required for the marking.

One compromise often struck between the chaos of irregularity and the constraining uniformity of large, even-aged stands is that which involves creating stand in which different categories of trees are systematically arranged in either the horizontal or the vertical dimension. It may help to have cuttings that advance across an area in sequential strips so that the removals are concentrated in narrow areas. With such procedures it is sometimes possible to extract the cut timber through uncut areas that do not yet have any regeneration that could be damaged in logging. Such strips can also be advanced toward the forest roads. If an even-aged stand consists of layers with fast-growing species in strata above slower-growing species, it may be possible to concentrate the removals in one stratum at a time. Some of these compromises about spatial arrangement of age classes and species will be discussed in subsequent chapters.

Administrative and supervisory activities of all sorts are difficult in heterogeneous stands and forests. While uniformity of stands may not be mandatory for sustained-yield management, it is at least helpful. It is much easier to keep track of uniform stands and to detect the places where protective measures and silvicultural treatments are necessary.

Heterogeneity of stands and forests is a characteristic that should not be encouraged or endured without good reason or because of vague naturalistic doctrines not known to apply to the circumstances. There are, of course, important arguments for heterogeneity; the logic for them is outlined in the preceding sections and in some subsequent chapters. However, none of them are reasons for purely haphazard or random scattering of species and age classes, except in situations where the forest is inherited in such condition and there seems no real cause for alteration.

There are strong natural tendencies for stands to become heterogeneous. The important decisions, therefore, relate to determining the degree of heterogeneity that is to be tolerated and whether to try to make stands more heterogeneous or more uniform.

It is often desirable to shape the boundaries of stands to improve the visual qualities of the landscape, especially where the terrain makes forests visible over long distances (Crowe, 1978; Bacon and Twombly, 1980).

Preexisting stand boundaries should not be changed without good reason. Where it is desirable to create new stand boundaries or variations within stands they should coincide with features such as changes in site, variations in accessibility, and barriers to damaging agencies. It is especially important to try to get stand boundaries to fit major differences in site conditions.

Resolution of Conflicting Objectives

It should be apparent that there is no inherent harmony among the various major objectives sought in managing forests. Such harmony can be created only by weighing the various objectives individually and inventing silvicultural systems that represent analytical compromises within forest management plans created by the same kind of procedure (Society of American Foresters, 1981).

The disharmony that has been deliberately painted in tones of sharp contrast is most evident if one considers all forestry in general. Fortunately these conflicting objectives need be resolved only for particular forests or stands. Analysis of each situation will usually reveal a few ruling considerations; the necessity of giving first attention to these will simplify and govern the solutions.

Sometimes one is faced with two or more problems, each of which is separately insoluble but which can be neatly combined into a single solution. The process starts with a consideration of the goals of the forest owner. Each of the remaining objectives must then receive some attention; it would almost invariably be a mistake to pursue any single one to the bitter limit. The analytical process generally works downward from the forest to the stand, but not without the formation of some preliminary idea of the range of treatments and results that is silviculturally feasible. Some immutable and absolutely restrictive natural factors are bound to exist and these must be recognized early in the process. However, the remaining latitude of silvicultural possibilities is likely to be broad enough that further efforts to develop some optimum silvicultural system is normally based on analysis of the effect of all factors, natural and social.

In the chapters that follow there are some examples not only of methods of reproduction but also of some particular silvicultural systems and the reasoning underlying their development. All of them are best regarded as generalized examples chosen mainly to illustrate specific points. Published accounts of complete silvicultural systems that have been developed and applied are few and usually generalized to the extent of referring to certain forest types in large regions rather than to specific forest properties. In application, silvicultural systems include a myriad of additional variations designed to accommodate differences in objectives of ownership, accessibility, site quality, and all the other factors that make every stand different from all others.

Silviculture by Stand Prescription

Silvicultural treatments are best prescribed, stand by stand, by foresters on the ground. However, the general forms of silvicultural systems that may be prescribed and the basic management policies ordinarily have to be determined for the forest ownership as a whole. Some degree of standardization is necessary to ensure uniformity and continuity of action. Too much standardization can lead to treatments that fit well in some stands but badly in others. Results can become especially bad if conformity to standard operating procedures causes field personnel to turn their minds either off or to non-silvicultural problems. Foresters on the ground should see things of silvicultural significance that are not obvious to others viewing stands from roadsides, distant offices, or university campuses.

Foresters at any vantage point are far more able to formulate and prescribe silvicultural treatments than such lay people such as legislators, corporation directors, accountants, or users of forests. However, when these other kinds of people represent the ownership of the forest, public or private, it is they who determine the objectives of management and policies of use of the forest. The foresters should advise and tell what the potentialities are, but they should then be content to execute the policies.

The need for quasi-independent stand prescription may be obvious if one reflects on the fact that one might logically clearcut an unhealthy 35-year-old pine stand yet thin a healthy one even though both are side by side on the same site within the same ownership. A standard procedure for *all* 35-year-old pine stands would dictate that the two be treated the same.

The Silvicultural System as a Working Hypothesis

The practice of silviculture must be conducted in the absence of complete knowledge about the immutable natural and changing social factors that affect each stand. Furthermore most of the treatments cannot be properly evaluated until many years after their application. Decisive action cannot await absolute proof of validity nor can it be evaded indefinitely by fence-straddling. The forester must, therefore, proceed as far as possible on the basis of proven fact and then complete plans for action in the light of the most objectively analytical opinions that can be formed. This combining of fact with opinion is treacherous, especially because the opinions to which one gets committed by actions based on them are easily mistaken for facts.

The soundest basis for action derived from a mixture of proven fact and unproven opinion is the *working hypothesis*, which is not truth but the best estimate of truth formed by analyzing all available information. It is not allowed to become a *ruling doctrine* but is, instead, constantly tested against new information and modified accordingly. It is not embraced so wholeheartedly that it cannot be discarded and replaced. One must always be ready to admit, at least inwardly, that an earlier decision was wrong and correct the procedures accordingly.

The silvicultural system is logically based on a working hypothesis and is altered as it becomes necessary to change the hypothesis. Lest they become ruling doctrines, existing procedures should be constantly examined to determine whether they have outlived their time or become inconsistent with new information. It is, for example, logical to consider whether silvicultural practices developed for the era of railroad logging remain valid. Similarly, it is well to question whether thinning policies should not be altered when it is found that the rate of diameter growth does not, as was once believed, control the properties of wood.

Radical changes and excessive fluctuations in silvicultural procedures lead to confusion and lost motion. If the silvicultural system for dealing with a particular local kind of stand has been kept in conformity to the circumstances by occasional modifications, disruptive reforms are unlikely to be necessary. It is prudent to avoid the petulant temptation to discard tested procedures in favor of those that are radically different and untested because of problems of the moment. If once promising plantations become riddled with root rots, one should not necessarily swear off planting. Neither should the lapse of 9 years between good seed crops cause a wholesale shift to artificial regeneration.

Forestry is almost unique in the extent to which the actions of the present govern those of future generations of practitioners. Any treatment that is applied to a stand now is likely to restrict the choices available in subsequent treatment. In a sense, the forester conducting a treatment in a stand is entering into a pact of mutual understanding with succeeding foresters about the stand. We of the present are entitled to expect that the plans we put into effect now will be given the benefit of all doubts by our successors, but not to the extent of unquestioning adherence.

The results of treatment that are most difficult to change are species composition and age-class structure of stands, so changes in these attributes should be approached with deliberation. It must be recognized that the period of regeneration is nearly the only one during which major changes can be made in stand characteristics and silvicultural systems. The period of intermediate cutting is more one of modification than of change. We do the best we can with stands inherited from the past; those that we start anew provide the best opportunity to inflict new ideas on the future (Fig. 12-3).

The silvicultural system should be built where it is to be used, not prefabricated and brought from some other kind of forest. This chapter has been an attempt to describe the considerations that enter into the construction; the remaining ones tell about most of the general categories of systems that foresters have devised.

FIGURE 12-3 One-and-a-half centuries of evolution in silvicultural practice in Bavaria illustrated in three stands at the Forest of the University of Munich. The Scotch pine stand in the background was established by direct seeding 150 years ago on soils degraded by long periods of overuse. On the left is a Norway spruce plantation established 70 years ago where the initial pines had been harvested and the soil had improved enough to allow the spruces to grow. On the right is a planted stand of mixed hardwoods, 30 years old, which represents the reestablishment of a forest resembling the original one and believed to be resistant to many of the maladies that pure spruce plantations sometimes suffer. (*Photograph by Yale University School of Forestry and Environmental Studies.*)

BIBLIOGRAPHY

Anman, G. D., *et al.* 1977. Guidelines for reducing losses of lodgepole pine to the mountain pine beetle in unmanaged stands in the Rocky Mountains. *USFS Gen. Tech. Rept.* INT-36. 19 pp.

Bacon, W. R., and A. D. Twombly. 1980. National forest landscape management. Vol. 2, Chap. 8, Timber. *USDA, Agr. Hbk.* 559. 223 pp.

Bailey, J. A. 1984. *Principles of wildlife management.* Wiley, New York. 373 pp.

Baker, W. L. 1972. Eastern forest insects. *USDA Misc. Publ.* 1175. 642 pp.

Bega, R. V. (Ed.) 1978. Diseases of Pacific Coast conifers. *USDA, Agr. Hbk.* 521. 206 pp.

Boyce, J. S. 1961. *Forest pathology.* 3rd ed. McGraw–Hill, New York. 572 pp.

Champion, H. G., and S. K. Seth. 1968. *General silviculture for India.* Govt. of India Press, Delhi. 511 pp.

Clutter, J. L., *et al.* 1983. *Timber management, a quantitative approach.* Wiley, New York. 333 pp.

Crowe, S. 1978. The landscape of forests and woods. *U.K. For. Comm. Booklet* 44. 47 pp.

Davis, K. P. 1959. Comments on "What is unit area control?" *J. For.* **57**:517–518.

Davis, K. P. 1966. *Forest management: regulation and valuation.* 2nd ed. McGraw–Hill, New York. 517 pp.

Dengler, A. 1972. *Waldbau auf ökologischer Grundlage.* Band II. *Baumartenwahl, Bestandesbegründung und Bestandespflege.* Revised by A. Bennemann and E. Röhrig. Parey, Hamburg. 264 pp.

Dinus, R. J., and R. A. Schmidt (Eds.) 1977. *Management of fusiform rust in south pines, symposium proceedings, 1976.* Univ. of Florida, Gainesville. 163 pp.

Doane, C. C., and M. L. McManus. 1981. The gypsy moth: research toward integrated peat management. *USDA Tech. Bull.* 1584. 757 pp.

Duerr, W. A., J. Fedkiw, and S. Guttenberg. 1956. Financial maturity: a guide to profitable timber management. *USDA Tech. Bull.* 1146. 74 pp.

Eyre, F. H., and P. Zehngraff. 1948. Red pine management in Minnesota. *USDA Circ.* 778. 70 pp.

Fedkiw, J., and J. G. Yoho. 1960. Economic models for thinning and reproducing even-aged stands. *J. For.* **58**:26–34.

Florence, R. G. 1979. The silvicultural decision. *For. Ecol. and Mgmt.* **1**:293–306.

Furniss, R. L., and V. M. Carolin. 1977. Western forest insects. *USDA Misc. Publ.* 1339. 654 pp.

Gunter, J. E., and H. L. Haney, Jr. 1984. *Essentials of forest investment analysis.* Oreg. State Univ. Bookstores, Corvallis. 333 pp.

Hallin, W. E. 1959. The application of unit area control in the management of ponderosa–Jeffrey pine at Black Mountain Experimental Forest. *USDA Tech. Bull.* 1191. 96 pp.

Hedden R. L., S. J. Barras, and J. E. Coster (Eds.) 1981. Hazard-rating systems in forest insect pest management: symposium proceedings, Athens, Georgia, 1980. *USFS Gen. Tech. Rept.* WO-27. 169 pp.

Hepting, G. H. 1971. Diseases of forest and shade trees of the United States. *USDA, Agr. Hbk.* 386. 658 pp.

Hewlett, J. D. 1982. *Principles of forest hydrology.* Univ. Georgia Press, Athens. 183 pp.

Knight, F. B., and H. J. Heikkenen. 1980. *Principles of forest entomology.* 5th ed. McGraw–Hill, New York. 461 pp.

Knuchel, H. 1953. *Planning and control in the managed forest*. Transl. by M. L. Anderson. Oliver and Boyd, Edinburgh. 360 pp.

Leuschner, W. A. 1984. *Introduction to forest resource management*. Wiley, New York. 298 pp.

Manion, P. D. 1981. *Tree disease concepts*. Prentice–Hall, Englewood Cliffs, N.J. 399 pp.

Mayer, H. 1977. *Waldbau auf soziologisch-ökologischer Grundlage*. Gustav Fischer, New York. 500 pp.

Myers, C. A. 1974. Multipurpose silviculture in ponderosa pine stands of the Montane Zone of central Colorado. *USFS Res. Paper* RM-132. 15 pp.

Nwoboshi, L. C. 1982. *Tropical silviculture, principles and techniques*. Ibadan Univ. Press, Ibadan, Nigeria. 333 pp.

Ohmann, L. F., *et al.* 1978. Some harvest options and their consequences for the aspen, birch, and associated conifer forest types of the Lake States. *USFS Gen. Tech. Rept*. NC-48. 34 pp.

Paulsen, H. A., Jr. 1975. Range management in the Central and Southern Rocky Mountains. *USFS Res. Paper* RM-154. 24 pp.

Sartwell, C., and R. E. Stevens. 1975. Mountain pine beetle in ponderosa pine, prospects for silvicultural control in second-growth stands. *J. For.* **73**:136–140.

Schemnitz, S. D. (Ed.) 1980. *The wildlife management techniques manual*. Wildlife Society, Washington, D.C. 686 pp.

Shaw, J. H. 1985. *Introduction to wildlife management*. McGraw–Hill, New York. 316 pp.

Society of American Foresters. 1981. *Choices in silviculture for American forests*. Society of American Foresters, Washington, D.C. 80 pp.

Spurr, S. H. 1956. German silvicultural systems. *For. Sci.* **2**:75–80.

Thatcher, R. C., *et al.* (Eds.) 1981. The southern pine beetle. *USDA Tech. Bull.* 1631. 266 pp.

Thomas, J. W. 1979. Wildlife habitats in managed forests. *USDA, Agr. Hbk.* 553. 512 pp.

Trimble, G. R., Jr., *et al.* 1974. Some options for managing forest land in the central Appalachians. *USFS Gen. Tech. Rept.* NE-12. 42 pp.

Troup, R. S. 1952. *Silvicultural systems*. 2nd ed. Edited by E. W. Jones. Oxford Univ. Press, London. 199 pp.

U. S. Forest Service. 1983. Silvicultural systems for the major forest types of the United States. *USDA, Agr. Hbk.* 445. 191 pp.

Verner, J., and A. S. Boss (Eds.) 1980. California wildlife and their habitats: western Sierra Nevada. *USFS Gen. Tech. Rept.* PSW-37. 439 pp.

Williams, M. R. W. 1981. *Decision-making in forest management*. Wiley, New York. 143 pp.

Worrell, A. C. 1953. Financial maturity: a questionable concept in forest management. *J. For.* **51**:711–714.

CHAPTER 13

Clearcutting Methods

Clearcutting is conceptually the simplest way of starting the replacement of old stands. It leads to the creation of even-aged stands that are often, but not necessarily, pure. It simulates some of the rather catastrophic events that have led to regeneration of certain species, especially the shade-intolerant and exposure-tolerant ones. Silvicultural systems involving clearcutting may range from the crudest of extensive kinds to the most intensive and expensive. Clearcutting is not the only way of creating even-aged stands; the shelterwood methods and others will be discussed in subsequent chapters.

Semantics

It might seem that no term about cutting practice could be less ambiguous than "clearcutting." Actually it is a semantic morass made even deeper by the negative connotation sometimes attached to the word. As a technical term of silviculture, clearcutting refers to treatments in which virtually all vegetation is removed and almost all of the growing space becomes available for new plants. Timber harvesting alone is seldom sufficient to achieve such complete removal of vegetation; that is really one of the main roles of the site preparation measures that were described in Chapter 8. If this kind of clearcutting is meant it is prudent to refer to it as "silvicultural," "true," or "complete" clearcutting; sometimes the term "clean cutting" is also used.

The use of "clearcutting" as a forester's technical term sometimes extends to operations in which heavy removal cuttings release new stands already established beneath them as advance growth. Use of the term "one-cut shelterwood cutting," discussed more in Chapter 14, has the virtue of focusing attention on the source of the regeneration. This is crucial in those situations in which new seedlings of the desired species are poorly adapted for establishment after the exposure resulting from true clearcutting. It is not merely a matter of academic semantics; many regeneration failures are caused by foresters who have gotten mixed up about the sources of regeneration associated with "clearcutting."

The simple coppice method, considered in Chapter 16, depends on regeneration from vegetative sprouting but the degree of removal of the old stand is so nearly complete that the operation looks very much like clearcutting. However, the term "coppice method" is sufficiently well recognized that this technique is seldom confused with true clearcutting.

One kind of semantic confusion is perhaps inevitable in this terminology. In some mixed stands it is entirely possible to carry out heavy removal cuttings that lead to regeneration arising from advance growth, sprouts, and new seedlings germinating after the cutting. In a certain sense these are combinations of the shelterwood, coppice, and clearcutting methods. Such regeneration is very common in mixed hardwood forests. The noncommittal term "heavy removal cutting" provides a useful solution to this dilemma.

The sloppiest usage of the term "clearcutting" is with reference to logging operations in which only the merchantable trees are cut. If only a few trees are cut from a stand, the result should often be termed selection thinning or regarded as some crude variant of the selection method for creating uneven-aged stands.

Basic Procedure

The true clearcutting method lays bare the area treated and leads to the establishment of an even-aged stand composed of naturally or artificially established trees not there before the cutting. Growing space is made vacant with the goal of refilling it promptly. As a method of regeneration it is logically applied, instead of those involving partial cutting, when residual trees are not worthy of retention for further increase in value, source of seed, protection of the new crop, wildlife habitat, amenity, or other useful purposes. It and the closely related seed-tree method can be used only with species that are capable of establishment, including that done by planting, in conditions of full exposure. There are some very intolerant pioneer species that not only endure such conditions but also, for practical purposes, require them.

Clearly justifiable application of clearcutting is also found in the harvesting of stands that are thoroughly mature or overmature. Clearcutting should also be seriously considered if the methods of partial cutting would be substantially more costly or if this method provides the most expeditious means of replacing a poor stand with a good one. If such conditions exist and the desirable species is ecologically adaptable to the conditions created, clearcutting may be distinctly superior to any other method of regeneration.

The points that have been mentioned so far apply nearly equally both to the intensive method of clearcutting with artificial regeneration and to that involving natural regeneration, which is more difficult yet is often associated with extensive silviculture. Even though artificial regeneration can be and occasionally is used to remedy failures of natural regeneration after clearcutting, it is useful to consider the two procedures separately.

CLEARCUTTING WITH ARTIFICIAL REGENERATION

In this method, simple in concept but sophisticated in detail, the vegetation of the previous stand is removed and replaced by seeding or planting. Many of the details have already been described in the sections on site preparation, nursery practices, planting, direct seeding, and operations for the release of established trees.

These techniques evade many restrictions on silvicultural treatment although the freedom conferred comes with much opportunity for making mistakes. The size and pattern of cutting areas are not limited by the necessity of reserving sources of seed. If the new crop is planted, there is no need to modify procedures to protect the seedlings from those hazards of the natural environment that are bypassed in the nursery. There are no residual trees or desirable advance growth to hamper logging or the use of fire, herbicides, or machinery during site preparation. It is usually much easier, cheaper, and more dependable to dispose of preexisting undesirable plants before the new stand is established than it is afterward.

The composition, arrangement, and density of the new stand are under close control. Opportunity exists for the use of improved genetic strains or of species not found on the site. The risk of delay in restocking of the site is small; in some cases it may even be possible to shorten the whole next rotation by planting trees that are older or larger than would be possible with trees that started from seed right on the site.

Clearcutting with Direct Seeding

The techniques of this procedure were dealt with in Chapter 7 about regeneration from seed and indirectly in Chapter 8 on site preparation. One way to consider its role in silviculture is to observe why it is used only in somewhat special cases. This is ironic because there would seem to be no way of creating new forests that was speedier, simpler, or grander than sowing tree seeds, especially from aircraft.

The chief difficulty is that direct seeding tends to be wasteful of seed. Tree seeds produced in intensive genetic improvement programs are not only costly but often in limited supply. Even with cheaper wild seeds it may be better to sow them in nurseries than out in the forest. As previously indicated, there are also problems with protecting seeds from birds and rodents; seed germination is also heavily dependent on the question of whether rain falls at the right times.

Successful direct seeding can also depend heavily on the right kind of site preparation or on careful placement of seeds in seed-spots. The amount and cost of such work is frequently the same as would be required for planting, which gives more dependable survival.

Because of these considerations, the use of direct seeding in combination with clearcutting is usually limited to rather unusual situations. Perhaps the most notable cases involve species with seeds that are small or easily obtained or both. Aerial seeding of eucalypts, which have very small seeds, is common after clearcutting and broadcast burning in Australia (Hillis and Brown, 1984). The same is true of jack and lodgepole pine, especially in parts of Canada (Smith and Brown, 1984). The same could also be true of such prolific seeders as the birches but known cases are few, perhaps because natural regeneration from seed of these species often appears almost miraculously where it is wanted anyhow, given the exposure of mineral soil.

There have been times when aerial seeding after clearcutting and site preparation was common in the culture of such species as loblolly pine and coast Douglas-fir. During the 1960s the advent of tree-improvement programs with their costly seed caused many public and industrial forestry organizations to substitute planting. Discouragement with the erratic patterns of stand density associated with seed regeneration also played a role in this shift. However, there are so many logistical problems involved in nursery operation and planting that some sort of continuing and developing use of direct seeding should be anticipated.

Direct seeding is also used sometimes as an emergency measure to reforest lands that have been devastated by wildfire, shifting cultivation, or other forest-killing disturbances beyond the control of foresters. This kind of emergency response does not have to wait upon the expansion of nurseries and other kinds of logistic support that would be necessary with planting.

Clearcutting and Planting as a Mode of Intensive Silviculture

One of the most important applications of clearcutting is in a kind of plantation silviculture that carries much of the technique of agriculture into forestry. For some foresters and in some circumstances this is the only kind of silviculture worth considering; in some naturalistic circles it is an anathema. Matters of opinion aside, the ideas underlying this approach must be recognized and understood. Neither this nor any brand of silviculture is a complete, unified system or ritual that must be applied as a whole or not at all. It is possible to apply the individual pieces selectively.

The establishment of new stands starts with complete removal of the previous crop. The growing space is made vacant by the use of machinery, herbicides, fire, or combinations thereof in site preparation. One of the common goals of the site preparation is reduction of physical impediments to planting. If the soil moisture conditions can be improved by reshaping the surface terrain, this may be done also. The trees that are then planted, usually in pure stands, represent the best genetic material available and are carefully fitted to the sites.

The growth of the planted crop of trees can often be improved by reducing any competing vegetation that survived the site preparation or came in after the planting. Deficiencies of chemical nutrients can be remedied by fertilization, although this is more often done after the stands have closed than at the time of planting. Thinning is an important part of any kind of truly intensive plantation silviculture, although the objectives and methods can vary widely. When the stand is deemed to be mature, ordinarily on financial grounds, it is replaced by a new planted stand and the whole cycle is repeated.

As little as possible is left to chance or nature. Planting plays a key role because it is a much more dependable kind of regeneration than that from natural or artificial seeding. Planting represents one way of shortening the period during which the stand does not occupy the site fully and has thus not resumed full production. The loss of even one year's growth can be a significant fraction of the total production of a rotation, especially if the rotation is short. Some gain in forest production, or shortening of the rotation, may be made possible if the seedlings are already one or more years old when moved from the nursery. The gain of one year with a 50-year rotation is a 2% increase in the production secured on each unit of land area. Prompt regeneration also ensures capture of the benefits of site preparation. Delay in regeneration increases the risk that the growing space will be usurped by undesirable vegetation.

Another major advantage is that with planting it is possible to govern the density, spacing pattern, species composition, and genetic constitution of the new stand more precisely than is possible with any other regeneration method. There is the greatest possible freedom for the use of various kinds of machinery, often powerful, in harvesting, site modification, and planting in ways that are similar to or borrowed from those used in agriculture. There can be a deliberate effort to reduce the need for tinkering with undesirable vegetation or erratic stocking after stand establishment. Sometimes there is even the hope that everything can be set up so well that the stand can be left alone, except for protecting it, until the clearcutting and replanting at the end of the rotation.

An effort is often also made to secure the closest possible approach to uniformity of species and tree size so as to reduce the cost and complications of harvesting and processing the timber crops.

Policies about Application

The whole approach is often associated with very forthright ideas about achieving definite improvement upon nature. One manifestation of this, though not a necessary part of the approach, is the tendency to plant exotics. The general approach is a very positive one with a powerful appeal. It seems to be the only possibility to those who believe that useful plants will not appear unless planted; there are certainly times and places where this view is correct.

The policy also has certain psychological dimensions. Anyone who tries to regenerate forests by planting trees cannot be accused of having been lazy or of leaving anything to the caprice of nature. Furthermore, there is something heart-warming about planting trees regardless of financial or ecological attitudes about the practice. The actions of people who are dedicated to planting can be ascribed to this attitude as much as to either cold calculation or sheer impatience with nature.

As the supply of wood decreases and as prices increase, the use of clearcutting and planting becomes greater and on a worldwide basis. It has long been the most common practice with most of the species of Europe and Japan. The use of the technique in growing loblolly and slash pine on industrial lands in the southeastern United States (Wahlenberg, 1960 and 1965) is the most important American manifestation of the general use of this silvicultural method with members of the hard pine group throughout temperate forests. In almost all Mediterranean climates, such as those of parts of the Southern Hemisphere and Spain, there is the plantation culture of the California *Pinus radiata* (Monterey pine) as an exotic. Clearcutting and planting has come to be the most common way of regenerating coast Douglas-fir in the Pacific Northwest (Fig. 13-1) (Cleary, Greaves, and Hermann, 1965). Disenchantment with efforts to control species composition by partial cutting in many forests of the humid tropics has led to much reliance on plantation silviculture.

Since the technique requires large investments right at the beginning of the rotation it is, not surprisingly, applied mostly by owners capable of long-term investments and with species that grow to merchantable size quickly. The owners are most likely to be forest industry corporations and public agencies that have a certain kind of immortality. Owners with shorter time outlooks ordinarily use this approach only if there is some direct or indirect public subsidy.

Even the more nearly immortal kinds of owners are usually sufficiently concerned about early return on their investments that they prefer to plant species that obviously grow quickly in height and diameter. For this reason there is a bias toward intolerant species of early successional status rather than toward the more shade-tolerant species that are adapted to start in the shade and grow slowly in the early stages.

With planting there is an incentive to use species that race for the sky and develop large stems quickly. So great is this incentive that it is sometimes pursued at the expense of mean annual production per unit of land surface. Those species that grow most rapidly in diameter and height, contrary to intuitive belief, often have comparatively low mean annual increments of total wood per acre. Sometimes both kinds of species can be planted together so that one gets the best of both worlds, a point that will be taken up in Chapter 17. There is no fundamental reason why plantations established after clearcutting must be pure; in fact, mixed plantings of this origin are fairly common in Central Europe.

The technique of clearcutting and planting originated as a means of

FIGURE 13-1 Clearcutting of over-mature old-growth and planting of Douglas-fir, Willa-mette National Foreset, Oregon. The upper picture was taken 4 years after planting when most of the trees were still overtopped by the herbaceous plants that appeared after the slash burning done just before the planting. The lower picture shows the same place when the plantation was 12 years old. (*Photographs by U. S. Forest Service.*)

replacing poor or unstable stands with new, good ones. This is still its most common role. Much of the plantation culture of southern pines is still the result of efforts to replace stands that had degenerated to mixtures of poor hardwoods poorly adapted to the soils and degraded by high-grading and fires. Much of the clearcutting and planting being done in the western United States involves stands that are too old and decrepit to endure regimes of partial cutting.

Sometimes it is significantly cheaper to clearcut and plant than it is to take various special measures to secure natural regeneration. The whole process of removing the previous crop and establishing a new one should be viewed as a financially integrated operation regardless of how it is administered. The exigencies of timber harvesting are most likely to favor clearcutting and planting if the machinery is ponderous, roads are costly, and the terrain is difficult. It is not purely a matter of harvesting and planting because the costs of site preparation usually loom large where the method of clearcutting and planting is used.

Even though this sequence of operations can seem very costly, there are still instances in which it may be the cheapest way of replacing an old stand. However, the financial analysis of the choice often depends heavily on how the cost of machinery is depreciated or whether the cost of regeneration is charged as an obligatory expense of removing the old crop or as an investment made in the new crop and carried at compound interest.

In many instances the benefit purchased by planting is the increased assurance that there will be regeneration of the right kind; this is as much a matter of human psychology as something that can be measured in terms of wood production or money. The opposing view that this mode of silvicultural procedure is unwise also comes from mixtures of psychological considerations with matters of natural science and economics.

CLEARCUTTING WITH NATURAL REGENERATION

It is possible to secure natural regeneration after clearcutting, although it is only under quite special circumstances that there is much assurance of success. This way of applying the clearcutting method has little in common with that involving planting. In both kinds of application one seeks to establish the regeneration as promptly as possible on some area that is nearly free of vegetation and exposed to sun and sky. Each also allows much latitude for disturbance of soil and vegetation during harvesting and site preparation. However, clearcutting with natural regeneration is usually associated with comparatively extensive, low-investment silviculture.

Real clearcutting exposes the site so much that it is compatible with natural (or artificial) seeding only when the species are capable of enduring exposure. This means that it is not suitable for certain shade-tolerant but exposure-intolerant species unless they are planted (and sometimes not even then). This attribute of clearcutting can, however, be of some advantage if the object is to encourage species of early successional status and discourage seed regeneration of shade-tolerant species that happen to be undesirable. This effect is usually perceived only when one examines clear-cut areas and observes what has *not* regenerated well; the unsuccessful species probably disappear when too young and small to be noticed.

The prospect of success with natural regeneration after clearcutting and the ways of securing it vary widely depending upon the source and means of dispersal of the seeds. These distinctions are so important that each will be considered separately.

Clearcutting with Seeding from Adjacent Stands

Sometimes it is possible to clearcut a stand and depend upon seeds dispersed from some nearby, untreated stand to provide the regeneration. With such an approach, the agency of dispersal is all-important. In the temperate zone, the pioneer species best adapted to this kind of silviculture are disseminated by wind. In the tropics, where it is less windy, birds, rodents, and bats may play a more important role but there are still a number of wind-dispersed pioneer tree species. There are even species of river floodplains, such as the tupelo gums of the bottomlands of the Southeast, that can be regenerated after clearcutting by floating seeds.

Wind, water, and animals can carry a few seeds over astonishingly long distances. However, in regenerating forests, it is usually necessary to remember that the density of seeds deposited on a unit of ground surface varies inversely as some power of the distance from the source. The rest of this discussion about dispersal of seed into clearcut areas involves wind, but some of the same ideas may apply to dissemination by other agencies.

Wind-dispersed seeds usually have wings and moving air is fortunately turbulent enough that some are wafted aloft rather than dropping only downward after they are released from the cones or fruits. Where surrounding trees are the only source of seed, the clearing must be sufficiently small (usually long and narrow) to allow for adequate dissemination to all points. Safe widths for clearings to be stocked by wind-disseminated seed are likely to range, depending on species, from *one to five times the height of the adjacent timber* from which seed will be obtained. Direction of the wind at the time of seed dispersal should be known, and the clearing should be so located that its long axis is at right angles to this direction. Unless the clearcut area is very narrow, the distribution of regeneration is likely to be uneven, as shown in Fig. 13-2.

Most dissemination occurs during dry, sunny weather when the winds are brisk and gusty. The most effective winds are frequently those that blow out of the dry interior regions of continents; they are not necessarily of the same direction as the so-called "prevailing winds." In rugged terrain, it is best to have the long axes of the openings run at right angles to the contour lines because the winds responsible for dispersal of seeds are usually altered in direction so that they blow up or down valleys. Where other considerations dictate that the clearings be oriented in a manner less

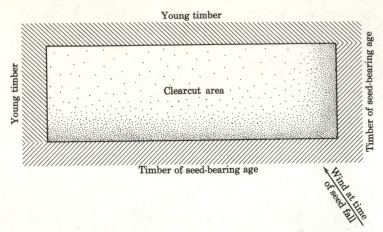

FIGURE 13-2   Clearcutting the whole stand, with reproduction secured by seed disseminated from trees located outside the cut stand. The density of the reproduction 5 years after the cutting is indicated by the dots.

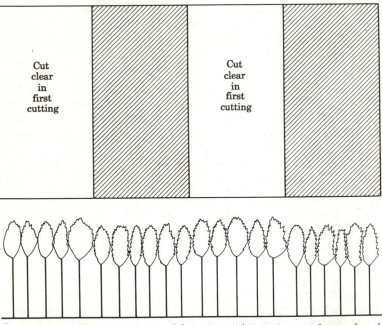

FIGURE 13-3   The arrangement of the strips within a pine stand reproduced by clearcutting in alternate strips 80 feet wide. The upper sketch shows the horizontal arrangement of the two sets of strips. The lower sketch shows a vertical profile of the new stand 50 years after the first cutting and 45 years after the second. The older trees are about 80 feet tall, whereas the younger have an average height of 75 feet; this difference would be of little significance in practice.

favorable from the standpoint of seed dispersal, the dimensions of the openings should be correspondingly reduced.

The width of the clearcut areas is often restricted by distances of effective seed dispersal but sometimes also by the desirability of retaining enough side shade to keep out pioneer weeds or to favor species not fully resistant to exposure. Marquis (1966) has shown how control over the width of openings can determine whether yellow, paper, or gray birch dominate regeneration in some northern hardwood forests.

Strip-clearcutting

If it is desirable to regenerate a stand by clearcutting, all within a period of several years, but the stand is too wide to clearcut all at once, it may be possible to do the clearcutting in alternate or progressive strips. With the **alternate-strip method,** the stand is clearcut in two stages (Fig. 13-3). During the first cutting, some strips are cut clear but intervening strips are left as sources of seed; the reserved strips are generally no wider than they absolutely must be. When seedlings have become established on the clearcut strips, the originally reserved strips are clearcut. The regeneration of these must come from seed on the trees that are cut or from artificial regeneration. The clearcut strips are best oriented so that they are at right angles to the direction of seed-dispersing winds or to that of the midday sun.

The **progressive-strip method** is a means of accomplishing the same objective but in three or more stages and with a lower proportion of the area in the awkward status of the final strip. Figure 13-4 is a schematic diagram of this kind of cutting sequence.

If, as is often the case, there is serious risk of wind damage to the residual strips, it may help to cut the strips in a sequence that advances toward the direction from which the most dangerous winds are likely to blow, that is, from leeward to windward. In this way the amount of edge of newly exposed but uncut timber is minimized. If, as is common in much of

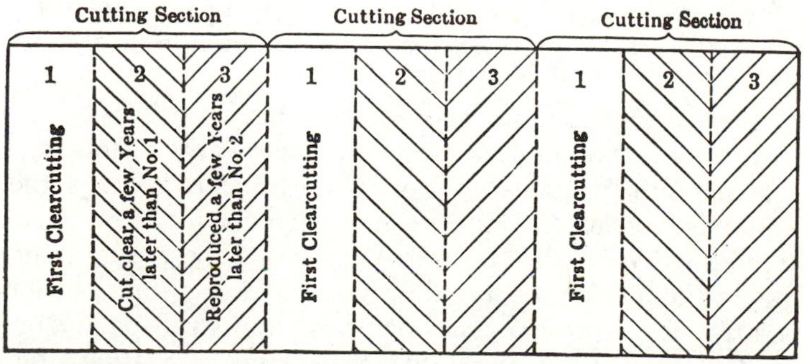

FIGURE 13-4 Clearcutting a single stand in progressive strips, using three cutting sections. The last strip (Number 3) in each section may be reproduced by a method other than clearcutting.

North America, the most dangerous winds come from the southwest, it becomes logical to advance the strips from northeast toward the southwest. This may also provide some partial protection against the slanting rays of the afternoon sun but does not necessarily coincide with the dry winds of seed dispersal. Along the Atlantic seaboard the most dangerous winds, those of tropical hurricanes, are from east or southeast and this would call for different strip orientations. In strip-clearcutting it is highly desirable that the windward edge of the last strip that is to be cut be composed of trees that are strong enough or deeply rooted enough to be unusually windfirm.

Clearcutting in Patches

On rugged and irregular terrain or in even-aged stands that lack uniformity, the regular arrangement of cutting areas employed in the strip methods is impractical. Some of the advantages of these methods may be retained, however, if the stand is removed in a series of clearcuttings made in patches.

In the first cutting, portions of the stand are selected that for some reason should be cut before the rest of the stand. In succeeding operations, these patches are enlarged or new patches are created elsewhere in the stand. The first areas to be cut may be ones with injured or overmature trees, places where previous disturbances have allowed the establishment of advance regeneration, or soils where the trees are shallow-rooted and subject to windfall if exposed by cutting of adjacent areas.

The advantage of these modifications of the clearcutting method is that they enable this method to be employed in regenerating even-aged stands that cannot be reproduced by natural seeding if all the timber is removed in a single cutting. The desirability of recognizing them as distinct methods of cutting rather than as series of unrelated clearcuttings of separate stands lies in the fact that, if conducted as outlined above, each leads to establishment of a single, relatively homogeneous stand. With the passage of time, the groups of new trees on each strip or patch tend to approach the same size and become indistinguishable. The difficulties of administering a forest consisting of a patchwork of minute stands are thus avoided.

Reliance on long-distance seed dispersal is such a risky proposition that it can seldom be counted upon to cause early restocking of large clearcut areas with desirable species. Unfortunately most of the species capable of this spectacular kind of reforestation are short-lived, weedy kinds of pioneers. With most desirable species it is best to think in terms of using this technique for regenerating narrow clearcuttings or merely the edges of large ones. One drawback of clearcutting in strips that are narrow enough is that most of the operational advantages of treating large areas at once are sacrificed.

Clearcutting with Regeneration from Seed on the Site

The restrictions set on the size of clearcut areas by seed-dispersal distances can clearly be avoided if one can rely on a supply of seed on the site. With almost any species, a substantial amount of seed may come from the trees removed in clearcutting, provided that the cutting is made during a good seed year. The seeds may still be on the trees at the time of cutting, or they may have fallen recently. With most species the only seed from the cut trees that is useful for regeneration is that produced in the current year. Certain softwoods, however, have stored seed in **serotinous** cones that open gradually over a period of years, as is true of black spruce, or that rarely open to any extent except under the influence of high temperatures. The most prominent species in the latter group are jack and lodgepole pine; some other less important hard pines that also reproduce in nature after crown fires have this characteristic. Seed stored in serotinous cones remains viable for many years, but special measures are often necessary to secure the release of this vast quantity of seed after cutting.

Regeneration arising from seed produced by the trees removed in clearcutting is uniform and complete if the conditions created by the cutting are subsequently favorable. For this reason it is distinctly advantageous to carry out any method of regeneration cutting when abundant crops of ripe seed are on the trees or have recently fallen.

Seeds Stored in the Forest Floor

Adequate amounts of seed of a *few* species may remain viable in the humus layers beneath uncut stands for periods longer than one year. One example is Atlantic white-cedar, the seeds of which are stored in large quantities in the poorly aerated peats on which this species grows (Little, 1950). Another is yellow-poplar (Clark and Boyce, 1964), which seems to have some sort of requirement for high temperature or exposure to light to germinate. The seeds of the ashes remain stored in the forest floor for one year because it takes them that long after they fall from the trees to mature. With such species it is often possible to expect seemingly miraculous regeneration after clearcutting the entire source of seed. However, it is well not to proceed without being sure that the seeds are there.

Unfortunately long-term storage of seeds in the forest floor is more characteristic of undesirable pioneer trees, annuals, and shrubs than of commercial tree species. Many of these weeds have seeds with hard, nearly impermeable coats that allow the embryos to survive for the many decades that may elapse between major fires or similar regenerative disturbances. They germinate or flourish long enough to produce seeds and then die; they may also lapse into the understory. Examples of such species are the genera *Rubus*, *Ribes*, and *Ceanothus*, as well as the short-lived pin cherry.

APPLICATION OF CLEARCUTTING METHODS

This technique of establishing new forests on areas that have been made nearly vacant of vegetation is best adapted for species that reproduce naturally after fire or other drastic disturbance. Such species are almost invariably airborne, being disseminated by wind or winged animals; they usually represent early or intermediate stages in succession and often form pure stands.

True clearcutting simulates the kind of fire that burns when large amounts of fuel are put onto the ground by windstorms or pest attacks. This kind of holocaust played a definite role in the regeneration of some species in a pure state of nature, although few are absolutely dependent on such events.

One species that is dependent is the mountain ash eucalypt of south-eastern Australia (Hillis and Brown, 1984). Investigations in Tasmania have shown that it is necessary to create substantial amounts of dry fuel for prescribed burning to prepare seedbeds after clearcutting. The fuel includes not only the logging debris from the ashes but also that from the deliberate felling of small non-eucalypts of the understory. The resulting hot fires induce mountain ash regeneration without destroying much of the organic matter incorporated in the mineral soil. Less severe fires induce regeneration of less desirable eucalypts.

Most regeneration associated with clearcutting simulates the effects of cooler fires that have far less fuel to burn and do not actually or immediately kill all of the trees on the burned areas (Fig. 13-5). The most common criterion of adaptability to clearcutting is the capacity of a species to grow faster in height than undesirable competitors. This is why clearcutting is not well suited to those kinds of shade-tolerant species that grow slowly in the juvenile stages, even if they are planted. They grow so much more slowly than the associated pioneer vegetation induced by clearcutting that they require much release work.

Evolution of Silviculture in the West Coast Douglas-fir Region

This provides a useful example of the various roles, advantages, and shortcomings of true clearcutting. Under entirely natural conditions, coastal Douglas-fir regenerated after lightning fires; the sources of seeds were scattered seed trees or patches of trees in swales that escaped fires or were not promptly killed by them (Fig. 13-5). Douglas-fir, since it is normally the fastest growing conifer of these forests, tended to dominate for the first century of stand development. The Douglas-firs then dwindled from the attrition of various lethal agencies and were gradually replaced by western hemlocks, true firs, and western red-cedar that became established in the understories at the time of the initiating fires or afterwards.

Early in this century these forests were being liquidated by railroad logging in combination with cable yarding. This meant that clearcut areas

FIGURE 13-5 Excellent natural regeneration of Douglas-fir 11 years after the Yacolt Burn of 1902 on the Gifford Pinchot National Forest in southwestern Washington. The seed came from scattered living trees like those in the background. A second fire at this stage would kill the reproduction and the old residual trees, allowing the area to revert to brush. (*Photograph by U. S. Forest Service.*)

were extended progressively along the logging railroads. Even though only the best trees were harvested, the shifting of the yarding cables usually pulled the rest over creating huge volumes of slash. Society then had no concern for forest regeneration but it did have the pathetic hope that laws requiring broadcast burning of slash might forestall the dangerous conflagrations of the time.

Foresters had little to say about what was going on and, even at that, tended to be complacent about the very large size of the clearcut areas. There was the mistaken view that the seeds of Douglas-fir from the old stands would remain stored and viable in the forest floor for many years and during the clearcutting and slash burning. This view was based on the observation of the fine Douglas-fir regeneration that had appeared after fires in uncut stands. Foresters had discounted the importance of wind dispersal of seeds from adjacent stands or from scattered survivors such as those shown in Fig. 13-5. They did correctly perceive that a quick repetition of tree-killing fires would eliminate the conifers because the seed supply, whatever its source, would be destroyed. Unfortunately the otherwise good research that had been done about the regeneration process had omitted any good test of the long-term viability of Douglas-fir seeds.

By the 1930s it was shown (Isaac, 1943) that retention of living seed sources was crucial. Foresters began to have some modest influence on harvesting practices and there was interest in regeneration provided that it cost little. For the first time there were tractors and trucks powerful enough to move the huge old-growth trees that could previously be logged only with cumbersome steam-driven machines. However, only the biggest and best trees were worth enough to be harvested.

These circumstances set the stage for an episode of ill-fated efforts to prolong the lives of old-growth stands by light partial cuttings referred to as "selective logging" (Fig. 13-6). The cuttings were intended to operate as selection cuttings to create uneven-aged stands. The only trees cut were commonly those roughly 3 feet in diameter; these were mostly the scattered surviving Douglas-fir, which, in spite of their great size and age, were still the most durable trees in the old stands. Unfortunately most of the remaining trees were western hemlocks and true firs already weakened by age and decay. The usual result was an acceleration of the process by which wind, fungi, and bark beetles break up ancient stands. By the 1950s the whole episode was recognized as a fiasco to be quietly forgotten.

During the next phase, clearcutting returned to fashion. By then it had been found, at least on the most common kinds of sites such as those in western Washington, that reasonably good natural regeneration could be obtained from windborne seed if the clearcuttings were less than about 100 acres in size. It was also found that if a few seed-trees were left within the cutting areas, the regeneration was better than when reliance was placed on adjacent uncut stands.

However, by the 1960s, the attitude about forest regeneration had changed to the point that state laws required it. The exigencies of cable logging in old-growth timber also made it desirable to clearcut areas too wide for natural seed dispersal. There was, therefore, a period during which aerial direct seeding was a common way of regenerating after clearcutting and broadcast burning of slash. Planting (Fig. 13-1) has subsequently become very common because of the willingness to invest money to be sure of prompt regeneration of properly spaced trees of the best available genetic qualities.

Nevertheless, the evolution of silvicultural practice does not halt. It was soon found that the routine of clearcutting, burning, and planting was not a universal solution for the silviculture in the West Coast Douglas-fir Region. It tended to fail when efforts were made to extend it to atypical sites.

The most complete failures were at high elevations in the Cascade Range, where there are severe microclimatic extremes and true firs abound. The most important failures were on certain sites in southwestern Oregon where the rainless summers are longer than they are farther north (Seidel, 1979). This silvicultural routine often produces eruptions of red alder and undesirable shrubs when it is extended westward into the narrow coastal fringe, where fog-drip supports luxuriant stands of western hemlock and Sitka spruce, which are advance-growth species.

FIGURE 13-6 "Selective logging" in old-growth stand of Douglas-fir, western hemlock, and true firs on the Olympic Peninsula, Washington, in the middle 1930s, before and after cutting. Unfortunately this alternative to clearcutting started the accelerated decline of most stands of this kind. (*Photographs by U. S. Forest Service.*)

Another general problem with clearcutting in the Northwest, especially on public lands, has been popular displeasure over the ugly appearance of clearcut areas. The result of these problems has been an upsurge in use of the shelterwood method (Figs. 14-1 and 14-9). This technique is difficult enough to apply that there is less risk that it will become a panacea.

Application for Southern Pines

Somewhat the same sequence of events has taken place in the development of the silviculture of southern pines on industrial lands of the southeastern Coastal Plain. Initially there was widespread devastation from the combination of clearcutting and annual burning to improve grazing. Most first attempts at long-term silviculture took the form of selection cutting. These proved to be fairly effective as a means of managing existing growing stocks but were not very compatible with the hardwood control measures necessary to establish pine regeneration.

During the period from 1950 to 1970 there was, on industrial forests, the same quick progression from one means of establishing even-aged stands to another. Initially the seed-tree method was the solution but this was generally replaced by aerial seeding when it was concluded that each technique required the same amount of site preparation. Then, when there was willingness to spend money for prompt regeneration with well-spaced pines of good genetic qualities, planting became the common technique. Since southern pines are comparatively resistant to exposure there has not been the necessity to use the shelterwood method or other kinds of partial cutting to protect regeneration. When such methods are used it is really for meeting management objectives other than or in addition to regeneration.

One useful illustration of the way silvicultural prescriptions should be fitted to the circumstances comes from observations (Lotti, 1961) in the well-watered loblolly pine stands of coastal South Carolina. Here it is possible for systematic programs of thinning and prescribed burning to enable natural regeneration by clearcutting throughout more than half of almost any year. The development of good seed-bearers by thinning provides ample seed for regeneration almost annually. Regeneration is then effective if litter and understory hardwoods are sufficiently reduced by prescribed burning. Harvests between April 1, the time of germination, and August 1, when the seedling stems harden, are the exceptions; during these 5 months it is necessary to reserve seed trees to allow for losses from logging damage to succulent seedlings. During the remainder of the year either the newly fallen seeds or the hardened first-year seedlings provide enough regeneration to allow clearcutting.

Actually it is possible to obtain regeneration of the southern pines by almost any method that provides for controlling competition from hardwoods if there are pine seeds or seedlings on the ground (Hu, Lillieholm, and Burns, 1983).

Closed-Cone Pines

The species of pines that have serotinous cones are the one important group that is generally more satisfactorily regenerated by clearcutting than by any other method. If the cones are truly serotinous the seed crops of many years are stored in the crowns of the trees and there is no necessity of reserving standing trees. The main problems are to get the cones to open and to expose enough mineral soil. In fact, if all of the cones on standing trees remain tightly closed, it may be futile to reserve any trees strictly for a source of seed. Since it is rarely feasible to duplicate the severe crown fires that bring these stands into being in nature, the effects of the fires must be simulated by other means.

Reasonably successful techniques for this purpose have been developed for jack pine in the Great Lakes Region (Cayford, Chrosciewicz, and Sims, 1967) and sand pine on old deposits in Florida. The lodgepole pine of the Rocky Mountains is quite variable in the extent to which it exhibits the closed-cone habit and must be handled accordingly (Lotan and Perry, 1983).

The first prerequisite is exposure of the mineral soil by mechanical scarification. If most of the cones are tightly closed, it is necessary that cone or cone-bearing slash be scattered over the area in such a manner that a sufficient number of cones lie exposed to the sun within 15 cm of the ground. Within this zone of sluggish air movement it is possible for temperatures to increase to about 50°C, which is the melting point of the resin that seals the cone scales. The necessity of such high surface temperatures poses a dilemma because they cause heat injury to the seedlings. Consequently it is important that some of the seeds fall into small spots shaded by debris.

It would be ideal if the scarification could be done before clearcutting, but it is usually more expeditious to do it afterward or by appropriate modifications of skidding procedures. The most common technique is to disk the areas after cutting and then lop and scatter the slash; if these steps are reversed the cones are likely to be buried unopened. Success can also be achieved if the slash is bunched with bulldozer root-rakes provided that enough cones are broken off and deposited on scarified soil. Sometimes the dense little clusters of seedlings that arise from the opening of cones on the ground pose a difficult kind of precommercial thinning problem. If slash burning is necessary it must be limited to the part of the slash that has been bunched into concentrated areas; broadcast burning destroys too many seeds.

These species display regional variation in the degree to which the closed-cone habit is manifest. If the cones do not remain closed and store seeds, it is necessary to employ some sort of partial cutting or to clearcut in narrow strips and thus depend on seeds from current production. The assumption that these species are always serotinous has led to regeneration failures in cases where reliance was placed on stored seeds that did not exist (Wyoming Forest Study Team, 1971). If some cones open and some do not, it is possible to compromise between the two approaches. Variations in cutting practice, degree of scarification, and slash treatment can be employed to reduce the risk of getting excessively dense natural regeneration that is likely to stagnate. However, if all the cones are truly serotinous, definite efforts must be made to open them and the difficulties are compensated only by the fact that the size, shape, and arrangement of clearcut areas can be governed entirely by considerations other than seed dispersal.

An important incidental advantage of clearcutting in the management of lodgepole pine is that it provides one of the surest means of eradicating infestations of the debilitating parasitic seed plant, dwarf-mistletoe (Hawksworth and Sharf, 1984). Since the seeds of the pathogen must come from living trees, it is necessary only to prevent slow reinfestation from adjoining stands adjacent to the clearcut areas. With partial cutting, the mistletoe is very likely to spread from the old stand to the new unless great care is taken to cut all the infected trees of the previous crop (Baumgartner, 1973).

The seeds of serotinous conifers rarely exhibit dormancy so there is little natural control over the season of germination. Therefore, it is necessary to time the release of seed so that the seedlings will germinate at the proper season. With jack and lodgepole pine, the seeds must germinate before summer so that the seedlings will harden before the first frost. With sand pine in Florida, however, it is best to schedule scarification and slash treatment such that seedlings commence development during late fall or early winter when rainfall is adequate to prevent drought and the risk of heat injury is lowest. It is indeed remarkable how much careful effort is necessary to obtain natural regeneration of these species that are adapted to reproduce themselves so easily and abundantly after catastrophic fires.

ENVIRONMENTAL EFFECTS OF CLEARCUTTING AND HARVESTING

No silvicultural practice excites more controversy than clearcutting (Hermann and Lavender, 1973; Roach, 1975). Its advocates are usually concerned about considerations such as achieving efficiency of the various operations involved in harvesting and silvicultural management, developing uniform stands, growing shade-intolerant species, and, where planting is involved, improving the certainty of regeneration. Its opponents often decry the problems that result from making the growing space vacant, the bad appearance of clearcut areas, the incompatibility of the method with growing some exposure-intolerant species, and the pure stands often created by this method.

Both groups tend to make much of certain carefully selected ecological principles (Barney and Dils, 1972). The advocates sometimes contend that many species are so intolerant of shade that their regeneration absolutely requires the exposure associated with clearcutting. The opponents often claim that, if the disturbances of silvicultural practice are kept small and light, natural tendencies will lead to some idyllic, intricate balance of species and age classes.

The term "clearcutting" has a bad connotation because of guilt by association with certain devastating practices and also because of the intuitive opinion that any practice that looks so bad must have many baneful effects. Even foresters share enough of these apprehensions that they tend to be hasty in blaming bad effects on clearcutting, especially that associated with planting, when the real causes may be something else.

Aesthetics

Some foresters also have a trained incapacity to perceive ugliness in logging. To most people, all cutting areas look bad and those created by clearcutting look worst of all because they are so highly visible. They are especially visible in mountainous or hilly terrain; it does not take much screening to hide them on flat land. People regard unutilized logging debris as more objectionable than the cutting itself. To many it is not only ugly but also evidence of slovenly wastefulness; no amount of learned talk about the economic factors that distinguish waste from utilizable material can altogether overcome this opinion. The problem is mostly an aesthetic one not limited to clearcutting.

Aesthetic problems have cosmetic solutions. Cutting areas can often be hidden by leaving screens of trees along roads and trails. It is especially useful to locate log-loading sites behind such screens. The screens can be removed when the new stands that have been created behind them become obvious. Where the clearcut areas cannot be hidden, as on slopes, it is well to apply the artistic principles of landscape architecture to the alignment of the edges of the cutting areas; the only common reason to have artificial-looking straight lines is the need to cut up to property boundaries (Crowe, 1978). Sometimes the shapes of cutting areas, which are stand boundaries with clearcutting, can be made to imitate some other pattern visible in the adjacent forests. It is worth noting that the boundaries between different site conditions, which are the ideal boundaries between stands, are seldom perfectly straight and almost never define perfect squares or rectangles.

Guilt by Association

Clearcutting is often associated with all sorts of abuse of land and vegetation. Paradoxically almost none of these things have anything to do with clearcutting as a conscious method of regenerating stands of trees.

It is a common belief that tree cutting caused the sad history of soil damage and erosion in the Mediterranean Region, East Asia, and many wet and dry tropical areas. In virtually all such cases the damage to the soil resulted from what people did *after* removing the forests. Grazing can be especially harmful, particularly along the dry fringes of forest areas where it is a characteristic way of harvesting vegetation and where forest vegetation has a tenuous foothold. Goats have probably destroyed more forests and caused more erosion than anything else. Repeated burning done to encourage grass often ensures deforestation after heavy cutting because that is the very purpose of the burning.

Vast areas of forest throughout the world have been clearcut not to regenerate forests but to use the land for agriculture. It is unfortunately true that most forms of agriculture cause some sort of degradation of the soil. It is counterproductive to be complacent about these effects, which really threaten civilization, while decrying actions taken to grow trees, which are the best soil defenders available. Silvicultural practices feed the soil with organic matter while clean-cultivation agriculture depletes the supply.

Overgrazing, overbrowsing, and overfarming of land are often associated with overuse of whatever forests may remain in overpopulated localities. These forests may be cut at very short intervals for fuelwood; this practice usually takes the form of coppice cutting and encourages increase in sprouting species, especially if the lands are burned frequently to favor grazing. The ultimate insult to the system is the gathering of forest litter for fuel, stable bedding, or agricultural mulches. The fact that such sequences of degrading events may start with clearing off the trees has nothing to do with clearcutting as a silvicultural system.

The kind of cutting done in the liquidation of forests without plan for future production has often been called clearcutting even when only the merchantable trees are cut. The expression "clearcut and get out" succinctly describes the policy followed in the United States while the lumber industry chopped its way through virgin forests from the North to the South to the West. It is still common for loggers and even foresters to refer to partial cuttings in which all usable trees are cut as "clearcutting." Regardless of whether this kind of cutting practice is wise or not, it seldom destroys forests if it is the only treatment that takes place. However, if this sort of timber harvesting is followed by repeated fires, as was often the case in the Great Lakes Region, the old longleaf pine forests of the South. or parts of the West, then the return of true forest vegetation can be long delayed. Such events were usually associated with the hope that the forests were being replaced by farms.

It is nearly impossible to eradicate forests by cutting alone. The exceptions exist on sites that verge on being too dry, too wet, or too cold for trees to grow; in such places, shrubs, grass, or grasslike plants may replace the trees after severe cuttings. Even on the susceptible sites it usually takes combinations of cutting and fire to produce this result. There are plenty of cases in which reckless cutting can degrade forests but few in which it really destroys the forest.

"Clearcutting" has such a bad connotation that it makes a handy whipping boy for those who want to publicize their unhappiness with any aspect of forest management. A 10-second television glimpse of an untidy clearcut area is often sufficient to convince people that the practice is devastating regardless of scientific evidence to the contrary. Often the issue is not clearcutting at all but the desire to have more forest land for primary or exclusive use for wildlife, wilderness, or similar uses that seem threatened by silvicultural treatments that involve cutting trees. Sometimes the critics are genuinely concerned about the possibilities of erosion, floods, landslides, loss of rare species, nutrient loss, and the violation of some hypotheses about ecology. There is chronic concern about the pure, even-aged stands often created by clearcutting. Much of this kind of fear is groundless; some is not, but only under certain circumstances.

True clearcutting sometimes causes regeneration failures by exposing seedlings to too much sun, frost, or wind. This is especially true with species that are adapted to start as advance growth beneath the old stands. These effects can often, but not always, be overcome by planting. The elimination of seed sources, possible with any mode of cutting but most likely with clearcutting, can also result in the failure of natural regeneration of desired species. The common consequence of regeneration failure after clearcutting is not only loss of production but also invasion by undesirable vegetation.

Possibilities of Soil Damage

The view that clearcutting causes physical erosion of the soil is almost groundless. The main defense of soil against erosion is the porosity of the surface soil. Water that sinks into the soil does not flow over the surface and cause the detachment of particles that is called erosion. Soil porosity is maintained chiefly by the burrowing activities of soil organisms that feed on dead organic matter, a material provided in large supply by the leaf litter of forest trees. The porosity of soil that has been built by the growing of tree crops is not destroyed by the clearcutting of the trees. It may diminish slowly if the supply of new organic matter is halted but this will not happen if vegetation regrows. Silvicultural clearcutting is aimed at establishing new vegetation and not at preventing it; vegetation of almost any kind will produce organic matter and maintain soil porosity regardless of whether it is the species envisioned in the silvicultural act of clearcutting.

Soil erosion in forest soils usually results from actions that *scrape*, *gouge*, or *compact* the mineral soil and thus eliminate or impair the porous surface layers. These effects are much more likely to come from road building, log transportation, or deliberate soil treatments than from clearcutting or any other mode of timber cutting. In one important sense, clearcutting causes *less* surface erosion and stream sedimentation than other methods of cutting because it covers the least amount of terrain per unit of volume removed and thus requires use of the minimum length of transportation surface.

Landslides on Steep Slopes

While clearcutting *per se* is not likely to induce surface erosion, it may induce various kinds of landslides on extraordinarily steep, geologically unstable slopes. Most of these kinds of sites in North America are found in small portions of the very steep and geologically very young Coast Ranges along the Pacific Ocean (Swanston, 1974); however, similar conditions can exist in other steep terrain. The causative conditions are not steepness alone but the existence of discontinuities of materials within a few meters of the surface that can become planes of slippage, especially when lubricated by water.

The action of webs of tree roots can bind the surfaces together well enough to reduce the tendency for lenses and slabs of material to slide or slip down the slopes. If the trees are killed by clearcutting, the web of tree roots is likely to decay after several years and cause delayed-action landslides. Whereas any surface erosion would be most serious immediately after soil disturbance, this kind has the disconcerting characteristic of occurring a few years later.

As in the case with surface erosion, these kinds of "mass-wasting" events are much more likely to be caused by roads and road building than by the cutting pattern. If landslides are likely to occur on a given kind of steep slope it is likely that there will be evidence of them having taken place without any human disturbance; however, good knowledge of the subsurface geological structures is often necessary to detect dangerous situations. Susceptible areas are best logged with helicopters or balloons, without roads, or not logged at all. However, with most such slopes it is not a question, from the geological standpoint, of whether there will be landslides but only of when and at what rate. If such areas are going to be subjected to cutting at all, it is highly logical to contemplate kinds of selection or shelterwood cutting that will encourage the permanent maintenance of a network of living roots.

Wetland Problems and Fog-drip

The most common case where clearcutting alone can cause true site degradation is on one of the flattest kinds of terrain. In many forested swamps and bogs transpiration by trees plays a major role in keeping water tables low enough for the roots of the trees to survive. If clearcutting turns off the "transpiration pump," the level of poorly aerated water may rise enough to retard or prevent regrowth of trees.

The intuitive belief that clearcutting somehow causes droughts or floods is largely groundless (Anderson *et al.*, 1976). Clearcutting actually increases the amount of water in the soil and the amount that flows to the streams (Lull and Reinhart, 1967). This is because of reduction of transpiration and of interception of raindrops and snow by tree cover. This increase of soil water almost never adds to the peak runoff associated with floods.

In fact, it is ordinarily beneficial because it is most important at the very times of year when transpiration is greatest and stream-flow the most feeble. However, if there was a flood-inducing rain soon after one of these times, the flood waters would be increased by the comparatively small amount of water that clearcutting had caused to be left in the soil detention storage capacity (Harr, Frederickson, and Rothacher, 1979).

The chief effect of clearcutting on the water regime is to provide more water for the vegetation of the site. Experimental tests of the maximization of stream-flow by mechanical or chemical soil treatments to prevent regrowth of vegetation have usually caused soil erosion or loss of nutrients to runoff (Bormann *et al.*, 1974). These effects and treatments are not those of silvicultural clearcutting, which is not aimed at denuding the landscape.

Fog-drip is one kind of precipitation that is reduced or prevented by clearcutting. This forms when the minute, suspended water droplets of clouds are blown through leaf canopies, swept out of the air, and coalesced into droplets large enough to fall to the ground. This kind of precipitation can be very significant in some "cloud forests" at high elevations and in some of the remarkable kinds of forests that occur along the foggy western shores of continents at middle latitudes. The coast redwood and western hemlock–Sitka spruce forests of the western edge of North America are heavily watered from this source. The extent to which clearcutting might harm such forests is not really known, perhaps because of the difficulty of measuring fog-drip precipitation accurately. Conventional rain gauges exposed beneath the open sky do not collect fog-drip. When supercooled cloud droplets impinge on solid objects and freeze they form a white ice called rime.

Possibilities of Nutrient Loss

If nearly all of the vegetation is temporarily eliminated, as with true clearcutting, the risk that soluble substances, especially chemical nutrients, may leach out of the forest system at inordinate rates must be considered (Bormann and Likens, 1979). The best defense against such loss is the continued maintenance of some sort of vegetative cover even if the trees or other woody plants are eliminated. The next best defense is prompt reestablishment of vegetation to continue the cycling of nutrients and other soluble substances on the site.

Living vegetation is not the only buffer against nutrient loss because the organic and inorganic colloidal surfaces that make up the cation exchange capacity of the soil also play a highly important role. Soils that are high in organic matter or clays (mineral colloids) may strongly resist loss of cations such as calcium, potassium, or ammonium nitrogen but this does not help retain anions such as nitrate or sulfate ions. The retention of nitrates (and the less important sulfates) depends mostly on living vegetation and conservation of soil organic matter. The question of whether phosphate anions leach away depends more on the vagaries of their binding with iron and aluminum compounds than on anything altered by cutting practices.

Most investigations of the matter have not shown major losses of nutrients in association with regeneration by clearcutting. There was, however, a significant case involving sandy soils and the northern hardwood forests of the granitic White Mountains of New Hampshire (Bormann *et al.,* 1974). Here the warming of the forest floor induced by clearcutting causes the quick decomposition of the thin leaves of the litter of maple, beech, and birch and, more importantly, nitrification. In this process nitrifying bacteria convert soil ammonium compounds, which can be retained on the cation exchange surfaces of colloids, to nitrate anions, which are free to move out in the soil solution.

If, as is the case late in the growing season, the pioneer vegetation does not develop fast enough to take up the nitrates and some other highly mobile nutrients, then there will be a period of some months during which there will be accelerated losses to stream-flow. The fact that very large losses were caused by repeated annual applications of a soil sterilant aimed at maximizing the yield of water in stream-flow has to do with the prevention rather than the establishment of regeneration by clearcutting. As far as regenerative clearcuttings in these particular forests are concerned, the important point is to keep their widths narrow and place them in areas already well vegetated with advance growth. Similar losses from nitrification have not been observed in experiments in conifer forests or in deciduous forests farther south or west.

Soil Disturbance

In thinking about effects of tree cutting on environmental factors, it is well to start from the premise that the most important, far-reaching, and enduring effects come from disturbance of the forest soil. Almost all such effects have to do with roads, log transportation, and mechanical forms of site preparation. Compared with these any effects of the spatial arrangement of cutting patterns are small, subtle, and involved with rather special cases.

A small opening created in the forest by what can be called a selection cutting is, in many respects, simply a little clearcutting. If the utilization is close, it may remove more chemical nutrients per square meter of land surface then a large clearcutting for sawlogs.

As far as nutrients are concerned, the objective of forest management cannot be that of keeping all of the substances on the site. So long as water falls on the site it is inevitable that some materials will be slowly eroded or leached away to stream-flow. In fact, if the streams flowed with chemically pure water, free of suspended colloids, in full shade, at temperatures slightly above the freezing point, neither fish nor water plants would survive in them. The object should instead be to provide life-supporting substances and energy in moderation. Some natural erosion is inevitable; it is the rapid, accelerated kind that should be forestalled. Ideally all soils should be managed such that nutrients and colloids are lost from them no more rapidly than chemical weathering and atmospheric fallout replace them.

Defense by Revegetation

There is no better means of conserving the renewability and productivity of terrestrial ecosystems than keeping the growing space full of vigorous forest vegetation. So long as the vegetation of a given forest age class continues to accumulate biomass it is in what has been called the aggrading phase of its development; this means that it is not only adding to the standing crop of organic substance but also accumulating nutrients, including carbon, on the site. When it is destroyed or reaches the stage at which decay equals or exceeds the accretion of biomass, it enters upon the degradation phase in which nutrients and carbon cease to accumulate and begin to be lost to the atmosphere or waters moving through the system. In other words, nutrients leak out of the forest system not only when stands are destroyed but also when they become biologically overmature.

If this be the case, the timber-production forest managed for sustained yield with all age classes from one year to that of rotation age would come close to the perfect system for keeping soil nutrients, nitrogen, and carbon cycling within the whole forest. The ideal rotation for this purpose would extend to the time when the rates of decay and accretion of total dry matter (not merchantable volume of wood) were equal. From the limited amount of evidence available (Satoo and Madgwick, 1982) it appears that such a rotation length is close to that of a typical sawlog rotation.

Since successful clearcutting is that which results in prompt revegetation it is not necessarily more harmful to the environment than other methods. What goes on in association with clearcutting or any other regeneration technique is as important as the kind of stand created. It is possible to do significant damage to soil and water during the regeneration of forests. It is to this and not to the pattern of cutting that primary environmental concerns should be addressed. Clearcutting in and of itself is suspect mainly because it creates stands that may be too large in area for some purposes, causes damage in a few sets of special circumstances, and looks bad.

BIBLIOGRAPHY

Anderson, H. W., *et al.* 1976. Forests and water: effects of forest management on floods, sedimentation, and water supply. *USFS Gen. Tech. Rept.* PSW-18. 115 pp.

Barney, C. W., and R. E. Dils. 1972. *Bibliography of clearcutting in western forests*. Col. State Univ., Coll. For. and Nat. Resources, Fort Collins. 65 pp.

Baumgartner, D. M. (Ed.) 1973. *Management of lodgepole pine ecosystems*. Wash. State Univ. Coop. Ext. Service, Pullman. 825 pp.

Bormann, F. H., *et al.* 1974. The export of nutrients and recovery of stable conditions following deforestation at Hubbard Brook. *Ecol. Monogr.* **44:**255–277.

Bormann, F. H., and G. E. Likens. 1979. *Pattern and process in a forested ecosystem.* Springer, New York. 253 pp.

Cayford, J. H., Z. Chrosciewicz, and H. P. Sims. 1967. A review of silvicultural research in jack pine. *Canada, For. Branch Publ.* 1173. 209 pp.

Clark, B. F., and S. G. Boyce. 1964. Yellow-poplar seed remain viable in the forest litter. *J. For.* **62:**564–567.

Cleary, B. D, R. D. Greaves, and R. K. Hermann (Eds.) 1978. *Regenerating Oregon's forests.* Oreg. State Univ., Ext. Service, Corvallis. 286 pp.

Crowe, S. 1978. The landscape of forest and woods. *U.K. For. Comm. Booklet* 44. 47 pp.

DeByle, N. V. 1981. Clearcutting and fire in the larch/Douglas-fir forest of western Montana—a multifaceted research summary. *USFS Gen. Tech. Rept.* INT-99. 73 pp.

Harr, R. D., R. L. Frederickson, and J. Rothacher. 1979. Changes in streamflow following timber harvest in southwestern Oregon. *USFS Res. Paper* PNW-249. 22 pp.

Hawksworth, F. G., and R. R. Scharf (Eds.) 1984. Biology of dwarf mistletoes. *USFS Gen. Tech. Rept.* RM-111. 131 pp.

Hermann, R. K., and D. P. Lavender (Eds.) 1973. Even-aged management, symposium, 1972. *Oreg. State Univ., Sch. For. Paper* 843. 250 pp.

Hillis, W. E., and A. G. Brown. 1984. *Eucalypts for wood production.* Academic, Orlando, Fla. 427 pp.

Hu, S. C., R. J. Lillieholm, and P. Y. Burns. 1983. Loblolly pine: a bibliography, 1959–82. *La. State Univ. Sch. For. & Wildlife Mgmt., Research Rept.* 2. 199 pp.

Isaac, L. A. 1943. *Reproductive habits of Douglas-fir.* Charles L. Pack Forestry Foundation, Washington, D.C. 107 pp.

Leaf, A. L. (Ed.) 1979. *Impact of intensive harvesting on forest nutrient cycling systems.* State Univ. of N.Y. Coll. of For., Syracuse. 421 pp.

Little, S., Jr. 1950. Ecology and silviculture of whitecedar and associated species in southern New Jersey. *Yale Univ. Sch. For. Bull.* 56. 103 pp.

Lotan, J. E., and D. A. Perry. 1983. Ecology and regeneration of lodgepole pine. *USDA, Agr. Hbk.* 606. 51 pp.

Lotti, T. 1961. The case for natural regeneration. *In* A. B. Crow (Ed.), *Advances in southern pine management.* Louisiana State University Press, Baton Rouge. Pp. 16–25.

Lull, H. W., and K. G. Reinhart. 1967. Increasing water yield in the Northeast by management of forested watersheds. *USFS Res. Paper* NE-66. 45 pp.

Marquis, D. A. 1966. Germination and growth of paper and yellow birch in simulated strip cuttings. *USFS Res. Paper* NE-54. 19 pp.

Roach, B. A. (Ed.) 1975. *Clearcutting in Pennsylvania.* Penn. State Univ., Sch. For. Resources, University Park. 81 pp.

Satoo, T., and H. A. I. Madgwick. 1982. *Forest biomass.* Nijhoff/Junk, The Hague. 152 pp.

Schmidt, W. C., R. C. Shearer, and A. L. Roe. 1976. Ecology and silviculture of western larch forests. *USDA Tech. Bull.* 1520. 96 pp.

Seidel, K. W. 1979. Regeneration in mixed conifer clearcuts in the Cascade Range and the Blue Mountains of eastern Oregon. *USFS Res. Paper* PNW-248. 24 pp.

Swanston, D. N. 1974. The forest ecosystem of southeast Alaska. 5. Soil mass movement. *USFS Gen. Tech. Rept.* PNW-17. 22 pp.

Wahlenburg, W. G. 1960. *Loblolly pine: its use, ecology, regeneration, protection, growth, and management.* Duke Univ., Sch. For., Durham, N.C. 603 pp.

Wahlenberg, W. G. (Ed.) 1965. A guide to loblolly and slash pine plantation management. *Ga. For. Research Council Rept.* 14. 360 pp.

Wyoming Forest Study Team. 1971. *Forest management in Wyoming.* USFS, Washington, D.C. 81 pp.

Shelterwood and Seed-Tree Methods

Clearcutting is not the only method of regeneration for creating even-aged stands. The shelterwood and seed-tree methods are ones in which some trees of the old stand are kept at least until the new stand is safely established. Any site preparation or other measures to induce regeneration are done before rather than after the final cutting. The transition from the old stand to the new is accomplished without losing command of the growing space. The regeneration period is short enough that the new stand is, for purposes of forest regulation, essentially even-aged. The advantages of the even-aged condition and those of partial cutting can be combined in varying ways and degrees when these two methods are applied.

The time sequences of cutting range from those in which old trees are left only until they have provided the seed for a new crop to those in which some of the trees of the previous crop are deliberately retained as a part of the new one. The patterns of cutting in space can include ones in which the residual trees are scattered uniformly throughout what will come to be one large, even-aged stand or sequences of frequent cuttings that advance across an area in strips or as gradually enlarging patches. The more complicated systems differ from the selection system mainly because the resulting stands can be counted as even-aged for purposes of forest mapping and in regulating harvests for sustained yield. The ensuing discussion starts with the simplest variant, seed-tree cutting, and proceeds to more complicated ones.

SEED-TREE METHODS

In this method the stand is cut clear except for a few **seed trees,** which are left standing singly or in groups to furnish seed to restock the cleared area naturally. After a new crop is established these seed trees may be removed in a second cutting or left indefinitely.

The chief distinction from shelterwood cutting is that the remaining crown cover is not enough to make the cutting area microclimatically different from a broad, open, clearcut area. The distinction from the clearcutting method is that the seed sources are retained within the cutting area and not off at the sides. Because of this, provision for a source of seed does not restrict the size of the cutting or regeneration area; this is the most common reason for using the seed-tree methods.

It is not really possible to distinguish seed-tree cutting from shelterwood cutting on the basis of some arbitrary number of reserved trees per hectare. It depends instead on whether enough crown cover is left to have both the purpose and the effect of providing a significant degree of shade. There are cases of old-growth stands consisting of very large Douglas-firs in southwestern Oregon where the temporary reservation of only about 6 trees per hectare provides enough shade to enable establishment of regeneration (natural or planted) on difficult sites (Fig. 14-1). Trees more than 1 m D.B.H. have wide crowns. With more ordinary stands it may take several times as many trees to create significant shade. In any event, the seed-

FIGURE 14-1 Large and very valuable old-growth Douglas-firs reserved in cutting 3 years ago to provide seeds and shade. On this particular site, on the Willamette National Forest in Oregon, seedlings cannot withstand direct sunlight even if planted. The old trees are not capable of increasing much more in timber value and will be removed as soon as the new stand is dependably established. (*Photograph by U. S. Forest Service.*)

tree method is a concept and not some number of trees per unit area; the concept is that of leaving just enough trees to provide seed to regenerate a stand.

It cannot be said that the concept works well enough that seed-tree methods are in common use or stay in use long when they are adopted. The reservation of seed trees is often a half-hearted gesture toward regenerating forests during the early stages of deliberate forest management. Frequently the number and fecundity of seed trees have been inadequate. In many instances, it was found that the method worked only when intensive site preparation made it possible to get adequate numbers of seedlings from limited seed supplies. In such cases, it was often soon concluded that it was more economical to use direct seeding to take advantage of the necessary site preparation. Another problem has sometimes been that the number of trees left were too few to justify returning to harvest them; leaving more trees in shelterwood cutting has often been the solution.

There have been misleading cases in which crops of new seedlings ascribed to reserved seed trees were, on closer examination, found to be from advance growth already present before cutting. If the circumstances are such that scattered seed trees cannot provide enough seed or shelter, these misinterpretations can lead to regeneration failures. Another kind of delusion is the reservation of a seemingly respectable number of trees that are too feeble to produce much seed.

In spite of its burden of guilt by association with self-deception, the seed-tree method can be useful and practical if intelligently applied under appropriate circumstances.

Arrangement of Seed Trees

Normally the seed trees are left isolated and rather uniformly distributed after the rest of the stand is removed, as shown in Fig. 14-2. Another option is that of leaving the seed trees in groups, strips, or rows rather than being scattered uniformly. The concentration of seed trees on restricted areas makes them somewhat easier to protect during the initial cutting and also to harvest when they have served their purpose. It is also possible that cross-pollination is increased, although seed dispersal would be less efficient and uniform. Sometimes it is hard to get new regeneration established under such concentrations of seed trees but this probably can be mitigated by applying localized shelterwood cutting to them at the time of their reservation. There is no evidence that groups of seed trees are significantly more wind-resistant than comparable trees left isolated.

Characteristics of Seed Trees

The primary considerations are windfirmness and seed-producing ability. The best trees to reserve are those members of the dominant crown class that have wide, deep crowns, relatively large live crown ratios, and stocky, tapering boles. Such trees are the ones most likely to combine

FIGURE 14-2 Two winter views of the same spot in a 95-year-old stand of loblolly pine at the Bigwoods Experimental Forest, North Carolina, immediately before and after a cutting in which eight seed trees were reserved on each acre. The larger hardwoods were killed with herbicides and prescribed burning was done in the late summer to kill small hardwoods, dispose of slash, and prepare the seedbed. (*Photograph by U. S. Forest Service.*)

windfirmness with the health and vigor that leads to good seed production. Trees with short, narrow crowns are, in spite of the smaller area on which the wind may act, vulnerable to breakage and uprooting because they have weak roots and slender boles without much taper. The occasional production of large "distress crops" of seeds by damaged trees is not a sufficiently dependable phenomenon to make it wise to use such trees for seed trees. The method is not really applicable to species that are character-istically weak in resistance to wind (or ice damage) or to soils that are conducive to shallow-rootedness.

Seed trees must be old enough to produce abundant fertile seed; the age at which seed bearing begins in closed stands is the safest criterion. Early and abundant production of seed can sometimes be encouraged by thinning out the crown canopy or fertilizing the soil around the chosen trees. Anything that can be done to protect seeds on such trees from insects, arboreal rodents, or similar pests will enhance production. The seed-producing ability of good trees is often quickly enhanced by their release in seed-tree cuttings. In fact, a few good seed trees sometimes produce as much seed per hectare as the full stand did before the cutting.

Because of both hereditary and environmental factors, the trees that have been good seed producers in the recent past are very likely to be the most dependable. The remains of cones or fruits on the trees or on the ground beneath them are often useful evidence of fecundity.

The kind of trees used for seed trees depends to some extent on the intensity of silviculture practiced. If a second cutting can be made within a few years to utilize the seed trees, there is little reason not to pick fine, healthy dominants, presumably containing valuable timber. When the seed trees must be abandoned after they have served their purpose, indi-viduals of lower commercial value may have to be chosen. In stands that are distinctly even-aged, the reservation of the smaller trees is likely to be a futile gesture. However, in uneven-aged stands, it may be possible to find relatively small trees that are sufficiently sturdy and vigorous to serve as seed trees.

The seed-tree method theoretically provides an opportunity for more rigorous selection of the parents of the new crop than is attainable with any other method of natural regeneration. As long as the patterns of inheri-tance of the characteristics of tree species remain obscure, it is most rational to assume that undesirable characteristics are heritable and to attempt to confine choices to the best phenotypes.

If there is some compelling reason why the most valuable trees cannot be reserved for seed, it is wise to try to evade the problem. The most important way that this can be done is to select trees that have developed imperfectly because of strong and obvious environmental influences that are likely to have operated at random. Trees that have developed poor form because of lack of competition would be logical choices in such cir-cumstances. Trees that are poor because of apparent susceptibility to dam-aging agencies that operate in immature stands should be rejected. There is probably no reason to avoid trees damaged by heart-rots, which character-istically infect trees only when they are older than the normal rotation age.

Number and Distribution of Seed Trees

The number of trees to be reserved depends on factors such as the amount of seed produced per tree, the prospective survival of the trees, and the number of seeds required to yield an established seedling under the surface conditions that will prevail.

If the seed trees are uniformly scattered, the prospective quantity of seed needed seems to be a more critical factor than the distance of dissemination. Satisfactory dissemination by wind can usually be counted upon for a distance of roughly three times the height of the trees. With species that are disseminated by winged or terrestrial animals, the distances depend upon the habits of these animals. Even these distances can be great enough that regeneration success depends more on the quantity of seed and the nature of the regeneration microenvironments than upon limitations of dispersal distances.

The number and distribution of seed trees influence the pollination of female flowers and stroboli and, therefore, the number of viable seeds produced. A few species, such as white ash and cottonwood, are actually dioecious; both male and female trees of these species must be left regardless of the method of regeneration employed. Furthermore, the male and female flowers of the same tree rarely become mature simultaneously. The possibility of self-fertilization is also reduced by the tendency for female flowers to develop at the top of the tree, above the male flowers; the wind is not often turbulent enough to cause much pollen to rise straight upward. Self-fertilization usually results in the production of a high proportion of inviable seeds. The concentration of pollen grains also decreases rapidly with distance from the source because of atmospheric turbulence; therefore, the proportion of fertile seeds probably decreases as seed trees become more widely separated.

It may, therefore, be assumed that isolated seed trees produce a lower percentage of *viable* seeds than they would if adjacent to other trees of the same species. Furthermore, since *total* seed production is usually increased by releasing trees from competition, it is clear that the fruitfulness of trees in closed stands is not a good criterion of the number of isolated seed trees required.

The proper number of seed trees to leave per hectare depends on knowledge of seed production and the way in which the young trees will develop. First, a goal must be set for the number and spatial distribution of established seedlings that constitute a satisfactorily stocked stand. For example, this minimum might, at the fourth year, be two thousand trees per hectare with 80 percent stocking of a random sample of 4 m² quadrats. The next step is to determine, after allowing for losses of seed before germination and of seedlings thereafter, the amount of seed per acre needed to secure established reproduction of the prescribed density. When the total requirement of seed is known the number of trees required per acre may be calculated, if the seed production of the trees can be predicted.

The risk of loss to wind or lightning may influence the number of trees reserved. Trousdell (1955) found that in northeastern North Carolina the annual losses to lightning occurred at the rate of 5 trees per 100 hectares, regardless of how many were reserved. An increase in the number of trees left would reduce the percentage, but not the number, that were struck.

Since the number of thunderstorms there is close to the national average, one might estimate prospective lightning losses in other localities from the relative frequency of such storms.

Sporadicity of Seed Crops

Success is likely to be achieved only in years of good seed crops. Except in the case of species with very small seeds, and sometimes even with those, almost all of the seeds produced by trees are consumed by rodents, birds, insects, fungi, or other pests. Only a small fraction remains to renew the forest and that small amount is enough to restock the forest only when the total seed crop is large.

Unfortunately the planning of natural regeneration is seriously hamstrung by our poor understanding of the factors that control seed production and its timing. If there is any important tendency toward regular periodicity of seed crops it is so obscured by disrupting influences that it is no basis for reliable predictions.

Sometimes there is evidence that suggests that the number or proportion of female reproductive buds, often a limiting factor, is enhanced by improved tree vigor or nitrogen nutrition. On the other hand, there is also evidence to suggest that the most critical question is whether various damaging abiotic agencies or biotic pests interrupt one of the many steps that lie between bud formation and seed production. Often the best crops of seed come after years of total failure. With such episodes it is suspected that the lack of seeds in one year may cause collapse of populations of seed-dependent organisms during the next year.

With most species it is possible to predict good seed crops only a few months in advance, when the fruits or cones have formed and the time during which damage can occur begins to dwindle. The earliest times for assessing the prospects come when, usually a year in advance, the buds first differentiate into vegetative or reproductive. If development of cones or fruits takes two years, as is true with most pines and red oaks, predictions with some degree of usefulness can be made rather far in advance. However, with a process that has many steps from bud formation to seedling establishment it is important to remember that failure at any one step can defeat the whole process.

In practice, it is normally best to reserve enough trees to restock the area in one moderately good seed year. Excellent crops can be depended upon only if already in being; meager crops are usually consumed by rodents or birds.

SHELTERWOOD METHODS

In the shelterwood methods of regeneration, one retains more old trees and for more reasons than is the case with seed-tree cutting. There is the same fundamental purpose of getting a new even-aged crop of trees established before the old one is completely removed (U.S. Forest Service, 1979). The establishment or use of advance growth is still the essential characteristic.

The shelterwood method involves the gradual removal of the entire stand in a series of partial cuttings that extend over a fraction of the rotation. The cuttings usually resemble heavy thinnings, and, under intensive practice, regeneration by the shelterwood method logically follows a series of thinnings. Natural reproduction starts under the protection of the older stand and is finally released when it becomes desirable to give the new crop full use of the growing space.

Within the framework of the shelterwood method, it is possible to achieve wide variation in the relative degrees of shelter and exposure in both space and time. Adjustments can be made to meet the environmental requirements of almost all species except those that are exceedingly intolerant of shade or root competition.

The term shelterwood denotes one purpose succinctly, although sheltering seedlings from damaging agencies is not the *only* one. The shelterwood method is also employed partly, even primarily, to get efficient use of the productive capacity of the growing stock. Trees that are no longer capable of much further increase in value are the ones cut to make space for regeneration. Those that are left are chosen not only to provide seed or protection for the new stand but also for their capacity to increase in value at attractive rates. The objectives of regeneration and growing stock management are readily combined because the trees of greatest vigor and quality are those that will provide good crops of seed and also grow most rapidly in value. In other words, the best trees grow best and the policy of cutting the worst first has much to recommend it.

If natural regeneration from advance growth tends to be too dense and thus to develop into overcrowded thickets, it may help to leave shelterwood trees to induce irregularity of height growth in the new stand. Such trees can also become an important part of the new stand. Residual trees are sometimes retained to favor wildlife species that depend on trees older than normal rotation age for food or shelter. Well-chosen residual trees also make a stand under regeneration look less devastated than is the case with clearcutting.

The shelterwood method is ordinarily aimed at securing natural regeneration from seed but there is no fundamental reason against full or partial substitution by artificial or vegetative regeneration. While the stands created by shelterwood cutting are usually even-aged there are variants, usually involving mixtures of species, in which the stands have two age classes or have trees ranging over several decades in age.

Basic Procedures in Shelterwood Methods

The basic ideas of shelterwood cutting are most easily described in terms of an application of the method in which the residual trees are uniformly distributed over the area of the new stand. This approach to shelterwood cutting is indeed sometimes called simply the "uniform method" (Fig. 14-3). The same steps and procedures are also followed with respect to any given spot in a stand when the spatial patterns of cutting are the more complicated ones described later.

The first cuttings create vacancies in the growing space of the stand in

FIGURE 14-3   A 74-year-old stand of eastern white pine in the Pack Forest in eastern New York being regenerated by the uniform-shelterwood method. The seed cutting was carried out 12 years previously and the partial shade of the overstory is being used to reduce damage to terminal shoots by the white pine weevil. (*Photograph by State University of New York, College of Environmental Sciences and Forestry.*)

which the new crop can become established. Besides furnishing seed, the old stand affords protection to the young seedlings. A time finally arrives when this shelter becomes a hindrance rather than a benefit to the growth of the seedlings. It is then necessary to remove the remainder of the old stand, giving the new stand possession of the area and opportunity to develop in even-aged form.

The whole process is usually accomplished within a relatively short period. Where adequate advance reproduction has become established entirely as a result of the natural opening-up of old stands, a single cutting may be sufficient. Normally, however, the shelterwood method requires a minimum of two cuttings. Under intensive practice, several cuttings are made in the gradual process of simultaneously freeing the reproduction and removing the mature stand. Regardless of the number of cuttings, the largest, most vigorous, and best-formed individuals of desirable species are normally retained until the final cutting. In this way the best trees of the stand are left to provide seed for the new crop; meanwhile they continue to lay down wood of high quality at a rapid rate. Trees of the poorer categories are gradually removed in successive cuttings with the least desirable usually being cut first.

The sequence of operations involves two or three different kinds of cutting applied in the following order:

1. **Preparatory cuttings,** which set the stage for regeneration.
2. **Establishment** or **seeding cuttings,** which induce the actual establishment of seedlings.
3. **Removal cuttings,** which release the established seedlings.

Preparatory Cuttings

If natural reproduction is to start under the old stand a supply of seed must be available and site conditions must be favorable for germination and establishment of seedlings. It is sometimes, but not always, necessary to conduct preparatory cuttings to improve the vigor of prospective seed bearers or to accelerate the decomposition of unfavorable humus layers.

If the unincorporated humus layers are too thick, their decomposition can be hastened by opening the canopy to admit more solar radiation and precipitation to the forest floor. This needs to be considered only with tightly closed stands in cool or dry regions favorable to the development of thick humus layers. When stands are so dense that the prospective seed bearers have short, narrow crowns, preparatory cutting may be necessary to induce enlargement of crowns, which is conducive to seed production and windfirmness.

Preparatory cutting involves objectives additional to those of low thinning but differs little otherwise. Most of the trees that are cut come from the lower crown classes. In principle, openings are made in the main canopy only to the extent necessary to permit the development of vigorous seed bearers. In practice, there is no reason why undesirable or defective trees of the dominant and codominant classes cannot also be removed. However, efforts should be made to avoid creating the kind of environment that will favor the establishment of undesirable species more shade-tolerant than those ultimately desired. If such species are likely to invade after light cutting and the objective is solely to develop the seed bearers, the preparatory cutting is logically confined to removing a few competitors of the chosen trees.

In the majority of cases no preparatory cuttings are necessary, particularly where their purpose has been largely accomplished by systematic thinnings prior to the reproduction period. In fact, reproduction of some species may become established as a result of thinnings or even in the absence of cutting. The various active measures of site preparation are usually more expeditious than preparatory cutting as a means of reducing the accumulation of humus and controlling undesirable vegetation beneath the old stands. For these reasons the establishment cutting is generally the first step in shelterwood cutting. Site preparation is very likely to be needed and is most logically done just before or simultaneously with the establishment cutting.

Establishment or Seeding Cuttings

The purpose of this cutting stage is to open up enough vacant growing space in a single operation to allow the establishment of regeneration. This is best done during a year in which the desired species bear seed in abundance. Otherwise undesirable vegetation may become established after the cutting. The desirability of executing cuttings to make way for regeneration during good seed years is not peculiar to the shelterwood method. A similar policy would be at least equally appropriate in stands managed under other methods of reproduction.

With the shelterwood system it is more nearly practicable to confine true regeneration cuttings to good seed years than with any other. In all other methods, with the possible exception of the selection system, an annual harvest of trees of the best quality is possible only if true regeneration cuttings are carried out somewhere in the forest every year. In the shelterwood method, the trees removed in the establishment cutting, which is the true reproduction cutting of the method, are not the best trees in the stand. During years of poor seed production, timber of high quality may be harvested by removal cuttings in those stands where regeneration has already become established as a result of previous seeding cuttings.

In any method of natural regeneration it is desirable to attempt to concentrate true reproduction cuttings in periods when good seed crops are in prospect or have recently fallen. When seed production is poor the problem is best evaded by emphasis on intermediate cuttings and the harvesting of timber in places where regeneration already exists. The shelterwood method provides a systematic means, under even-aged management, for reconciling the need for a steady flow of products with the natural tendency for reproduction to become established at erratic and infrequent intervals.

The trees removed in the establishment cuttings are the least desirable remaining in the stand. They include any remaining intermediate and overtopped trees as well as all or part of the codominants. It is particularly important that trees of undesirable species be cut (or killed) regardless of crown class. Any trees too poor to retain for further growth should be removed, although the temptation to eliminate all coarse, branchy dominants should be restrained if these are the only good seed bearers.

The establishment cutting is best confined to a single operation so that the new crop will be reasonably uniform in age and size. In this way the severity of the cutting can also be adjusted so as to create a range of environmental conditions that is restricted as closely as possible to that which is optimum for germination and establishment of the species desired. A series of cuttings of ever-increasing severity might involve the reservation of a shelterwood that was at first too dense and later too thin to provide the proper environment.

Where moisture deficiencies are likely to limit regeneration, an excessively dense overwood may cause too much root competition and interception of precipitation. If heat injury to succulent young seedlings is the primary difficulty, the establishment cutting should be regulated so as to increase the amount of diffuse light admitted from the sky as much as possible without allowing much direct sunlight to reach the forest floor.

The appropriate density of the overwood can be determined by observation and experimentation. It will differ within wide limits depending on the requirements of the species sought and the site factors; it may also be as important to make conditions unfavorable to undesirable species as to create those favorable to the desirable. If the species that appear are more shade-tolerant than those desired, it is usually necessary to increase the severity of the establishment cutting. If, for example, the existing species to be reproduced is pine and some more tolerant species burgeons after shelterwood cutting, it is probably a sign that the cutting was not heavy enough or site preparation was inadequate.

Various indexes can be used to regulate and guide the severity of establishment cuttings and thus enable transfer of experience from one case to another. The amount of ground left covered by tree crowns is the best index, but it is difficult to measure. Therefore, indexes such as basal area, sum of stem diameters, or even timber volumes per hectare are generally used. If these parameters are expressed as percentages of the total before cutting, there is an approximate indication of the amount of growing space that has been made vacant. In view of the prospect that the extraction of trees felled in the subsequent removal cuttings can cause damage to the new seedlings, it is often desirable to take off as much of the old stand as possible in the establishment cutting.

It is not necessary that the distribution of reserved trees be absolutely uniform. An uneven distribution may occasionally be favorable to the seedlings in their competition with the older trees for soil moisture. Advantage is taken of all groups of advance growth that may have started naturally or as an unintentional result of preparatory cuttings or thinnings. Over such groups the cutting is heavy and may remove all the old timber. In fact it is possible to find stands in which reproduction has started so well after thinnings that neither preparatory nor establishment cuttings are needed. Here the first reproduction cutting will be a removal cutting.

If satisfactory regeneration does not become established as a result of the seed crop present at the time of cutting or from seed produced in the next few years, two courses of action are available. The inadequate natural reproduction may be supplemented by planting, or measures of site preparation may be applied during a subsequent seed year. If good reproduction is secured without delay, 3 to 10 years usually elapse before the stand is ready for the first removal cutting.

Removal Cuttings

Removal cuttings have, in varying degrees, the objectives of gradually uncovering the new crop and of making best use of the potentialities of the remnants of the old crop to increase in value. There may be one or, in intensive management, several removal cuttings, the last of which is called the **final cutting.** The ideal objective of this process is to proceed so that the new crop fills the growing space as fast as the old crop is caused to relinquish it. Although perfect achievement of this goal is impossible, the procedure enables regeneration to be accomplished without reducing total wood production as far below the capacity of the site as would be the case if a new crop of seedlings, natural or planted, were left to regain full occupancy alone.

The largest and most vigorous trees usually have the greatest capacity to increase in value and so are the ones most logically reserved until the final cutting (Fig. 14-4) unless they happen to be ones that interfere unduly with the new crop.

After the reproduction is established it is watched for signs of poor growth. If the young trees develop unhealthy foliage, bend aside toward the light, or fail to maintain a satisfactory rate of growth in height, the competition of the overwood should be eliminated or reduced. If the development of the new stand is not uniform and if there is more than one removal cutting, it may be desirable to vary the severity of a given cutting operation in different parts of the area. Some patches may need full release, others may need partial release, and still others that remain unstocked may need a repetition of the seeding cutting.

It is important to distinguish between the conditions suitable for the establishment of regeneration and those necessary to make it grow satisfactorily in height. It is frequently the case that establishment cuttings provide ideal shelter for new seedlings but much heavier removal cuttings are necessary to get them to grow well.

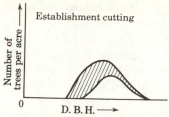
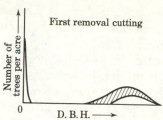

FIGURE 14-4   Diameter distribution curves showing (by cross-hatching) the relative numbers of trees of different diameters removed in a typical establishment cutting and the first of two removal cuttings. In the establishment cutting (*left*) the smaller and less desirable trees, representing about a third of the volume, are cut. In this example, the first removal cutting (*right*) takes place after the trees left in the establishment cutting have increased substantially in diameter and after some of the reproduction resulting from that cutting has grown to measurable D.B.H. The trees reserved in the first removal cutting are the biggest and best in the stand; they will all be harvested in the second and final removal cutting.

The new regeneration may consist of a mixture of species. If the desirable trees are the least tolerant members of the mixture, it is important that subsequent removal cutting opens the stand rapidly to allow them to outgrow more tolerant competitors. Advantage is thus taken of the fact that intolerant species tend to increase in height growth with each added increment of light right up to full sunlight whereas the tolerant ones attain a modest but maximum height growth at some intermediate light intensity.

If the purpose is to discourage the intolerants and encourage tolerant species or ones that represent an intermediate condition, the rate at which the new vegetation is given command of the growing space should be regulated accordingly. There are many situations in which over-exposure calls forth invasions of grass or other weedy pioneers while excessive shade allows usurpation by shade-tolerant species. It is very important to know where each of the interacting species stands in the spectrum of microenvironmental interactions discussed in Chapter 7.

The intervals between removal cuttings and the periods over which they are extended are highly variable. Except for the need to regulate the composition and rate of development of the new crop there is little reason why the timing cannot be fitted to any management considerations. There is a convention that if the new stand is to be counted as even-aged, the period of removal cuttings should not exceed 20 percent of the rotation length. If it does, the program can be termed the irregular shelterwood system as described in the next section and some of the characteristics of even-aged management will be lost or blurred, presumably to gain other advantages.

The removal cuttings, if uniformly applied over the regeneration area, are almost certain to cause some injury to the new stand. This can be reduced, but not eliminated, by care in logging. The least injury is likely to result if the overwood is harvested while the seedlings are still flexible.

Greatest difficulties result when it is necessary to fell trees with broad crowns into stands of saplings. The damage usually appears more serious than it really is and it is sometimes even a benefit in disguise. Shelterwood cutting often leads to the development of grossly overstocked patches of reproduction that look handsome when young but are apt to stagnate in the sapling stage. These clumps can be crudely thinned by the skidding or felling of trees across them. In fact, it is usually best to direct the inevitable logging damage toward the densest parts of the new stand or those that are entirely unstocked and away from the sparsely stocked portions.

Variations in Shelterwood Methods

Shelterwood cuttings can be arranged in time and space by many different patterns that can be categorized as follows:

1. **Uniform method**—application uniformly over the entire stand.
2. **Strip-shelterwood method**—application in strips.
3. **Group-shelterwood method**—application in groups or patches.
4. **Irregular shelterwood method**—long-extended regeneration period with or without spatial variations in cutting patterns.

The previous section was deliberately focused on the uniform method but the same principles apply in varying degrees to the other variants.

Strip-Shelterwood Method

In this modification the stand is cut in strips that are advanced gradually across the area. At any given time cutting is concentrated in certain strips and the rest of the stand remains temporarily untouched.

Starting on one side of the stand, a seeding cutting is conducted on the first strip. After a few years a removal cutting is made on the first strip, and the next strip is given a seeding cutting. A few years later the first strip is subjected to a final cutting, the second strip to a removal cutting, and a third strip to a seeding cutting. In this way (Fig. 14-5), the series of cuttings progresses strip by strip across the stand. It may be necessary to divide a large stand into cutting sections, as shown in Fig. 14-5, if the whole stand is to be even-aged in the next rotation.

The strip-shelterwood method can be applied in a variety of other ways. In the simplest application, the only kind of cutting involved would be clearcutting in very narrow strips so that the side shade of the trees on the adjacent strip of uncut timber takes the place of the protection of an overwood. This effect can seldom be achieved if the strip is wider than half the height of the adjoining trees, but much depends on the latitude and the sun angles.

In the most intensive applications, such as some common in central Europe, the cuttings are applied at short intervals so that there is a gradual

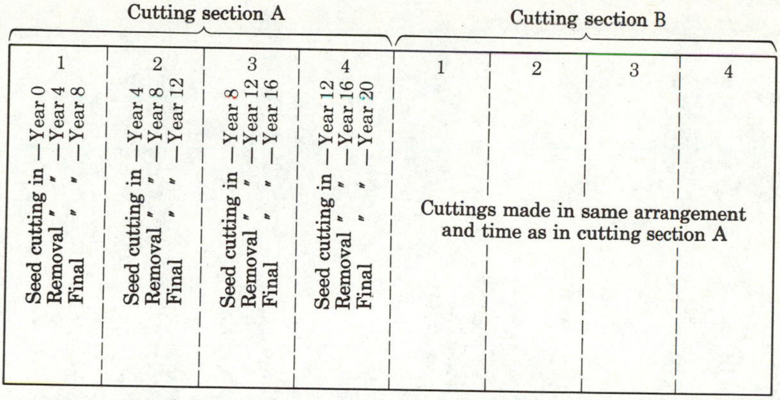

Cutting section A				Cutting section B			
1	2	3	4	1	2	3	4
Seed cutting in —Year 0 Removal " " —Year 4 Final " " —Year 8	Seed cutting in —Year 4 Removal " " —Year 8 Final " " —Year 12	Seed cutting in —Year 8 Removal " " —Year 12 Final " " —Year 16	Seed cutting in —Year 12 Removal " " —Year 16 Final " " —Year 20	Cuttings made in same arrangement and time as in cutting section A			

FIGURE 14-5 Diagram of a stand reproduced in two cutting sections by the strip-shelterwood method. The period of regeneration is about 20 years long, and, if the rotation is 100 years, the stand is kept in even-aged form. Preparatory cutting is unnecessary in most cases and hence is omitted from the diagram.

transition from final removal cutting to light preparatory cutting across strips about 30 meters wide (Fig. 14-6). Within these strips there are preparatory, establishment, removal, and final removal cuttings, but each in zones that are scarcely distinguishable from each other. The whole strip is advanced slowly across the stand, at average annual rates of 2–6 meters. Several years normally elapse between harvesting operations in a given stand; any advantages that might be gained from annual fine tuning of the regeneration process would be outweighed by the nuisance and disruption of very frequent cuttings.

It is not necessary that the strips be straight unless this would be of real advantage in systematizing or concentrating operations. An irregular or undulating line may even expose a longer frontage for regeneration and thus hasten the advancement of the strip cuttings.

The strip-shelterwood methods can have certain advantages over the uniform method. Most of these result, in one way or another, from the distinction between having a diffuse cover dispersed over large areas and being able to adjust conditions within systematically arranged transition zones. With the dispersed cover it is almost inevitable that degrees of shade and root competition will vary over wide ranges from spot to spot. With the advancing strips there is much more opportunity to use walls of standing trees to provide side shade and create microenvironmental zones that will fall in belts that can be predictably located on the ground. This approach is the most dependable way to regulate the proportions of direct and diffuse radiation and amounts of root competition depicted in Fig. 8-1.

The positive purpose of cutting in advancing strips is to provide the initial conditions necessary for establishment of desirable regeneration and subsequently the appropriate degree of release from overhead competition. There is also the negative purpose of accomplishing these steps with-

FIGURE 14-6 Regeneration of spruce, fir, and beech obtained by use of the strip-shelter-wood system, with cutting advancing from north to south (left to right) in a forest in southern Bavaria. The timber is extracted to the right. If the cutting is done before the advance growth is established, dense grass invades and hampers tree seedlings. (*Photograph by Yale University School of Forestry and Environmental Studies.*)

out favoring too much vegetation that is more or less tolerant than that desired. One of the objectives of strip-shelterwood cutting in Central Europe is prevention of invasion by coarse grasses that thrive on over-exposure and hamper seedlings of spruce, fir, and beech.

If the strips are advanced toward the direction from which dangerous winds come, with their long axes approximately at right angles to that direction, there may be some reduction in the risk that residual trees will blow down. The trees that are left isolated stand in the lee of closed stands while their own enhanced stem growth improves their windfirmness.

Perhaps the greatest advantage the advancing strips have over the uniform method is the opportunity to extract the felled timber through the old stand rather than through the regenerated belts. The possibility of doing this is greatest if the stands have been appropriately thinned before the start of regeneration cuttings so that there is enough space between the old trees. Both the strip cutting and the thinning require that the stand be served by a good transportation system. If logs are best skidded downhill, it is logical that on hillsides the strips advance either downhill or along contour lines but not uphill.

Group-Shelterwood Method

Sometimes it may be useful to apply shelterwood cutting in a pattern of expanding groups or patches. This can be done to induce patches of advance regeneration or to take opportunistic advantage of those that have arisen from thinning or natural disturbances. While there may be a certain appeal to the notion that such groups be expanded uniformly in all directions as shown in Fig. 14-7, it would usually be more logical to advance the cuttings mostly in one direction and turn the patches gradually into advancing strips. The chief advantage of such a shift would be the systematization of timber harvesting and administration; the direction of progression can also be determined by the need to protect seedlings from direct sunlight or residual stands from wind damage.

The little openings may become frost pockets and the question of whether this is undesirable depends on one's attitude about whatever frost-sensitive species might appear. Regeneration groups or patches are sometimes deliberately started in conjunction with shelterwood cutting as a means of enriching mixed stands with intolerant species or in giving slow-growing species a head start over more aggressive ones.

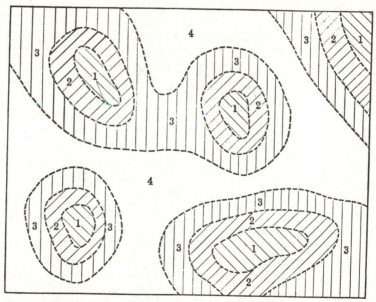

Kind of Cutting	Areas Marked:-			
	1	2	3	4
Preparatory and Seed Cutting combined	Received Cuttings in Years as follows:-			
		0	5	10
Removal Cutting		5	10	15
Final Cutting	0	10	15	20

FIGURE 14-7 Arrangement of cuttings in a stand regenerated by the group-shelterwood method. Advance reproduction was present before cutting on areas marked 1.

Irregular Shelterwood Method

This differs from other variants of the shelterwood method in that the regeneration period is extended so long that the new stand is not really even-aged. This does not mean that it has the three or more age classes that denote the uneven-aged condition. It does, however, mean that the stand will include two age classes for long periods and sometimes even for a whole rotation. The adjective "irregular" refers mainly to the variations in tree heights within the new stand.

As far as spatial arrangement of cuttings are concerned, this technique can be applied in uniform arrangement or in the strip and group patterns just discussed. In fact, if the advancement of successive extensions of the strip and group arrangements proceeds slowly, as it often does when these methods are feasible, the regeneration of an entire stand may well take more than a fifth of the rotation. The stands thus created may, therefore, have irregularities of height but ones arranged in patterns that show gradational change in the horizontal dimension.

The irregular shelterwood method is sufficiently similar to the selection methods that any one harvesting operation, if viewed close up, may seem to be a selection cutting. The difference lies in the goal and the ultimate age-class structure of the stand, which can be perceived only by looking at the whole stand and not just part of it. In the irregular shelterwood system variations in age are more nearly accepted than deliberately created. The stands are fully as attractive as truly uneven-aged ones.

The irregular shelterwood method is often but not necessarily associated with the maintenance of mixtures of species. This is because almost every species has a different regime of growth in height and diameter such that they seldom reach maturity at the same stages of stand development.

The simplest kinds of irregular shelterwood cutting are those done in pure stands and typified by Fig. 14-8. The most common purpose is to take advantage of the growth potential of some scattered trees of outstanding vigor or quality. In pure, even-aged stands such trees are likely to be good dominants, although sometimes small-crowned codominants with fine stems can be revived. Occasionally the purpose of retaining scattered trees of certain sizes is to make up for shortages of trees in these sizes in the overall diameter distribution of whole forests being managed for sustained yield on an even-aged basis. In still other cases, one purpose may be to provide some sort of long-lasting shade protection to saplings of the new stand.

Irregular shelterwood cuttings become more complicated when they are applied in stands that are already irregular because of variations in development of individual trees due to differences in species, age, or both. These circumstances provide increased latitude for retaining small trees, especially those of shade tolerant species, and deliberately using them as a part of the new, essentially even-aged crop. The small, reserved trees are apt to develop into excessively large-crowned wolf trees unless the sur-

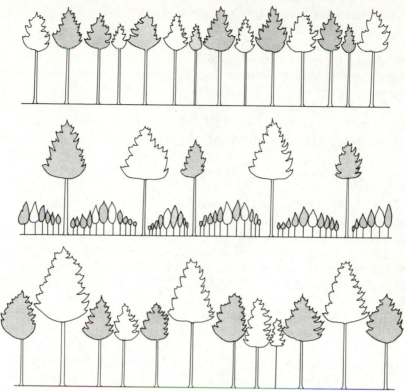

FIGURE 14-8   A stand of a moderately tolerant conifer at three stages in the application of the irregular shelterwood system. Each sketch shows the stand before a cutting in which the trees with the dark crowns are reserved. The *upper* sketch shows the initial even-aged condition just before trees still capable of increase in value are reserved in a seed cutting. The *middle* sketch shows the stand one-third of the way through the next rotation and ready for the *second* of two removal cuttings. A few merchantable trees are thinned from the new stand and three original trees are left to grow until the end of the second rotation. The tree on the left illustrates the use of the method to grow a few trees to great size and value; the main justification for leaving small-crowned trees like the two on the right would be to remedy a deficiency, on the forest as a whole, of good stems in their diameter class. The *lower* sketch shows the stand after additional thinnings, at the end of the second rotation, and ready for seeding cutting as well as for harvest of the three original trees. These three trees could have been cut earlier; there is no necessity that they be carried through two whole rotations.

rounding smaller trees catch up with them in time to maintain adequate training effects. This works best if the previously subordinate trees are of species that are more geotropic than phototropic and thus tend to remain straight when growing under taller trees. Discussion of irregular shelterwood cutting in mixed stands will resume in Chapter 17, in connection with consideration of stratified mixtures.

APPLICATION OF
SHELTERWOOD AND SEED-TREE METHODS

These methods can be adapted to such a wide range of ecological conditions and management objectives that there are many variants. As is the case with all other methods, there is no such thing as *the* shelterwood method or even *the* uniform method. About the only things that all the variants have in common are (1) reliance on getting regeneration established before the old crop is removed and (2) creation of stands that either have the even-aged condition or approach it. Ordinarily they involve natural regeneration but there is no rule against planting beneath the cover of a shelterwood.

The degree of exposure and treatment of the forest floor can be varied within wide limits to favor the regeneration of species of different shade tolerance and successional status. In fact, these methods can be used to regenerate almost any species if the treatments are suitably modified. The chief exceptions are those extremely intolerant species that are more harmed than helped by overhead shading. It is also true that there may be no purpose in employing these methods for serotinous-coned conifers or species that reproduce vegetatively.

At one end of the range of adaptable species are those shade-tolerant ones of late successional status. These are adapted to persist as advance growth under old stands and to respond to release as overstory trees succumb to agencies such as wind and biotic pests; these are the agencies previously described as killing old stands from the top downward. Under natural conditions these lethal disturbances often operate slowly enough to give rise to stands of irregular or uneven age-class structure. In the shelterwood or seed-tree systems, the old stands are removed rapidly enough that the new stands are even-aged or irregular and not uneven-aged. Species in this category often require shade in the early stages and, furthermore, cannot be made to grow very rapidly in height until they become well established.

For such species it may be necessary to have a shelterwood overstory not only to protect the new regeneration but also to have some trees that can make economically effective use of the growing space while the new trees are developing to the stage when they can do so. Under the circumstances one is usually just accelerating natural tendencies to a modest extent; the process is usually easy and low in cost but cannot be hurried.

At the other extreme are those species of low or intermediate shade tolerance that are adapted to regenerate after the kind of surface fires that rather closely simulate the simpler kinds of seed-tree and uniform shelterwood cuttings. For these species one cannot be content to drift along with natural tendencies but must usually employ site preparation measures or post-regeneration release cuttings to maintain the desired species. Since intolerant species seldom require shade after the succulent germination stage and usually have rapid juvenile height growth, it is common, under more intensive practice, to use clearcutting and planting to regenerate them.

There are, of course, many species of intermediate status but these modes of partial cutting can be modified in ways to regenerate them in the essentially even-aged condition. The common observation that many tree species are "tolerant in youth but intolerant after reaching the sapling stage" is evidence of the broad applicability of this mode of regeneration cutting.

Safety Factors

Both seed-tree and shelterwood regeneration cutting are procedures that have an inherent safeguard. The seed source is retained on the site until the new stand is established. If regeneration fails after one good seed crop, it is possible to wait for the next with no cost other than loss of time and the need to repeat any site preparation that may be necessary.

In the case of most shelterwood cutting, there are some other potential advantages that can be used in varying degrees. When one seeks natural regeneration from seed with clearcutting, seed-tree cutting, or even some forms of group-selection cutting it is too easy to say that the methods lead to regeneration *if* one cuts in a good seed year. This is not much help for the normal situation in which good seed years come sporadically and seldom. Most kinds of shelterwood (and selection) cutting provide a basic way around this dilemma.

The significant point is that the main harvest cuttings follow rather than precede the establishment of seedlings. The ecological adaptations of most species are flexible enough that shelterwood cutting can usually be applied with some degree of latitude about timing. Even a seeding cutting can often be made in any year with the idea of simply waiting for the next seed crop. The procedure that is really most crucial is any site preparation for regeneration that may be necessary; this may call for an extraordinary campaign in some year when a good crop of seed is due to fall.

Most species also have seedlings tolerant enough of shade that their release in removal cuttings does not have to be done on some precise time schedule. Because of this it is sometimes possible to store the seedlings that become established in one spectacular seed year and then release them, first in one stand and then in another, over a period of years or even of decades. One does not always have such latitude but it can be very advantageous if it does exist.

With most shelterwood cutting there is also the very important advantage that, if there are enough residual trees, the old stand can continue to make economically effective use of the growing space while regeneration is awaited. This eases the cost of waiting for good seed crops. Even more importantly it can provide for the "telescoping" of one rotation into another. If the new stand is, in effect, 6 years old at the time of the final removal of a rather full shelterwood that is 60 years old at the time, the average rotation may, for practical purposes, be shortened by 10 percent. The gain could be greater yet if the alternative is some sort of clearcutting procedure in which seeding or planting is, by accident or design, delayed for a year or more after cutting.

If the new stand gets well started before the old one is removed, the inevitable loss of production during the transition is kept under some control; this is because the desired vegetation comes close to retaining full command of the growing space. With true clearcutting, for example, there is an inevitable hiatus between the time the old vegetation is cleared away and that when the new trees have developed sufficient foliage and roots to make full use of the growing space. This hiatus is ordinarily several years long. It is also a time when unwanted vegetation can usurp some of the temporarily vacant growing space.

It is, in other words, highly advantageous to start the new forest under the old one if it is possible. The less than ideal ways of starting over again with a slate clean of vegetation are resorted to mainly as matters of necessity. The risk of logging damage to residual trees and established regeneration is a chronic problem with any method of partial cutting. Repetitive cuttings can have costs that can outweigh the value of their advantages. It may also be desirable to replace old stands with ones of different species or genetic constitution.

The presence of a shelterwood can make it difficult to burn slash; conversely, the shade that it casts could, in some climates, restrict drying or speed decomposition enough to make slash disposal unnecessary. The presence of shade often eliminates necessity for exposing mineral soil as a means of getting light-seeded species to germinate.

Because of the principle of leaving the best and most vigorous trees until the end, the shelterwood and seed-tree methods usually provide the best way of applying the concept of financial maturity to harvesting even-aged stands. Efficient use can thus be made of the capacity of the better trees to increase in value without sacrificing the advantages of concentration of operations associated with even-aged management. The sequence of distinctly separate operations is more visible, more systematic, and simpler to administer than with the selection systems.

The shelterwood method can be applied intensively only if there are markets for trees of small size or poor quality, unless such trees can be killed rather than harvested. The cost of logging is greater than when virtually all the trees are cut in a single operation. The otherwise sound principle of cutting the poor trees before the best can make the first cuttings of the shelterwood sequence financially unattractive, especially if the cost of making the stands accessible is high. Some of these difficulties are partially evaded in the modifications associated with extensive practice. They can be largely overcome once the forest is well served with roads and so organized that both seed and removal cuttings can be carried on simultaneously in different places.

The method is superior to all others except the selection method with respect to protection of the site and aesthetic considerations.

Applications in Extensive Silviculture

With the seed-tree method the cruder applications usually involve compromises about the quality of seed trees or associated site preparation. With the shelterwood method the number of steps may be reduced and the interval between them lengthened; efforts to secure complete restocking or to remove all undesirable vegetation may be relaxed.

One common extensive application of shelterwood cutting involves reliance on advance growth that has started by accident or without deliberate effort to establish it. Fires or other damaging agencies as well as the canopy gaps that develop from tree-sway and other effects in aging stands often allow advance growth to become established. If soil moisture conditions are favorable, advance growth of species that are only moderately shade-tolerant can appear as a result of canopy opening that may seem insignificant. It is entirely possible for stands of some early successional species to have desirable advance growth of a latter successional stage beneath them. In the Great Lakes Region, for example, red pine may follow jack pine in this manner.

If satisfactory advance growth is already established there is no particular reason against releasing it with a single removal cutting. This sort of "one-cut" shelterwood cutting, often done unintentionally, has been remarkably effective in the early stages of forest management in many places, especially in eastern North America. It is a useful procedure for rehabilitating many kinds of degenerate old-growth stands or ones that have been high-graded. The best results in such circumstances have often been associated with the development of good markets for trees that are defective or small. One drawback of the "one-cut" shelterwood method it that it gives little control over the nature of the regeneration. Furthermore, substantial losses of volume are sometimes sustained from the processes that led to establishment. The one-cut shelterwood system has a place in the early development of silvicultural practice but it does not lend itself well to regenerating stands that are kept well stocked and on shortened rotations.

Because of the importance of keeping attention focused on the sources of regeneration it is better to call this the "one-cut shelterwood method" than "clearcutting." Some foresters object to dignifying such a blunt approach with a sophisticated term. However, the reasons for not calling it "clearcutting" are not just pedantic semantics.

Dangers of Ignoring Advance Growth

Some of the worst blunders in American silviculture have resulted from misconceptions that heavy cuttings called "clearcutting" were leading to new regeneration when they were only releasing advance growth. It is easy to overlook small seedlings or to underestimate the age of small trees a few years after a harvesting operation. The blunders come about when the "clearcutting" is shifted to stands that are too young or too tightly closed to have advance growth or when the logging machinery is changed to kinds that destroy it.

One instance of this was in northern hardwood stands on the Allegheny Plateau in northwestern Pennsylvania. These had been subjected to successive high-gradings during the late nineteenth century. Early in the twentieth century a wood-distillation industry developed in the locality and all remaining trees larger than about 6 cm D.B.H. were removed. Fine new stands burgeoned and were dominated by good, fast-growing black cherry along with sugar maple, beech, and other species. In the keen vision of hindsight, it can now been seen that this all came from the advance growth that had been unintentionally established during the era of exploitive high-grading. What foresters mistakenly thought they knew was that the fine black cherries had germinated *after* "clearcutting" and possibly even depended upon the exposure.

It also happened that the heavy chemical-wood cuttings had taken place during the short span of a couple of decades. A large forest mostly of one age class was created. Not surprisingly there was eagerness to start regenerating some stands very early in order to get started with readjusting the age-class distribution for sustained yield. Since "clearcutting" seemed to have worked before some of it was tried in stands only 40 years old. The new vegetation commonly consisted of expanses of fern and a grass with a few beech root-suckers. Initially the blame was put on the excessive deer populations that are common in the locality but it soon became apparent that the lack of advance growth was the real reason. Shelterwood cutting has now been successfully substituted (Marquis, 1979).

Evolution of Application

Shelterwood management of the sort that involves sequential harvest cutting ordinarily does not develop until conditions favor silviculture of at least moderate intensity. There is a classic history (Eyre and Zehngraff, 1948) of such an evolution of practice leading to the formulation of a silvicultural system for red pine in Minnesota. The point of departure was the seed-tree method, which was found wanting because of the inadequacy of the seed supply. As markets improved, it became obvious that the shelterwood system had important advantages in enabling good regeneration, reducing invasion by shrubs, and allowing retention of the best trees for growth to optimum size and quality.

Some form of shelterwood cutting often comes into use when forestry has developed to the point where it becomes necessary to regenerate stands that have thinned and were otherwise well cared for. However, this development usually depends on the existence of other conditions. Among these are the desirability of establishing even-aged stands by natural seeding and the existence of terrain and machinery suitable for partial cutting. Another common and important impetus is the opportunity to reap financial advantage from the continued growth of the trees retained as the shelterwood (or as seed trees).

Matters of aesthetics and public relations also play an important role. The residual trees of a shelterwood are not only attractive but also provide, at a glance, evidence of concern for the future. These considerations had much to do with increasing use of the shelterwood method, especially on public forests, in the United States during the 1970s. The tendency for it to become a panacea, as had been the case with "selective logging" around 1940 and clearcutting around 1960, usually seems to have been avoided.

The use of shelterwood and seed-tree cutting to secure natural regeneration is common among owners who have the long-term outlook but wish, for any reason, to avoid the large out-of-pocket investments that artificial regeneration requires. The validity of this policy depends on the extent to which the gains may be counteracted by the potentially higher costs of logging or post-regeneration treatments. As is the case with other matters, many forest owners will follow courses of action that require little investment even if the results are less than ideal.

The situation in which this policy of seeking low-cost regeneration fits best is with those species, usually rather tolerant of shade, that grow slowly when young but faster later in life. With these the rapid compounding of high regeneration costs can be a more serious financial impediment than that of costs of delayed precommercial thinning of overdense shelterwood regeneration.

Examples of Application

The most common American use of uniform shelterwood cutting for the regeneration of pure stands is for many pines and other conifers. It has been used successfully for loblolly, shortleaf, and pitch pines if coupled with adequate control of hardwoods (Brender, 1973). It is now regarded as the best way of solving the difficult problem of regenerating longleaf pine, which is difficult to plant and has very slow-growing seedlings (Balmer, 1979; Boyer, 1979). When ponderosa pine is grown under even-aged management, the regeneration is often accomplished by shelterwood cutting. In the case of the Black Hills of South Dakota, which have summer rainfall and abundant regeneration, this is fairly easy; in other parts of the rather arid range of the species it may be necessary to wait patiently or plant (Barrett, 1979).

With lodgepole and jack pines shelterwood cutting is often used in stands where the cones are not closed or serotinous; even if they are, the reservation of a shelterwood may provide some shade for seeds from cones on the slash and also impose the kind of variations in seedling height that mitigate overstocking.

In the cases of the moderately shade-tolerant eastern and western white pines the shelterwood system is advantageous for pest management as well as being ideal for natural regeneration. The main problem with growing eastern white pine is usually the white pine weevil. This insect often kills the terminal shoots of trees with tops in the sunshine. The replacement branches are usually the laterals of the previous year and, when they turn upward, they form crooks and forks that badly degrade the stems. One solution is to to use the irregular shelterwood method to keep the young trees shaded until at least one log-length has formed. Older white pines provide the best kind of shade because the weevils are lured away by their sunlit tops. The overstory must remain for such a long time that it should be composed of pines that can produce value rapidly and thus compensate for the slow growth of the deliberately stunted regeneration.

With western white pine, for which the introduced blister rust fungus is the major problem, shelterwood cutting plays a role in keeping the understory *Ribes* shrubs, the alternate hosts of the fungus, in check. The partial shade that is conducive to pine seedlings may be too much to allow the *Ribes* to survive and produce seeds. Recent evidence also suggests that there may be possibilities of building rust-resistant natural populations if rust-resistant parent trees are retained in the shelterwood seed source. With both eastern and western white pines, it is necessary that shelterwood cuttings be severe enough to allow the pine seedlings to grow faster than those of hemlocks and other associated species that are more shade-tolerant than the pines.

Douglas-fir and many other conifers of western North America are ecologically well adapted to shelterwood cutting (Fig. 14-9) (Williamson, 1973; Ruth and Harris, 1979). In the past, the use of the method did not develop because of the difficulties of doing partial cutting, especially in ancient unstable stands and on the steep terrain that is so common. In such places as western Washington the climate and soils are favorable enough that clearcutting and planting is a dependable solution; shelterwood cutting may be used to meet various management objectives but partial shade is not necessary for regeneration.

At the other extreme are at least two categories of sites where Douglas-fir seedlings require shading even if they have been planted. Most of the region is so mountainous that elevational differences and rain-shadow effects create an intricate pattern of variations in site factors to which silviculture must be adapted.

One large locality where shelterwood cutting often seems necessary is southwestern Oregon, which has rainless summers that are longer than those farther north. The problem sites there include south-facing slopes, very coarse-textured soils derived from physically weathered granite, and low elevations where the rainfall is small. On such sites partial shade often appears to reduce direct evaporation and extremes of surface temperature enough to allow seedlings to become established. It is also possible that the retention of some shade reduces the development of ceanothus and manzanita shrubs, which are very prevalent and can easily exclude trees.

FIGURE 14-9   Regeneration, 7 years old, of Douglas-fir from uniform shelterwood cutting on a forest of the U. S. Bureau of Land Management in the Oregon Coast Range. (*Photograph by Yale University School of Forestry and Environmental Studies.*)

Sometimes it is necessary to leave very large, valuable, old-growth Douglas-fir to provide shade; even though they are harvested later, their subsequent growth in value is so small that they are costly umbrellas. Sometimes adroit use can be made of madrone or other trees of low value, as well as nonliving shade, to alleviate the problem. It is interesting to note that shelterwood cutting has often worked in this locality even though the transpiration of the residual trees must reduce the supply of water in the soil strata where most rooting occurs. This suggests that the regeneration bottle neck is at the soil surface or just beneath it rather than at the lower depths.

Somewhat similar problems have been encountered at the high elevations as has been the case in the Alps and other high mountain areas throughout the world (Tranquillini, 1979). In the Cascade Range of Oregon and Washington the problem has been most acute in the zone where Douglas-fir tends to give way to true firs and mountain hemlock, species which, among other adaptations, have saplings that restraighten themselves better after annual bending by massive snowfalls. At these high elevations there is so little atmosphere and water vapor above the terrain that seedlings in the open are defeated by the effects of heat injury by day, frost by night, and probably direct evaporation from the uppermost soil strata. The true firs are even more unsuccessful at surviving in the open than Douglas-fir. Shelterwood cutting has come to be a common solution to the problem but imaginative use has been made of ponderosa pine as a nurse crop. Although ponderosa pine is not adapted to these elevations it survives for some years after planting and tends to exclude shrubs and allow natural invasion of the local conifers beneath it. The planted pines dwindle away after they have served their purpose, which is often that of providing a remedy for misguided clearcutting.

On large areas of public forests in the western United States the shelterwood system plays a major role because of aesthetic considerations, policies of even-aged management, and the high cost of artificial regeneration, especially for sites with low or mediocre productivity.

FIGURE 14-10 Very dense natural regeneration of oak and beech from use of the irregular shelterwood method in the Spessart District of West Germany. (*Photograph by Yale University School of Forestry and Environmental Studies.*)

The approach can have some drawbacks. One common one is the prevalence of dwarf-mistletoe, a parasitic seed-plant that can spread from shelterwood trees to their progeny if they are left there too long. The topography can also be such that high investments are necessary to reach previously inaccessible stands. If this is the case, there are tendencies to clearcut to be sure to pay off the road costs promptly. Any scheme of cutting in which the poorer trees are harvested first is difficult to start in the early phases of deliberate programs of long-term forest management.

The seed-tree system can be useful for hardwood species that are very intolerant of shade and are small-seeded, such as some birches and eucalypts, provided that site preparation is adequate. More generally, however, the shelterwood system is the common mode of even-aged management of hardwoods, at least in temperate climates (Roach and Gingrich, 1968; Kelty and Nyland, 1981; Kellison, Frederick, and Gardner, 1981). In Western Europe oaks are often grown in pure stands on the uniform shelterwood system; very dense regeneration may be secured by planting or direct seeding with the overstory being kept mainly for shade and for additional growth (Fig. 14-10). The shelterwood cutting applied to them can be uniform or irregular. Advance growth of oaks usually develops best in partial shade and can then be stored for some years until it is well enough established to release (Carvell and Tryon, 1961; Holt and Fischer, 1979).

Most hardwood stands are complex mixtures of species of the kinds that will be discussed in Chapter 17 on mixed stands.

BIBLIOGRAPHY

Atkinson, W. A., and R. J. Zasoki (Eds.) 1976. Western hemlock management conference. *Univ. of Wash. Inst. For. Products Contrib.* 34. 317 pp.

Balmer, W. E. (Ed.) 1979. Proceedings, Longleaf Pine Workshop. *USFS Tech. Publ.* SA-TP-3. 119 pp.

Barrett, J. W. 1979. Silviculture of ponderosa pine in the Pacific Northwest: the state of our knowledge. *USFS Gen. Tech. Rept.* PNW-97. 106 pp.

Boyer, W. D. 1979. Regenerating the natural longleaf pine forest. *J. For.* **77**:572–579.

Brender, E. V. 1973. Silviculture of loblolly pine in the Georgia Piedmont. *Ga. For. Res. Counc. Hept.* 33. 74 pp.

Carvell, K. L., and E. H. Tryon. 1961. The effect of environmental factors on the abundance of oak regeneration beneath mature oak stands. *For. Sci.* **7**:98–105.

Eyre, F. H., and P. Zehngraff. 1948. Red pine management in Minnesota. *USDA Circ.* 778. 70 pp.

Holt, H. A., and B. C. Fischer (Eds.) 1979. *Regenerating oaks in upland hardwood forests.* Purdue Univ., Dept. of For. and Nat. Resources, W. Lafayette, Ind. 132 pp.

Kellison, R. C., D. J. Frederick, and W. E. Gardner. 1981. A guide for regenerating and managing natural stands of southern hardwoods. *N. C. Agr. Exp. Sta. Bull.* 463. 23 pp.

Kelty, M. J., and R. D. Nyland. 1981. Regenerating Adirondack hardwoods by shelterwood cutting and control of deer density. *J. For.* **79**:22–26.

Marquis, D. A. 1979. Shelterwood cutting in Allegheny hardwoods. *J. For.* **77**:140–144.

Roach, B. A., and S. F. Gingrich. 1968. Even-aged silviculture for upland central hardwoods. *USDA, Agr. Hbk.* 355. 39 pp.

Ruth, R. H., and A. S. Harris. 1979. Management of western hemlock–Sitka spruce ecosystems for timber production. *USFS Gen. Tech. Rept.* PNW-88. 197 pp.

Sander, I. L. 1977. Manager's handbook for oaks in the Central States. *USFS Gen. Tech. Rept.* NC-37. 35 pp.

Tranquillini, W. 1979. *Physiological ecology of the alpine timberline.* Springer-Verlag, New York. 137 pp.

Trousdell, K. B. 1955. Loblolly pine seed tree mortality. *Southeastern For. Exp. Sta., Sta. Paper* 61. 11 pp.

U.S. Forest Service. 1979. *The shelterwood regeneration method.* USFS Div. Timber Management, Washington, D.C. 221 pp.

Wenger, K. F., and K. B. Trousdell. 1957. Natural regeneration of loblolly pine in the South Atlantic Coastal Plain. *USDA Production Research Rept.* 13. 78 pp.

Williamson, R. L. 1973. Results of shelterwood harvesting of Douglas-fir in the Cascades of western Oregon. *USFS Res. Paper* PMW-161. 13 pp.

CHAPTER 15

Selection Systems and Uneven-Aged Management

The term **selection system** applies to silvicultural programs used to *create or maintain* uneven-aged stands; the **selection method** is employed *to regenerate* such stands. An uneven-aged stand contains at least three well-defined age classes; "well-defined" means differing in total height and age, not just in stem diameter.

The selection systems have been subject to so many different interpretations that certain qualifications are here placed on definitions to clarify discussion.

The first qualification is that the constituent even-aged aggregations involved in this interpretation of the selection method are small. The second is that regulation of the cut by volume within a stand is *not* regarded as an essential characteristic of the selection method.

The need for a restricted definition arises from the fact that there is not, in the strictest sense of the term, any such thing as an uneven-aged stand. Even when a single large tree dies it is replaced not by one new tree but by many that appear nearly simultaneously. This is true even if the new trees are from advance regeneration.

The uneven-aged stand is an artificial entity required for the comprehension of what might otherwise be a chaos of little "stands." The question of how large an even-aged aggregation must be to represent an individual stand depends entirely on the purpose for which one happens to be looking at the forest at the moment. Someone concerned about silviculture or logging may see many little even-aged stands. Someone concerned about forest administration, wildlife, or watershed management may see the same thing as a homogeneous entity characterized by much internal variation. The first party would map the little units separately but the second would regard that as useless complication.

In this chapter an essentially ecological definition will be used for the minimum size of a stand that might be called even-aged. This size is taken to be that of the largest opening that could be fully under the microclimatic influence of adjacent mature trees. The opening of this critical size would, at the very center, have the same temperature regime as that which prevailed over almost all of a large clearcut area. The critical opening is probably about twice as wide as the height of the mature trees. One might, for administrative reasons, set a larger critical size but it is hard to see any reason for using a smaller area as the criterion.

The view is sometimes taken that application of the selection systems necessarily requires that each stand be made into some sort of self-contained sustained-yield unit. This condition is almost never attained in practice and would be difficult to maintain even if achieved. In fact, efforts to mold stands into sustained-yield units often produce results so illogical in the light of other considerations as to give the whole idea a bad name. Nevertheless, the selection systems and uneven-aged stands can have other important advantages. Sustained yield is a noble objective best pursued on the scale of forests of hundreds of hectares and not on that of single stands. However, the procedures for developing sustained yield are basically the same whether it is sought in stands or in whole forests.

In the selection method of reproduction, the mature timber is removed either as single scattered trees or in small groups at relatively short intervals; such cuttings are repeated indefinitely with the deliberate purpose and effect of creating or maintaining an uneven-aged stand. This process depends on the establishment of reproduction at intervals and making it free to grow so that the continuing recruitment of new age classes is achieved. The method is usually associated with natural reproduction but there is no reason why planting or artificial seeding could not be used. True reproduction cutting is typically accomplished by cutting the oldest or largest trees. The seed and any protection necessary for natural reproduction come from the trees that remain around the openings.

Intermediate cuttings may be made among the younger trees at the same time that the older or larger trees are removed. Each immature even-aged aggregation is treated essentially as if it were a stand by itself. Within the whole uneven-aged stand the periods of regeneration and intermediate cutting may thus extend through the entire rotation and be indistinguishable from one another.

If the stand is to have three age classes, all that is required is that there be at least three successful regeneration cuttings, separated by major intervals of time, during the equivalent of one rotation. It should not be inferred that either the cuttings or the age classes need to be equally spaced in time. No age-class structure should be an end in itself. There should be some objective in developing the uneven-aged condition because it is best from the administrative standpoint that a stand have no more age classes than necessary.

Reasons for Uneven-aged Stands

There are two common reasons for developing such stands. The first is the existence of some management objective requiring that a stand always have some large trees. The second is simply that the stands were inherited in that condition and cannot be rendered even-aged without prematurely cutting too many young trees. The first reason requires more explanation than the second even though the second involves a condition common in forests and often easily accepted.

There can be many reasons for maintaining stands that always include some large trees. Not the least important are aesthetic considerations, especially in roadside strips and parts of forests in full public view. Any approach, even a crude one, to achieving sustained yield within small holdings necessitates a rather steady source of large trees. The juxtaposition of old trees and ones of younger ages is frequently important in wildlife management or for other purposes that require diversity of habitats and the species of plants and animals wanted in them. If natural regeneration is difficult, as on adverse sites, there may be reason to maintain a permanent source of seed. Uneven-aged stands may be essential on steep slopes that are either geologically unstable or subject to avalanches.

It has often been stated that the selection method favors regeneration only of shade-tolerant species or cannot be used for intolerant ones. This really is not true. As has been pointed out before, tree seedlings are small and their survival is governed by the characteristics of microenvironments with dimensions measured in centimeters. The shelterwood methods can be modified to regenerate even the most shade-requiring species in even-aged stands and the selection methods to accommodate intolerant ones in large openings. However, the single-tree selection method, next described, is associated mainly with shade-tolerant species.

Variations in Spatial Patterns of Selection Cutting

Single-Tree Selection Method

In this form of the method each little even-aged component of the uneven-aged stand occupies the space created by the removal of a single mature individual or exceedingly small clumps consisting of several such trees (Fig. 15-1). The development of reproduction in the very small, scattered openings thus created is the main characteristic. The species most likely to be perpetuated are those that are very tolerant, although the opening left by the removal of even a single large tree will, if site conditions are otherwise favorable, allow the establishment of a few trees representing early successional stages. The species composition of the reproduction depends as much on the competitive influence of adjacent trees and the way that they are subsequently treated as it does on the size of the holes initially created.

The space left by the removal of each mature tree must be filled by numerous seedlings, which are reduced to roughly a dozen saplings as the clump develops. The number of trees is gradually reduced either by thinning or through natural competition until but one tree is left at the end of the rotation for each that was initially cut.

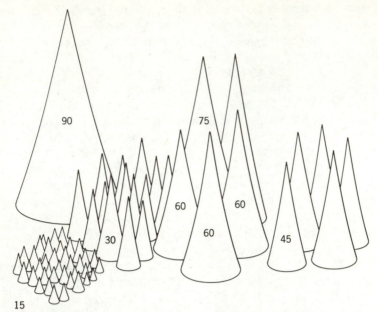

FIGURE 15-1   Schematic oblique view of a 1/10 hectare segment of a balanced uneven-aged stand being managed by the single-tree selection system on a 90-year rotation with a 15-year cutting cycle. Each tree is represented by a cone extending to the ground; the numbers indicate ages. Each age group occupies about 1/60 hectare. The 90-year-old tree is now ready to be replaced by numerous seedlings while the numbers of trees in the middle-aged groups are appropriately reduced by thinning.

One of the problems associated with single-tree openings is the risk that they will be closed by the expansion of the crowns of adjacent trees before the crowns of the new trees can grow to their place in the sunshine. Frequent light cuttings are, therefore, required to keep the system working. Logging is thus difficult and costly; the protection of young trees is a chronic problem because many kinds of operations are continually scattered throughout the stands.

Group-Selection Method

The selection system usually works best, from the procedural standpoint, if the final age-class units consist of more than single mature trees (Fig. 15-2). If the regeneration openings are made larger it becomes possible to accommodate the ecological requirements of almost any tree species. As previously stated, if the matter is viewed from the standpoint of ecological site factors, the maximum width of what are called "groups" could be set at approximately twice the height of the mature trees. With such a classification limit, the same environmental conditions that exist in a huge clearcut area could be made to exist but only near the center of the open-

FIGURE 15-2   A stand of ponderosa pine in the Fort Valley Experimental forest, Arizona, managed under the group-selection system. The saplings form a group that resulted from one unusually favorable regeneration year. The group that occupies the foreground has been thinned twice to favor trees of best bole form such as the ones now remaining. The small cards mark the stumps of medium-sized trees cut 32 years earlier; the large card marks the position of a large member of the group harvested 2 years ago. (*Photograph by U. S. Forest Service.*)

ing. Regeneration of some very intolerant species might thus be impractical.

As indicated earlier, this semantic matter can be clarified by viewing it from the standpoint of someone making administrative maps of forests for such purposes as determining areas of different age classes. Under this interpretation, the smallest areal unit of a given age class that one was ready to delineate on a map would be the smallest recognizable "stand." Anything smaller would be deemed a "group" and part of a larger "stand" with two or more age classes. With aerial photographs (Paine, 1981) it is possible to identify "stands" very much smaller than was the case when this terminology was first used and foresters could map stands only by measuring distances on the ground. In other words, it is now much easier to recognize small even-aged stands and there is less reason to lump them into larger uneven-aged stands.

The group-selection method has certain definite advantages (Roach, 1974) over single-tree selection cutting. Harvesting of the older trees can be carried out more cheaply and with less damage to the residual stand. There is greater latitude for creating the kind of environmental conditions necessary for reproduction. The trees develop in clearly defined even-aged aggregations; this is of substantial advantage in developing good form in many species, especially in hardwoods. This modification is also more readily applied to stands that have become uneven-aged through natural processes because such stands are more likely to contain even-aged groups of mature trees than a mixture of age classes by single trees. Furthermore, the openings created are large enough that the progress of regeneration and the growth of younger age classes are more readily apparent.

It is important to consider the significance of the effects involved in having trees growing on the edges of even-aged aggregations. The total perimeter of such edges is, of course, at a maximum with the single-tree selection method. The effects are both beneficial and harmful but vary according to the circumstances. Influences on reproduction are beneficial to the extent that side shade protects young seedlings. However, the competition from the older trees for soil moisture can be serious. The roots of the older trees are capable of spreading out into the newly created openings; they may even do so automatically if they are already attached to the roots of the cut trees through intraspecific root grafts. On the other hand, this lateral expansion of the roots and crowns of older trees along the edges of groups is precisely the same effect that makes it possible to allocate more area to the growth of large trees than is the case with forests of even-aged stands (see Fig. 15-7).

The juxtaposition of different age classes provides an opportunity to release the young trees along each edge once each rotation when the older adjoining groups are harvested. This takes the place of one heavy thinning and is of some advantage when ordinary thinning must be long delayed or is not feasible. The effects of this phenomenon on the form of the edge trees vary considerably depending on the species and circumstances. The development of large branches is forestalled as long as there are larger trees at the side but the effect is reversed when the larger trees are removed. Distinctly phototropic species, such as most of the hardwoods, do not fare well because of the tendency of the stems to bend outward toward the light or to become crooked if the terminal shoots are battered by the crowns of taller trees. This is not so true of distinctly geotropic species like conifers, which tend to grow straight under all circumstances. In climates subject to wet, clinging snows, any trees growing beside taller conifers are likely to be broken or ruined when accumulations of snow suddenly slide off the higher crowns.

Many species of wildlife profit from the combination of environmental conditions existing along the boundaries between very young groups and older trees. Several different kinds of protective cover are available in close proximity to the variety of food plants that may be fostered by the broad spectrum of microclimatic conditions existing between the edges and the centers of the young groups.

In the application of the group-selection method it is logical to weigh the relative advantages and disadvantages of growing trees along the edges of groups. If the advantages outweigh the disadvantages, an effort may be made to increase the length of group perimeter per acre by creating smaller groups.

The simplest kinds of group-selection cutting create definite gaps in the forest canopy. Some ecologists regard "gap-phase" regeneration as the common way that forests are renewed without departing from the "climax" or late successional kind of species composition. The gaps are construed as being caused by wind or maladies associated with old age. If the gaps are large enough, they will, for the reasons described in Chapter 7, provide microenvironmental conditions suitable for the regeneration of species from almost the whole spectrum of local vegetation. This may help explain why climax vegetation is regarded as having much higher species diversity than that of earlier successional stages. If so, this is because the environment of the gap is not uniform but covers a wide range of microclimates. Neither the forester nor the ecologist should regard the entire area of small openings as conducive to only one category of vegetation.

Another ecological attribute of small openings that must be anticipated is their tendency to become pockets of hot air by day and of frost and dew by night. This is because they are not well ventilated by the wind but are open to the sky and, if large enough, to the sun. At night plant surfaces cooled by radiational heat loss may not be adequately reheated by contact with warm air wafted down from above; dew or frost may form on the plants. This may cause risk of damage by frost or increase that of infection of leaves by fungus spores, such as those of the stem rusts of conifers. The frost effect has been suggested, along with deer browsing, as a means of controlling unwanted true firs in the regeneration of pine forests in the Sierra Nevada (Gordon, 1970).

The effects of hot-air accumulation in sunlit openings on plants are more obscure; humans find them very pleasant in winter and oppressive in summer. Studies in German spruce forests have shown that the risk of frost is most extreme in openings that are $1\frac{1}{2}$ times the height of the surrounding trees (Geiger, 1965). Wider openings are better ventilated and smaller ones more shaded, especially at high latitude. If these effects are undesirable, one can wholly or partly substitute the effects of more diffuse cover of the kinds associated with, but not limited to, shelterwood cutting. Those combine some shade with more opportunity for ventilation by the wind, which may be thought of as entering the forest from above in the degree that canopy openings allow.

There is no reason why all the mature trees of a group must be cut at a single time. If desirable because of the ecological requirements for reproduction, need for application of the financial maturity concept, or some other valid consideration, the groups can be reproduced by small-scale applications of the shelterwood and seed-tree methods. The group-selection method is not one procedure but a category including many variations that could be separated only with difficulty and by the use of complicated terminology. Figure 17-5, in Chapter 17 about mixed stands, shows how the group-selection method might be applied to stands of stratified structure. This is a sample of one of many possible variations and is probably more complicated than most.

There is no necessity that the individual groups in one stand be uniform in size, shape, or arrangement, nor is it necessary that one age class be represented by a single group. In fact, if proper advantage is taken of the existence of groups of different ages when an irregularly uneven-aged stand is placed under management, such regularity is unlikely. A systematic arrangement of individual groups would be convenient but is usually almost impossible to secure. If the total area occupied by different age classes can be determined with a fair degree of accuracy, the cut under the group-selection method can be regulated by area as well as by volume.

Strip-Selection Method

In both the single-tree and group-selection methods, the timber of the oldest age class is scattered through the stand and must often be extracted across areas under regeneration. The attendant difficulties of logging and administration may be alleviated by bringing all the trees of each class together in a separate part of the stand. This is best done by the strip-selection method in which each age class is concentrated in a long, narrow strip. The adjoining age classes on the two sides are respectively younger and older than the strip between. A complete series of age classes from reproduction to mature timber may thus be formed in consecutive order. This kind of arrangement is similar to the strip-shelterwood method, except that it has the objective of securing a complete range of age classes across the whole series of strips.

This arrangement enables the transportation of logs through the next strip to be harvested. There is ample opportunity to obtain reproduction under an overwood as in the shelterwood method. Successive cuttings should progress toward the south or against the direction of the most dangerous winds, as in the strip-shelterwood method. If protection against wind is desired, the strip-selection method has the additional advantage of creating a stand that is streamlined so that the main force of strong winds is diverted up and over the stand (Fig. 15-3).

Strip and patch selection cutting have sometimes been effective in helping increase the yield of water from snow melt in the mountains of the western United States (Anderson, 1960 and Leaf, 1975). Although the increases come mainly from reductions of transpiration caused by tree removal, there are some advantages in having snow concentrated in drifts that form in the openings. If it is desirable to have these drifts shaded so as to alter the period during which the snow melts, this can be done by having successive strips advance toward the south.

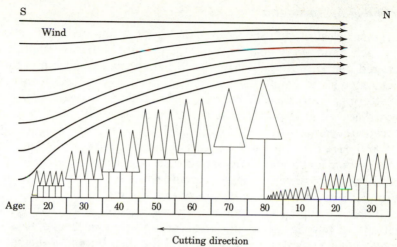

FIGURE 15-3   Cross section along a north-south axis through a portion of a stand reproduced by the strip-selection method and managed on an 80-year rotation with a cutting cycle of 10 years. By this means the stand is stream-lined against southerly winds and regeneration is protected from the direct rays of the sun. Cutting progresses from north to south; the 80-year-old strip is about to be cut.

The outstanding drawback of the strip-selection arrangement is the difficulty of creating it. The sequence of age classes is highly artificial and radically different from any likely to be found in nature.

Balanced Uneven-aged Stands

Ideal Status

One of the most attractive and idyllic concepts of forestry is that of the all-aged stand; it continually yields benefits and regenerates itself steadily; it is a dynamic system that is always the same and always in balance. It would be a perfect sustained-yield unit and would also approximate some state of ecological equilibrium, however that might be defined.

Even though this condition has a powerful naturalistic appeal, it is not one that would come into existence in nature but would have to be an essentially artificial creation. One would have to regenerate the same amount of forest area each year for about a whole rotation to create the condition exactly. It is a simple encapsulation of good ideas toward which foresters strive by various means but should not expect to achieve per-fectly.

The theoretical all-aged stand, the ideal sustained-yield unit, would have trees of every age class from 1-year seedlings to veterans of rotation age, with each age class occupying an equal area. Since it is virtually impossible to achieve this condition or even to carry out the measurements of area to determine whether one had, the idea must be modified if it is to come close to applicability.

Simplifying Approximations

The simplest modification has to do with the impracticality of operating in each stand every year. Instead it is presumed that the stand would be treated periodically with each period being called a **cutting cycle.** The uneven-aged stand would have as many age classes as there were cutting cycles during the rotation. If the result were a stand with an arrangement of age classes that approximated the all-aged stand it could be called a balanced uneven-aged stand. Figure 15-4 is a schematic map of such a stand and Fig. 15-5 shows how such stands might be combined into a whole forest managed on some perfect selection system. An uneven-aged stand that did not have age classes equally spaced and equally represented in terms of area would be an irregular uneven-aged stand.

The most complicated modifications are attempts to avoid direct determinations of tree ages and the areas occupied by each age class. Tree diameter is taken as the index of age and the distribution of diameters in the stand is used to assess and control the allocation of area of growing space to each age class. Because of the impracticality of logging in each

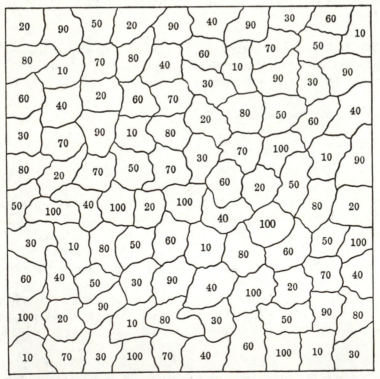

FIGURE 15-4 A 1-acre portion of a fully balanced selection stand managed on a rotation of 100 years under a 10-year cutting cycle. Ten age classes are represented, each occupying approximately one-tenth of the area. The numbers indicate the ages of the individual groups of trees.

Stand 1 contains age classes: 1, 11, 21, 31, 41, 51, 61, 71, 81, and 91	Stand 2 contains age classes: 2, 12, 22, 32, 42, 52, 62, 72, 82, and 92	Stand 3 contains age classes: 3, 13, 23, 33, 43, 53, 63, 73, 83, and 93	Stand 4 contains age classes: 4, 14, 24, 34, 44, 54, 64, 74, 84, and 94	Stand 5 contains age classes: 5, 15, 25, 35, 45, 55, 65, 75, 85, and 95
Stand 6 contains age classes: 6, 16, 26, 36, 46, 56, 66, 76, 86, and 96	Stand 7 contains age classes: 7, 17, 27, 37, 47, 57, 67, 77, 87, and 97	Stand 8 contains age classes: 8, 18, 28, 38, 48, 58, 68, 78, 88, and 98	Stand 9 contains age classes: 9, 19, 29, 39, 49, 59, 69, 79, 89, and 99	Stand 10 contains age classes: 10, 20, 30, 40, 50, 60, 70, 80, 90, and 100

FIGURE 15-5 Diagram of a selection forest managed on a rotation of 100 years with a cutting cycle of 10 years. The forest contains ten stands, one of which is cut through each year, thus giving equal annual cuts. Each stand contains ten age classes, and together the age classes in the ten stands form a continuous series of ages from 1 to 100 years.

stand each year, it is regarded as sufficient that there be three or more equally represented age classes with equal intervals of time between each, provided that they collectively have the proper diameter distribution.

Regulation of Cut by Diameter Distribution

The characteristics of the proper diameter distributions are depicted in Fig. 15-6. Since it requires many saplings to cover the space eventually occupied by a single mature tree, the distribution should approximate the smooth, "reverse-J-shaped" curve of Fig. 15-6A. This curve really represents the collective total of the diameter distributions of a series of little even-aged, pure groups of trees covering equal areas and separated by equal intervals of age, as shown in Fig. 15-6B. The objective is to harvest and replace trees in such patterns that this diameter distribution is the same either just before or just after each harvesting operation.

This procedure guides what is called "volume regulation for sustained yield" (Davis, 1966). This term probably refers back to a time when the ordinate of the graphs was timber volume rather than numbers of trees. "Regulation by diameter distribution" could be a better term. However, as is shown in Fig. 15-6B, this approach is really only area regulation by an indirect route that is necessitated by the problems of determining tree ages and land areas within stands of complex structure.

The proper J-shaped curve is often assumed to be convertible to a straight line if the logarithms of the numbers of trees per unit area are plotted over the arithmetical diameters. This relationship was first hypothesized by a French forester, deLiocourt. Diameter distributions defined by

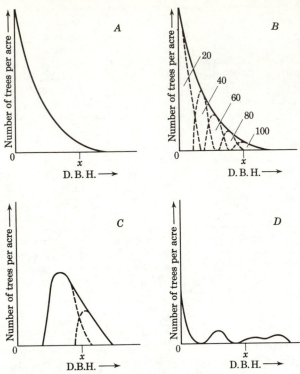

FIGURE 15-6 Several types of diameter distributions found in uneven-aged stands. *A* shows the distribution curve of an all-aged stand of 1 acre containing sufficient trees in each diameter class to produce an unvarying number of trees of optimum size (of diameter *x* or larger) at rotation age. *B* indicates how a balanced uneven-aged stand of the same diameter distribution may be composed of five age classes, each occupying an equal area, with a 100-year rotation. *C* represents a stand with two closely spaced age classes and no advance reproduction. Uncritically supervised selection cuttings in stands of this kind will produce abnormally high yields of timber until all existing trees reach optimum size; thereafter they will yield nothing until the new growth reaches merchantable size. It is difficult to create the balanced distribution in such stands in one rotation period, especially if the smaller trees do not respond to release after selection cuttings. *D* shows one of the many kinds of irregular uneven-aged distributions which may be found in virgin stands. This one contains four well-distributed age classes, one of which is well beyond optimum age for a commercial rotation. Such a stand could be gradually converted to the balanced form provided that the intermediate age classes did not deteriorate after partial cutting.

this relationship do not really fit those deduced, as shown by Fig. 15-6*B*, from yield tables for pure even-aged stands. The discrepancy lies in the fact that, if the relationship specifies logical numbers of trees in the larger sizes, as it often can, then there seem to be far too few in the smaller diameter classes. For some, this casts doubt on the validity of using this mathematical relationship in managing stands or forests for sustained yield. For advocates of the concept, the difference in numbers of small trees is often regarded as evidence of some advantages deemed to be inherent in the uneven-aged arrangement.

Possibilities with More Large and Fewer Small Trees

It certainly would be advantageous to have forests that provided a sustained yield of timber but did not have substantial areas that were stocked with trees that were too small to be merchantable. This is because a given amount of wood is more valuable if laid down on tree stems of merchantable size than on small ones. However, trees have to be small before they can be large. There are various ways, including thinning, for shifting production onto large stems by diminishing the proportion of small ones, but there are limits on what can be done.

If the regeneration units of a selection stand are small, the area that was previously occupied by one or two mature trees is not recolonized fully by seedlings. Some is taken over by the horizontal expansion of the crowns and roots of adjacent trees. When they are cut in their turn, the regeneration group can, in its turn, take over part of the newly vacated growing space. In this way, as shown in Fig. 15-7, the amount of area used by a given age-class group increases each time an adjacent older group is removed. This effect must have some reality in stands with small age-class groups and, therefore, much area of boundary zones between groups. The effect would be negligible if the groups are large.

Another potential source of gain of efficiency lies in the "advance-growth effect" described in connection with shelterwood cutting in Chapter 14. If small trees can be grown under large ones, it is not necessary to allocate to them the space that would be required if they were grown in the open. This advantage cannot be obtained unless the species are sufficiently shade-tolerant and the site factors permit the existence of advance growth.

If this effect were perfectly achieved, it might seem that one could simply allocate all of the "space" in the regulated diameter distribution to the released trees and none to those too small to release. However, it would have to be kept in mind that trees of sufficient size to be released had to be available and that they would not grow to this size anywhere near as fast as they would in an open-grown even-aged stand. This source of efficiency, to the extent that it exists, is equally attainable with shelterwood cutting. The point is that being able to give small trees a head start under big ones affects the shape of the appropriate curves of diameter distribution.

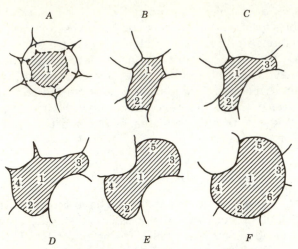

FIGURE 15-7 The initial contraction and subsequent expansion of area occupied by the trees of one small age-group in an uneven-aged stand. *A* shows the circular outline of the area previously covered by a single mature tree that has just been removed; the smaller cross-hatched area 1 is that to which the new age-group of seedlings is confined by expansion of the older adjacent trees. Sketches *B* to *F* show how this group of trees would reexpand as periodic, sequential cutting of areas 2 to 6 successively removed adjacent trees during the remainder of the rotation.

In stands on very dry sites, there can be another effect that diminishes the required numbers of small trees. In such cases, as in some ponderosa pine stands in the interior western United States (Fig. 15-2), groups of trees can have root systems that fully occupy the soil but crowns that do not come close to touching because the possible amount of foliage is so restricted. In such situations, it may not take many small trees to restock an area and often the effect of root competition seems severe enough to restrict branch size and stem taper without any obvious crown competition (Pearson, 1950). If this is so, it may take only a few seedlings to replace one mature tree. This is not really a source of efficiency but it does explain why a seemingly inadequate number of small trees may suffice. More importantly it is a phenomenon that is solace to foresters in places so dry that forest regeneration borders on the impossible. Suitable training effects of root competition presumably depend on having older trees close by.

It should also be noted that a series of well-thinned even-aged stands, representing a full sequence of age classes up to rotation age, would have more trees in the larger sizes and fewer in the smaller ones than the densely stocked unthinned stands depicted in the kind of "normal" yield tables that used to be developed in the United States.

Adjustments of q-Factor

There has been much American effort to use the deLiocourt hypothesis to develop balanced uneven-aged stands and to regulate cuttings in them by it. This means that the shapes of these curves of numbers of trees over D.B.H. must be determined for specific kinds of stands and management regimes (U. S. Forest Service, 1978). The choices to be made are of the slopes and intercepts of the kinds of semilogarithmic relationships shown in Fig. 15-8. The slope of each line is called the *q*-factor and defines a geometric progression of increasing numbers of trees with decreasing diameters. In customary usage, if the *q*-factor is 1.4 the number of trees in a given 2-inch D.B.H. class is 1.4 times that of the next larger 2-inch class.

The intercept of the line is the number of trees per acre at some given D.B.H. The most important thing about this intercept value is that it is defined by the interaction of (1) the total basal area chosen for the whole

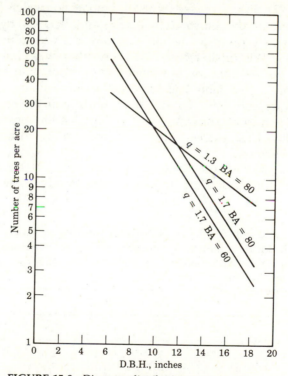

FIGURE 15-8 Diameter-distribution curves converted to straight lines by plotting logarithms of numbers of trees over arithmetic values of D.B.H., showing the effects of some changes in values of the *q*-factor and stand basal area (sq. ft. per acre) that are also shown in Table 15-2. The values of *q* are for 2-inch classes and the stands defined would have no trees larger than the 18-inch class.

stand and (2) the largest diameter class that will be grown. This also means that the resulting diameter distribution is for the stand just before one of the periodic harvesting operations. Table 15-1 by Tubbs and Oberg (1978) can be used to guide the calculations, which are based on the fact that basal area, tree numbers, and diameters are all perfectly related.

Table 15-2 shows the effect of some of the choices on diameter distributions. In addition to some examples of the deLiocourt relationship (with different values of q, basal area, and maximum D.B.H.), it also shows some diameter distributions formulated under various assumptions for certain hardwood stands.

The first four columns have examples in which two attributes are kept nearly the same. The stand basal area for each of the four cases is about 80 ft^2 per acre and the 18-inch class is the largest or nearly the largest to be grown. The first column is not based on deLiocourt's hypothesis but is constructed by Gingrich (1978) from the normal yield tables of Schnur (1937) for a balanced series of even-aged mixed hardwood stands.

The second column represents a stand model constructed for a certain kind of northern hardwood stand by Adams and Ek (1974) by a simulation based upon some growth-prediction equations derived from observations of many plots in Wisconsin forests. This simulation was aimed at optimizing financial returns from multiproduct timber production but within the tight limitations of sustained yield. Both of these distributions specify fewer large trees and usually more small ones than the semilogarithmic ones that happen to be shown. However, the distribution resulting from use of a q-factor of 1.7 is close to that derived from the normal yield tables for unthinned stands.

On the other hand, the distribution based on a q-factor of 1.3 typifies those often recommended for such hardwood stands (Crow *et al.*, 1981).

Table 15-1 Table from Tubbs and Oberg (1978) showing values of a coefficient, K, for determining the number of trees to be grown in uneven-aged stands with diameter distributions with different amounts of basal area per acre and various q-factors.

Maximum D.B.H. (in.)	q-Factor						
	1.1	1.2	1.3	1.4	1.5	1.6	1.7
24	18.5	24.8	34.3	48.1	68.0	97.3	139.3
22	14.0	18.2	24.0	32.1	43.4	58.9	80.1
20	10.3	13.0	16.4	21.1	27.2	35.2	45.6
18	7.4	9.0	11.0	13.5	16.7	20.6	25.5

Divide the chosen value of stand basal area by the K-value for the desired combination of q-factor and maximum D.B.H. to determine the number of trees in the largest diameter class. To determine the number of trees in each 2-inch class, multiply the number of trees in the next class with larger D.B.H. by the q-factor, as has been done for Table 15-2.

Table 15-2 Diameter distributions for eastern hardwood stands exemplifying the results of different criteria[a] used to determine these distributions for various kinds of uneven-aged stands.

D.B.H. (in.)	Yield table (1)	Adams and Ek (2)	BA 80 $q = 1.3$ (3)	BA 80 $q = 1.7$ (3)	BA 60 $q = 1.7$ (3)
6	73.4	103.6	35.1	75.7	56.8
8	50.5	43.1	27.0	44.5	33.4
10	32.7	27.6	20.8	26.2	19.7
12	18.7	22.1	16.0	15.4	11.6
14	9.2	6.7	12.3	9.1	6.8
16	4.1	2.3	9.5	5.3	4.0
18	1.0	1.1	7.3	3.1	2.4
20	0.2				
Total	189.8	206.5	128.0	179.3	134.7

(1) Based on a series of even-aged stands in a normal yield table for eastern oak stands, site index of 70 feet; total basal area of this hypothetical stand would be 82.3 ft² per acre.
(2) Based on simulation model aimed at securing sustained yield and optimizing financial returns; total basal area of stand would be 80 ft² per acre.
(3) Derived entirely from chosen values of q, basal areas per acre, and maximum D.B.H.

[a] See discussion in text for explanation of these criteria.

Reference back to Fig. 15-8, which shows some of these same relationships graphically, indicates the way that the diameter-distribution curve has been rotated so that it specifies more large and fewer small trees. It is very important to note that this kind of alteration in the distribution is usually made on the basis of observations that stands with many large trees grow more in merchantable volume than those defined by larger values of the q-factor. However, this advantage prevails only for part of a rotation because there are enough small trees in the whole system to continue the sustained yield; too much of the production has been shifted into the large diameter classes.

In this case involving hardwood stands, it appears that use of a q-factor of 1.7 would ensure sustained yield but too much lowering of the q-factor might cause the well-spring of sustained yield to run dry. In other words, sustained yield is not guaranteed by developing stands that conform to just *any* value of the q-factor. Figure 15-8 and Table 15-2 can also be used to examine the effect of changing the basal area and maximum diameter while the q-factor is constant.

The reason why there is so much confusion about appropriate diameter distributions is that no balanced uneven-aged stands are positively known to exist anywhere. Even if they did, it would take decades of close observation to prove that they conformed. Virgin forests of climax vegetation do not necessarily conform. The perfection of a given diameter distribution could be demonstrated only if one cut back to it at the end of each cutting cycle and the stand returned to the same preharvest distribution by the end of the next cycle.

Even if the proper distribution is not known it is still possible to keep track of how it changes. In this connection it is important to count the seedlings and saplings as well as the trees of larger D.B.H. Ordinarily there will be changes in numbers of trees in different classes from one inventory to another. These changes should be analyzed as objectively as possible. If they seem to be departing too much from one's best estimate of a logical diameter distribution, the silvicultural management plans should be altered accordingly. If the situation is kept under thoughtful surveillance, it is often possible to detect problems and correct them before they become serious.

At this point it is well to reiterate that the use of a selection method does not require that diameter distributions need to be adjusted to make stands into any sort of sustained yield units. Most of the benefits of the selection system really lie in other considerations. The method probably works better if stands are managed without great attention to diameter distributions, provided that it is recognized that concern about sustained yield within the stand has been relaxed. The main reason why foresters need to know about this mode of analysis lies in its utility in dealing with sustained-yield policies for whole forests and regions.

Application on Broad Geographical Scales

The forests of some large ownerships, localities, or regions often include polyglot assemblages of irregular uneven-aged stands, even-aged stands, understocked areas, and mixed stands. Not many forests are all neat arrays of nice, pure, even-aged plantations that lend themselves to easy analysis by area regulation. The mode of analysis by diameter distribution that has been described can also be used, if only for want of better methods, to analyze the status of many complicated forest situations with respect to sustained yield.

An example of the result of such analysis is the revelations of serious shortages of pine regeneration in the pine forest economy of the southeastern United States (Boyce and Knight, 1979). These are deemed to be the result of the tendency to cut pines and let hardwoods take over on small ownerships. An additional cause is the cessation of at least a century of abandonment of agricultural land that led to recolonization by old-field stands of pine. The mode of analysis has proven to be useful in this and similar cases regardless of whether any sort of uneven-aged selection management was contemplated.

Creation of Balanced Uneven-Aged Stands

It has been necessary to dwell at some length on the principles involved in maintaining the balanced distribution of age and diameter classes in stands that have already been brought to this condition. However, such stands exist far more often in print than in nature so the procedure for creating reasonable approximations of them is really more important.

Theoretically an absolutely even-aged stand can be replaced by a balanced uneven-aged stand during the period of one rotation. If the new stand is to have five age classes with a rotation of 100 years, all that is necessary is to conduct cuttings leading to the establishment of reproduction on one-fifth of the area at 20-year intervals. This necessarily means that some parts of the stand must be cut before or after the age of economic maturity. The disadvantages of such awkward timing of the harvests can sometimes be mitigated if any parts of the stand that mature early are replaced first and those that continue to increase in value longest are cut last. The conversion will be successful only to the extent that the cuttings are appropriately timed and heavy enough to lead to the establishment of the age classes and species of reproduction desired. It is usually a fallacy to pretend that the large trees of essentially even-aged stands are older than the small ones and to attempt to make the conversion by successive cuttings of dominants.

The creation of balanced uneven-aged stands can be carried out more rapidly and easily in stands that already contain several age classes in irregular distribution. It will still be necessary to make the sacrifice represented by cutting some trees earlier or later than might otherwise be desirable. Sometimes small, relatively old trees of the lower crown classes can be substituted for nonexistent dominants of the same size and younger age, thus inventing "age" classes that might otherwise take decades to grow from seedlings. This shortcut is useful to the extent that the trees involved can withstand exposure and return to full vigor. However, it is a procedure that sometimes becomes so attractive that it is indulged in to the exclusion of the creation of the new age classes of seedlings that must be continually recruited and made free to grow if the balanced distribution of age classes is to be completely developed.

The readjustment of the age-class distribution of a stand is likely to be accomplished with least sacrifice of other objectives if it is prosecuted slowly. Several rotations might well be required to transform essentially even-aged stands to the balanced uneven-aged form.

Procedures and Precautions Necessary for Sustained Yield

If the stand is to be treated as a sustained-yield unit and the whole forest were to consist of similarly treated stands, any errors that lead to the overcutting of stands would endanger the continuity of harvests in the whole forest.

The basic procedures for regulating the cut are the same whether the sustained-yield unit is a 50-hectare stand or the 200,000 acres ordained by the U. S. Congress in 1976 as the *minimum* size for such a unit in management of National Forests. The main essential is to develop a normal distribution of age classes, that is, an arrangement in which all classes, from the youngest to those of rotation age, are represented by equal areas. If, as under even-aged management, the areas and ages of all stands can be readily determined, this can be done by area regulation. Under complete uneven-aged management the areas of individual age classes are so small and scattered that the allowable cut must be determined in relation to the annual (or decennial) growth of the stand. This depends, in turn, on knowing and regulating the distribution of numbers, volumes, or basal areas of trees of different diameter classes.

Under either kind of management, an attempt is made to arrive at an average distribution of diameters that will coincide with the appropriate J-shaped curve (Fig. 15-6A). Once this perfect balance is secured the total annual cut of the forest can be equal to the annual growth without endangering future yields. If there is a surplus of trees of mature size, the annual harvest may actually exceed the growth until the balanced distribution is created. If there is a deficiency of trees of optimum size, the annual cut must be less than the growth so long as this condition prevails. *A surplus in one part of the diameter distribution can exist only if there is a compensating deficiency in another.*

The degree of precision of regulation of the cut necessary for the maintenance or creation of any real approximation of a balanced, uneven-aged stand cannot be achieved without making an inventory of the stand before marking it for cutting. Usually this involves determining the distribution of diameter classes by basal area or numbers of trees per acre. The prospective cut is then allocated among the various diameter classes on the basis of a comparison between the actual distribution and that which is assumed to represent the balanced condition.

Pitfalls

This means of regulation for sustained yield is not a technique that works in the absence of an understanding of the numerous sources of error. The risk of mistakes is greatest when individual unbalanced stands are being converted to the proper distribution of age and diameter classes.

The most important kind of error arises when stands are regarded as balanced and uneven-aged when they are not. A stand can have a diameter distribution that plots as a J-shaped curve on arithmetic axes or as a straight line logarithmically and still be far from balanced. Not just any slope of these relationships will give sustained yield.

It has sometimes happened that even-aged stands are treated as if they were balanced and uneven-aged. This folly ordinarily starts with the misconception that stands are uneven-aged if they merely exhibit a broad *range* of diameter classes. It is especially easy to be misled in mixed, even-aged stands composed of species that develop at different rates. This is part of the reason why such stands are discussed in the separate, final chapter of this book. A balanced uneven-aged stand cannot have a canopy that is smooth on top; it cannot be uniformly open and park-like underneath; the trees of pole size cannot outnumber the saplings nor they the seedlings.

These characteristics are associated mostly with even-aged stands and they tend to remain even-aged regardless of treatment.

The practice of ignoring small trees in timber inventories leads to misapprehensions about the diameter distribution of uneven-aged stands, especially if the proper distribution is not well known. If no real account is made of small trees and almost anything that looks like a J-shaped diameter distribution of larger trees is accepted, it becomes all too easy to ignore the necessity of continuing recruitment of seedlings and saplings. The greatest danger of misapprehension exists with distributions such as those shown in Fig. 15-6, *C* and *D*.

If several species are present, there is a strong tendency for the more tolerant ones to predominate in the reproduction, thus endangering the sustained yield of any desirable intolerant species present in the larger diameter classes. The tendency can be combatted only by attempting to make the cuttings heavy enough to provide the proper environmental conditions for the desired species of reproduction and, if necessary, by release cutting. Advance reproduction has many desirable attributes but begins to make a positive contribution to sustained yield only when it is made open to the sky and becomes free to grow.

The most serious mistake that can be made in regulating timber harvests by these ideas results from casually assuming that the allowable annual harvest is equal to the periodic annual increment. This assumption is valid only if the stand (or forest) is in the balanced uneven-aged condition.

Problems arise most often with high-graded forests that are irregular enough to bear superficial resemblance to uneven-aged ones, especially if the trees are essentially middle-aged. Someone makes a passing show of fitting a *q*-factor (usually a low one) to the diameter distribution and pronounces the stand all-aged. This can be rather easy to do if there are some shade-tolerant species such as spruces, firs, or certain hardwoods. The next move is determining the periodic annual increment of the stand. This is usually done by the technique of stand-table projection in which some future stand volume is determined by extrapolation of diameter growth rates of different D.B.H. classes.

While it may be useful for other purposes to know how much a stand will grow in the next decade or two from determining its periodic (or current) annual increment, this does *not* define the long-term allowable cut. If there is a preponderance of middle-aged trees of the larger submature sizes and a shortage of small trees, the periodic annual increment will be higher than the sustainable cutting rate.

If one cuts at this rate too long in a whole forest of stands that have thus been mistaken for all-aged, one will run out of timber to cut. This will not happen in a few stands but in all. It will not happen all at once but the bankruptcy will develop slowly. One generation of foresters may not even be aware that the average D.B.H. of the trees has shrunk because of the high-grading done by their predecessors, especially if they ignore the old inventory data.

Precautions

One way to guard against overcutting is to bear in mind that the allowable cutting rate for any conglomeration of trees under sustained yield cannot significantly exceed the peak mean annual increment for an even-aged stand of the same species composition on land of the same site quality. Even then it is not correct to calculate that the allowable annual harvest is x units of cubic volume and then remove it all from the biggest and best trees of the choicest species, all close to the road system. The harvests must instead be properly distributed among the mature (and thinnable) sizes and kinds of trees on all kinds of sites, if sustained yield is to be obtained. One must, in other words, pay heed to how the allowable cut in volume is distributed over the whole diameter distribution and over the land.

Failure to do this can lead to the kind of forest that some Scandinavian foresters have called the Green Lie. This is a high-graded kind of forest that is full of trees but looks good only if viewed from across a valley through the wrong end of a telescope. Scandinavia now has laws against forestry practices that create such stands, but they can be found in North America and other parts of the world. It is not necessarily sinful to create them but it is folly to do so and fail to perceive what is happening.

In schemes of partial cutting, there can be "undercutting" as well as overcutting. Without the assurance that comes from quantifying the amount of forest growing stock and predicting its growth, the conscientious forester is very likely to leave too much standing. If one is following the selection system there can be a nervous tendency to want to see all future harvests from one spot. Unless there is some quantitative guide, this often leads foresters to leave stands overcrowded with middle-aged trees and deficient in growing space for regeneration.

In this connection, it is also well to be aware that the desirability of retaining trees in sizes, age classes, or other categories needed for a well-balanced growing stock for a stand or whole forest can lead one into patterns of marking for cutting that may seem irrational from any other standpoint. One may be led not only into retaining rather poor trees in deficient categories but also into seemingly reckless removals in ones that have surpluses. Most matters of silviculture are questions of how much of the fixed growing space of forests is allocated to what.

Analysis of Stand Increment

The bookkeeping of the change of amount of wood in stands and forests can be very tricky. In the first place there is the problem that the change or "growth" is determined usually from measurements made at two different times. Each measurement itself is subject to errors, so this "growth," the difference between two measurements with high variance, is an estimate with even higher variance.

Second, it is important to subdivide the "growth" into its components. **Survivor growth** is that on trees present at the beginning and end of the period. Another component is the **volume harvested** during the period. Then there is the **mortality** that took place but went unsalvaged. The growth that took place during the period on trees that died during it cannot be determined by this method, but it presumably competed with survivor growth.

It may be noted that the growth loss from unsalvaged mortality counts as a heavy tax against the survivor growth of the period. This is because the loss is not merely of what grew during the period but the loss of all the production that had previously accumulated in those trees.

However, this is counteracted in some degree by another important artifact of the bookkeeping, **ingrowth**. Ingrowth takes place when trees pass some arbitrary minimum D.B.H. and are counted for the first time; this contribution to production is magnified because these trees suddenly acquire a substantial volume that was really laid down before the measurement period started.

In a perfectly balanced uneven-aged stand (or forest), these components would theoretically be the same in each successive period. Since most stands and forests are unbalanced, the magnitude of these components fluctuates markedly from one measurement period to another. However, by keeping track of the separate components one can more dependably analyze the growth of forests and reduce the risk of large errors in prediction.

There is even more opportunity for error if volume estimates are made in board feet rather than cubic volume. The ingrowth, a component always difficult to assess properly, is usually very large when measured in board feet. The board-foot volume of a tree increases so rapidly with diameter that small fluctuations in the diameter distribution cause remarkable differences in the apparent growth. This phenomenon is accentuated by the use of misleading log-rules; with the Doyle Rule, for example, there are overestimates of growth in diameter classes less than 22 inches.

Recapitulation

Before leaving this confusing topic, it may be well to refer to the simple beauty of the basic quantitative relationship, shown in Fig. 15-9, difficult though it may be to create or to define perfectly. Once the appropriate distribution of age classes has been converted to a diameter distribution and created in a stand (or larger forest component), it is necessary to consider the small but crucial changes that will be made in the distribution at the time of each cutting. Selection cuttings in such stands involve successive harvests of the *largest* trees in a stand; the age of these trees relative to that of their neighbors is actually a somewhat secondary consideration.

A diameter limit (point *x* in Fig. 15-9) is established as an indication of age, with the understanding that trees below this size are in general to be

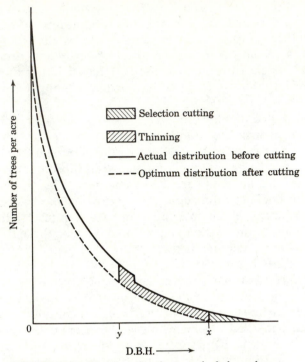

FIGURE 15-9 Diameter distribution of a balanced uneven-aged stand under intensive management, indicating
the number of trees of different diameter classes theoretically removed in a single cutting. All trees larger than
diameter *x*, which has been set as the index of rotation
age, are removed. The trees of smaller diameter may also
be reduced in number by thinning, provided that they are
larger than the tree of lowest diameter (*y*) which can be
profitably utilized. Under extensive practice there would
have been no cutting of trees less than diameter *x* and
most of those represented above as being cut in thinnings
would have been lost through natural suppression.

reserved and those above cut. This diameter limit should be regarded as a
flexible guide rather than as a rigid dividing line. Depending on their
silvicultural condition, especially capacity for further increase in value, a
few trees above the limit should be left and some below the limit should be
removed.

Under intensive practice, true thinning may be conducted simultaneously with the harvest cutting. Such thinning must be carefully distinguished from the harvest of the oldest trees and should take place in
clumps of trees too young for harvest cutting. The methods of thinning
followed are the same as those employed in even-aged stands. Unlike the
harvest cuttings, the thinnings should not remove the largest trees in the
various clumps, unless they are of undesirable form or species. The pri-

mary objective of the thinnings should be to anticipate the inevitable reduction of numbers denoted by the steep slope of the "J-shaped" curve (Fig. 15-9), so as to salvage prospective mortality and allow the remaining trees space for more rapid growth.

The whole program of thinning and stand-density regulation is the key to the crucial question of how rapidly the growing and aging trees flow through the diameter distribution. The whole process also depends on keeping up with the continual regeneration of stands. It has been said that one cannot cut more than is grown or grow more than is cut.

Irregular Uneven-Aged Stands

The only real objective of the balanced arrangement is that of making the stand a sustained-yield unit. There is no need to apply the regulatory procedures that have just been discussed to a stand unless it must provide a sustained yield. Usually the whole forest rather than the stand is the unit from which sustained yield is necessary. The objective can be attained far more easily and quickly by integrating many stands in such manner that the deficiencies of essential age or diameter classes in some stands are compensated by the surpluses in others. In this way a balanced condition can be obtained for the forest without attempting thorough readjustment of any individual stands.

In other words, the problem can be solved faster by working from the top down than from the bottom up. It should also be remembered that the diameter distribution is only a diagnostic tool and not an end in itself. The computer simulation of future production by trees and stands, as well as various techniques of operations analysis for comparing different courses of future action, may provide solutions without so much need for rigid regulation of diameter distributions.

If an entire forest ownership consists of one stand of about 20 acres or less and is required to produce a sustained yield, it would probably be essential to develop a balanced uneven-aged stand. While such small ownerships are numerous, it is seldom realistic to assume that the individuals likely to own such tracts are in an economic position to achieve close adherence to any program of sustained yield.

There are numerous valid reasons other than sustained yield for deliberate maintenance of uneven-aged stands. These other objectives can be achieved without the balanced arrangement. Long-continued application of the concept of financial maturity in efforts to extract optimum value from the growing stock is very likely to cause the development of uneven-aged stands. If this concept is the best key to intelligent management of an uneven-aged stand, it is as logical to use the selection system as it would be to employ the shelterwood system under similar circumstances in an even-aged stand. However, the sequences of cutting necessary for sustained yield and those guided by the concept of financial maturity are usually compatible only in uneven-aged stands that are already balanced and composed of good trees. If the diameter distribution is unbalanced, it can be brought into balance only by cutting some trees long before or long after they are financially mature.

Most reasons that exist for having uneven-aged stands and using the selection system are not concerned with manipulating the growing stock of stands in keeping with the concepts of either sustained yield or financial maturity. They may have to do with nothing more than the preexistence of the uneven-aged condition. The purposes may be wholly or partly matters of beauty (Fig. 15-10), wildlife, seedling ecology, or protection against damage. Such objectives may require uneven-aged stands but not necessarily balanced ones. In fact, the warping of patterns of harvest removals to fit some preordained diameter distribution commonly interferes with fulfillment of many other logical purposes. If an uneven-aged stand is to be

FIGURE 15-10 An uneven-aged stand consisting of a complex mixture of conifers, managed under the selection system, in a scenic strip along a highway leading through the Rogue River National Forest to Crater Lake in Oregon. (*Photograph by Yale University School of Forestry and Environmental Studies.*)

allowed to remain irregular and not be refitted to some diameter distribution, it is not necessary to take the costly step of determining the diameter distribution before the stand is marked for partial cutting.

In other words, adoption of the selection system does not necessitate development of the balanced arrangement. It is feasible to maintain an unbalanced uneven-aged stand indefinitely and reap any benefits, other than sustained yield, that may be induced by the uneven-aged arrangement. The distribution of age and diameter classes can be allowed to fluctuate almost at random, except to the extent that the cutting or reservation of particular classes must be adjusted to balance the books of sustained yield for the whole forest. In this kind of management, the allowable cut and its approximate distribution among diameter classes are determined for the entire forest and then harvests are made in various stands on the basis of other considerations, silvicultural and otherwise, until the scheduled volume and kinds of trees are removed.

Choice of Trees to Be Harvested

Theoretically the true reproduction cuttings of the selection system involve removing the oldest and largest trees with some predetermined **diameter limit** being taken as a definition of the smallest trees old enough for such cutting. However, this should be only a first approximation of the characteristics of trees selected for harvest.

Trees above the limit may be left because they are increasing in value with unusual rapidity or for reasons such as their capacity to provide seed and protection for reproduction. The trees designated for cutting even though they are below the limit are those that increase in value too slowly because of low quality or poor growth, those that interfere with logging or the growth of better trees, or those that are likely to be lost after the cutting. The guiding diameter limit is completely ignored when it comes to determining what trees are removed in the true thinnings among the younger age groups (Fig. 15-9). The diameter limit is also abrogated to whatever extent may be necessary to correct and allow for the effects of imbalances in the diameter distribution of stands or forests being managed as sustained-yield units.

The guidance of such complicated timber marking is sometimes facilitated by tree classifications. These are especially necessary if the various age groups are small, because the standard crown classification is most useful in rather large, single-canopied, even-aged aggregations. In other kinds of stands it must be supplemented by other information although the matter of position in the crown canopy is always important. The other characteristics important in deciding whether trees should be cut or left are (1) age or size, (2) quality, and (3) vigor.

Most ways of assessing tree vigor are based on observations of the foliage, which is the most crucial productive machinery of the tree. The condition of the foliage can be categorized in terms of its amount, color, density, and leaf size. If attention is given to the quality of the foliage, the live crown ratio is a very useful index of tree vigor. Sometimes these diagnostic attributes of trees can be expressed in numerical scoring systems. However, the tyranny of numbers can be great enough that it is well to remember that such numbers are good only to the extent that they were based on measurements of real processes. They should not relieve the user from responsibility for thought.

It is highly desirable to adapt tree classifications so that they can be used in assessing the degree of financial maturity of individual trees. This requires a means of estimating the rate of diameter growth from external characteristics. The extraction of increment cores is so slow and sometimes so harmful that it is done only to provide occasional checks on estimates. The best evidence is provided by the size and vigor of the crown but the characteristics of the bark are useful and convenient (see Fig. 6-3).

A tree classification, as with any classification, can be no better than the assumptions and principles on which it is based. It often conveys an air of authority that causes this point to be overlooked. One of the pitfalls is the common error of classifying trees on the basis of *past* growth or *present* characteristics without allowing for the effects of release from competition in the projected cuttings. Another is the possibility of arbitrarily combining promising trees with poor ones in some single category.

Evolution of Selection Methods

The first timber harvests in most forests typically involve crude, unconscious applications of the selection method in which only the biggest and best trees are removed. At the other end of the spectrum are some of the most intensive and intricate kinds of silviculture. Although one may gradually develop from the other, it is most common that the early crude selection cuttings are superceded by some other method of cutting.

The extensive kinds of selection cutting are often diameter-limit cuttings. The diameter limits can be set with varying degrees of sophistication. There may, for example, be benefit in setting a low limit for species of little future promise but higher limits for the more valuable species. Such extensive applications may be practiced indefinitely or used as temporary expedients until conditions become more favorable. The cutting cycles are usually long and there is seldom anything akin to intermediate cutting; there is little control over stand density or species composition.

In a pure stand or in a mixed stand where all species are desirable, an extensive application of the selection method, particularly group selection, may prove reasonably successful. Best results are obtained when the desirable species represent a physiographic or climatic climax. In the ordinary mixed stand, containing one or more inferior species, form and composition usually degenerate under such treatment. The desirable trees and species are cut; the inferior are left. The undesirable species multiply rapidly if they are tolerant. This type of treatment soon results in building up a growing stock of cull trees and poor species at the expense of the better ones.

If the undesirable trees are left in command of the growing space and regeneration of good ones is prevented, there can be a deleterious effect on the vegetation more subtle and enduring than any of the most destructive kinds of clearcutting. In fact, it may take something akin to clearcutting to set the stage for rehabilitation of the vegetation. On the other hand, the kind of high-grading that is deleterious to the species composition at least leaves the soil covered.

Economic Selection Methods and "Selective Cutting"

In situations where landowners have no interest beyond liquidation of existing merchantable timber, foresters have used one important principle of extensive application of the selection method as a device to discourage heavy cutting. There is always a limiting diameter below which trees cannot be profitably utilized for a given purpose, because the handling costs exceed the sale value of the small material produced. The tree that can be harvested with neither profit nor loss is referred to as the **marginal tree** of the stand or forest in which it occurs. A cutting that removes all trees that can be utilized without financial loss is called a **zero-margin** selection cutting. If a landowner can be induced to recognize the existence of the marginal tree, it becomes clearly desirable to leave smaller trees in logging. If the presently unmerchantable trees are sufficiently large and numerous to provide the basis for another harvest in the near future, there is an additional incentive to protect the smaller trees and shift away from a policy of short-term liquidation.

This approach is effective in getting the practice of forestry established, but is hardly an appropriate basis for long-term practice. In fact, as economic factors become more favorable, the size of the marginal tree often drops to the point where virtually every tree in a stand can be utilized at a profit. Under these conditions zero-margin selection cutting may become synonymous with clearcutting. The primary drawback of basing the selection method on the concept of the marginal tree is that the potentialities of different trees for future growth are not taken into account. Although a tree may be logged profitably at a given time, it often can be harvested at a greater profit at some later date.

Once a landowner is committed to retaining small trees for growing stock it becomes desirable to know which trees are best reserved to the next cutting cycle. This can be done by using the concept of financial maturity to detect those trees that will earn an attractive rate of compound interest on their own present value if left to grow. This procedure usually results in the setting of an approximate diameter limit substantially higher than that which would apply to zero-margin cutting. If the total return to be obtained during another cutting cycle is high enough, the practicability of continuing operations at least to the end of that cycle is demonstrated.

The practices involved in this procedure are often referred to as "selective cutting," although **economic selection method** is a more precise term. The line of financial logic is often sufficient to guide forest practice up from mere mining to a level of extensive practice likely to ensure continuity of management. However, if it is feasible to intensify practice still further to secure optimum production and sustained yield, the suitability of the selection method should be reexamined.

During the period 1930 to 1950, extensive applications of the selection method, described by the vague term "selective cutting," were regarded by many as the panacea of American forest management. The times were peculiarly appropriate for such proposals. The development of trucks and tractors suitable for use in logging big timber had commenced to displace railroad logging thus making partial cutting possible in old growth. Lumber markets were so badly depressed that only the biggest and best trees could be harvested profitably.

These developments played a highly important role in demonstrating that partial cutting was feasible and that residual stands of usable timber could be profitably left for future harvests. Unfortunately this campaign implanted certain misconceptions in the minds of some foresters and the public that have persisted. Selective cutting is a tool that can be used for good or evil. All too often it has resulted in neither maintenance of the uneven-aged form, establishment of desirable reproduction, nor preservation of residual stands.

Although selective cutting is frequently conducted in the name of the selection method, it more commonly bears closer resemblance to other kinds of silvicultural treatment. Many of the best selective cuttings are aimed exclusively at the most effective management of existing stands without provision for regeneration; such operations are definitely in the category of thinnings. Those that are conducted primarily to remove merchantable trees of undesirable form or species from the stand are best termed improvement cuttings and represent advantageous treatment preliminary to the initiation of more standard intermediate and reproduction cuttings. Selective cuttings that remove mostly trees threatened or already damaged by injurious agencies should be regarded as salvage cuttings.

Many selective cuttings, especially those conducted under extensive practice, remove such a high proportion of the merchantable volume that they destroy rather than accentuate any unevenness that may have existed in the age distribution of the stand. This development has also resulted from the deterioration of residual stands left after cutting. Treatments of this kind tend to create even-aged stands and should be regarded as crude variants of the shelterwood method. Finally, there are some selective cuttings that create or maintain the inherent characteristics of the uneven-aged stand, and may be correctly referred to as selection cuttings.

"Selective cutting" began to fall out of fashion in the United States about 1950. This was partly because of poor results in some kinds of forests but also because of rapid increases in willingness to invest money in forest regeneration. The most harmful results tended to be from high-grading of unstable old-growth forests in the more humid parts of the West. In the Coast Douglas-fir type, for example (Isaac, 1956), the trees most commonly cut were the few large, ancient Douglas-fir scattered above shade-tolerant species that were more decadent even though often younger (Fig. 13-6). The very trees being cut were actually often the ones most likely to endure. In the southern pine forests, at least on industrial lands, it was recognized that the hardwood control essential to pine regeneration was very difficult in uneven-aged stands. The baneful effects of high-grading in mixed hardwood stands throughout the East became more painfully obvious.

During the two decades that started about 1950, there was a major reaction in favor of clearcutting of the kinds in which regeneration comes either artificially or from released advance growth. It should not have been altogether surprising that by 1970 this rather unattractive form of silviculture evoked protests based upon grounds both sound and imaginary. The resulting shifts have been more toward variants of the shelterwood methods than to selection methods, although interest in those has revived also.

Application of Selection Methods

Southern Pines

Some useful applications of the selection method have often been sequels to high-grading of the sort that produces irregular stands. This is especially true of conifers or other species that are geotropic and tend to remain straight when growing in the shade. This is not so true of hardwoods because most are phototropic and become crooked when growing under other trees.

One well-documented case history (Reynolds, Baker, and Ku, 1984) involves selection cutting that was started on high-graded industrial ownerships in the western South with loblolly and shortleaf pine. These species characteristically grow together in even-aged stands so their successful culture in uneven-aged stands is proof of the versatility of the method. The basic objective of the procedure is to develop good sawtimber growing stocks as swiftly as possible from the remnants of high-graded even-aged stands. Good new stands could be created by liquidating the remnants and replacing them with even-aged regeneration but this might cause a temporary suspension of harvests when the old trees were gone. Furthermore, there would be needless sacrifice of many small- or medium-sized trees of good potentialities.

The guiding principle followed in this kind of cutting is the concept of financial maturity, which is applied with careful attention to the quality of increment as well as to volume. Full advantage is taken of the excellent natural pruning of some of those remnants of earlier stands that are capable of regaining full vigor (Fig. 15-11).

FIGURE 15-11   A stand, predominantly of loblolly and shortleaf pine, being managed under the selection system in southern Arkansas. The large-branched pine in front of the man is to be cut to foster the growth of the small-branched one behind him. (*Photograph by U. S. Forest Service.*)

This procedure rehabilitates the stands and also preserves enough of the irregularity of the initial stands that some intermingling of age classes remains. However, there were enough problems with controlling hardwoods and getting pine regeneration that most of the industrial forestry programs shifted to clearcutting and planting.

During the 1970s interest in selection (and shelterwood) management of southern pines revived because of need for low-investment silviculture for the small, private, nonindustrial ownerships that comprise almost two-thirds of the forest land (Williston, 1978). The chief emphasis tends to be on manipulating existing growing stocks to extract optimum advantage from them over the longest possible time. Although regeneration is seldom perfect, it is often possible to get sufficient natural regeneration if fire, herbicides, browsing, and harvesting of hardwoods are used to keep the hardwoods under some degree of control.

Central European Selection Systems

The selection system was devised by foresters in the mountainous parts of Europe as a means of growing mixtures of Norway spruce, European beech, and European silver fir. The fir is especially difficult to regenerate and requires partial shade. Another purpose has also been to imitate presumed natural tendencies so as to increase the resistance of the stands to damaging agencies. The most common reason has been to make some partial approaches toward the sustained-yield objective on small holdings.

Except for the small holdings, true selection stands are uncommon show-places, usually maintained in remote stands that had become irregular because of earlier high-grading. On the larger forest holdings, it is more common to maintain the mixtures by the irregular shelterwood system.

However, on small holdings (Fig. 15-12), many of which are typically 10 m wide and several hundred meters long, the sporadic requirements of owners for income dictate occasional selection cuttings. The ownerships stay within families and the forests are managed as a kind of savings account drawn upon mainly during periods of special need. The amount of growing stock goes up and down, but there is always something left in the bank. The owners have much discretion about how much income to re-

FIGURE 15-12 Uneven-aged stand, mainly of Norway spruce, under single-tree selection system on a very small private forest in West Germany. (*Photograph by Yale University School of Forestry and Environmental Studies.*)

move at any time, but public foresters have some control over how it is done.

There is seldom any sophisticated regulation of the cut in terms of diameter distributions. If there is, it usually involves deliberate maintenance of the overabundance of large trees that enhances current annual increment of sawtimber but is not conducive to the amount of regeneration ultimately needed for sustained yield. It is often recognized that the growing stock will ultimately dwindle and then attention will have to be given to regeneration and rebuilding growing stock.

There is a dictum of Bavarian farmers about their forests that epitomizes the informal approach to sustained yield. This is that one may cut sawlog-sized trees at the average annual rate of one per *"Tagewerk."* This archaic unit of land measure, literally "day's work," is about the size of an English acre and refers to what two oxen could plow in a day. While the approach to the regulation of the cut may be approximate, the silvicultural treatment of the stands is highly intensive.

The earliest selection forests are the village protection forests that were set aside in the Swiss Alps in medieval times. These are triangular stands, pointed uphill, which are maintained to divert snow avalanches from the villages in their lee. The purpose is to maintain a perpetual and continuous forest cover but to use the selection method to replace old trees before they inevitably die. Most of the adjoining slopes were deforested by heavy grazing that still continues. Nowadays the effect of the trees in these kinds of crucial protection forests is supplemented by steel barriers. Such forests were the source of the idea of creating uneven-aged stands to solve other problems at lower elevations and of their regulation by diameter distribution (Knuchel, 1953).

The classic selection system originated in the various European countries that surround the Alps. The purpose was to solve some specific problems in certain kinds of stands, mainly comparatively simple mixtures of spruce, fir, and beech. The assortment of species was, because of the accidents of glacial history, so limited that it was desirable to build rather than fight complexity of stand composition.

Attempts have been made, with varying degrees of understanding, in many parts of the world to apply the concepts that were developed to complex stands in general. These attempts have often failed because of at least two reasons. First, economic and site conditions do not always permit the intensive control of species composition and stand development that is routine in Europe. Second, the concept of the uneven-aged condition is not sufficient or even necessarily correct as a basis for understanding mixtures of species; this is why Chapter 17 on such stands is separate, even though managed stands can certainly be both mixed and uneven-aged.

Selection Systems on Dry Sites

There can be reasons for maintaining uneven-aged stands on sites that restrict regeneration if the stands have already become irregular. If climatic conditions are seldom conducive to regeneration it is usually desirable to maintain uneven-aged stands to have trees that can produce seeds whenever circumstances allow. In such cases tree growth is usually too slow to justify the high cost of planting, so one is restricted to natural regeneration. Another reason for uneven-aged stands in these circumstances is the desirability of trying to keep nearly every seedling that becomes established. If they come in sparsely and infrequently, the inevitable result is likely to be an uneven-aged stand.

This approach has long been used in the dry forests of the interior of the western United States, especially those composed of ponderosa pine, pure or in mixture with other species. Many of the stands are composed of small even-aged groups of trees that have arisen when older trees were killed by bark beetles, lightning, dwarf-mistletoe, or the irregular lethal effects of fires.

The modes of application of the selection system to the ponderosa pine type have varied significantly in time and place. The species has a very broad geographical range and the patterns of seasonal dryness to which it is exposed differ considerably as do the management problems.

The only part of the range in which regeneration is easily secured is a region of summer rainfall that lies east of the Rocky Mountains and includes the Black Hills of South Dakota. Most of the stands are even-aged and, because of the relative ease of regeneration, they are typically managed that way.

The widest part of the ponderosa pine range lies between the Rocky Mountains and the mountain chain formed by the Cascades and Sierra Nevada to the west. At the north there is a long, dry summer but with just enough extension of the winter rains into the spring that it is not extremely difficult to obtain regeneration. The selection system is commonly used but has evolved as conditions have allowed the intensity of practice to increase.

Before the 1940s the old-growth forest was still ruled by *Dendroctonus* bark beetles and the limitations imposed by railroad logging. It was anticipated that this cumbersome mode of log transportation would dictate a long cutting cycle of about 30 years. Only fine old trees more than about 20 inches D.B.H. were worth cutting. The Keen Tree Classification (Fig. 6-2) was devised as a basis for predicting prospects of survival of various categories of trees for the long cutting cycles that were envisioned. It was also used as the basis of financial maturity analysis, which also guided the choice of trees for harvest or retention. In fact, this approach, termed the **maturity-selection system** (Munger, 1941), was perhaps the most successful of the "selective logging" schemes of the time because it closely fitted existing developmental processes in truly uneven-aged stands.

With the advent of tractor logging and much more favorable markets, it became possible to shorten the cutting cycles and cut to produce thinning effects through a broader range of diameter classes. Problems with the bark beetles subsided as large, old, susceptible trees were eliminated. It has gradually become more customary to prescribe treatments for individual stands on the basis of their prevailing condition rather than in terms of some standard system. At one time the term **unit area control** (Hallin, 1959) was used to denote this shift in emphasis; the "unit" was any homogenous part of some large, heterogenous stand, but gradually came to be recognized as a small, separate stand in its own right. As increased attention became necessary to problems such as dwarf-mistletoe and the need for site preparation to deal with brush competition, even-aged systems, including those involving planting, have become more prominent among the alternatives that are used.

In Arizona and New Mexico severe drought in spring makes regeneration of the forest a rare event. There are summer showers and thunderstorms but they do not come in time to induce seedlings to germinate early enough to harden before frost. In fact, in northern Arizona there was abundant natural regeneration in 1919 and but little at no other time so far in this century. This came from the combination of an abundant seed crop and unusual spring rains. The 1919 class filled so much growing space that for some decades it was logical to be concerned about other silvicultural management problems. Since bark beetles were not a major problem, it was possible to aim the selection cutting at developing trees with good bole form and natural pruning. A program for doing this, called **improvement-selection cutting** (Fig. 15-2), and many other aspects of ponderosa pine silviculture were described by Pearson (1950). It is noteworthy that the kinds of pines with small crowns and branches that he found capable of responding to release and turning into fine trees were among those deemed to be highly vulnerable to beetle damage farther north.

Ponderosa pines grow well on the west slopes of the Sierra Nevada but they must share this habitat with sugar pine, incense-cedar, Douglas-fir, true firs, and other members of a mixed conifer type. Water stored in the soil from the deep winter snows carries over into the long, rainless summers well enough to support a magnificent forest. However, the surface soil becomes so dry that it is difficult to get regeneration established. Many species of brush are apt to usurp the growing space after fire or cutting; fire, herbicides, and mechanical site preparation must be used to reduce this competition enough to get regeneration started, often by planting. The group-by-group prescriptive approach of unit area control has become prominent, although the units tend to come to be viewed as small stands rather than parts of larger uneven-aged stands.

In most of the ponderosa pine types, the selection system has been used partly to take advantage of rare regeneration episodes. There is a different general case in which the selection system is often used to take advantage of easy regeneration coupled with minimal competition from undesirable species. Such circumstances come about when there is good rainfall at the season of seedling establishment but such poor moisture supply at other times that only drought-tolerant tree species can endure. The pines are the species that most often have this adaptation. Under just the right soil moisture regime, the regeneration phase of silviculture may require little more than drifting with the tide of natural processes.

This situation can develop on soils of deep sand, such as glacial outwash or old coastal beach deposits, in humid climates. If the stands are already uneven-aged, there may be even less reason than usual to convert them to the uneven-aged condition. The dry soils usually make the logging easy and reduce problems with undesirable vegetation. Any large vacancies created in the growing space by cutting usually fill up promptly and almost automatically with dense regeneration. The main problem is ordinarily that the natural regeneration is far too dense and precommercial thinning is required.

"Continuous Forest"

One famous and influential case involved Scotch pine on some deep sands south of the Baltic Sea. Because of the almost effortless regeneration, a mode of management called the "Dauerwald" or "continuous forest" was developed. This was a revolt against policies of strict regimentation of the forest, fixed rotation lengths, and efforts for securing reproduction. Most efforts were concentrated on tending individual trees and harvesting them only when they interfered with better trees. This policy tended to accentuate any variations in age-class structure that already existed in the stands.

This general approach is so easy that it is useful to detect the narrow range of site conditions that make it feasible. They occur in the eastern and southern parts of the United States only on deep, sandy soils, which are often among the poorer soils for tree growth. They are, for example, the easiest places to grow eastern white pine and some other pines but the production rates are not high. The main virtue is that hardwood competition is feeble. Such sites are more common in the western interior; examples there are the kinds of ponderosa pine sites where regeneration is easy but brush competition is not serious. Another important case is the mountains of northern Mexico, which have some uneven-aged, easily regenerated forests of *Pinus durangensis* and *P. arizonica* in a summer-wet, winter-dry climate.

Whether this dry-site regeneration is difficult or easy, it is often associated with some peculiar phenomena of root competition. The supply of available water is often small enough that the invisible root systems can be fully closed but unable to support a closed canopy of foliage. The stands easily develop persistent gaps that appear to be unstocked but are actually being fully utilized by the roots of adjoining trees. If root-grafting is well-developed, as it often is with pines, partial cuttings may merely donate the use of living roots of the cut trees to the remaining ones. This is good for the remaining trees but can defeat efforts to create real soil vacancies for regeneration.

With deep sands in humid climates there can be an additional peculiar phenomenon. The problems with moisture deficiency may be mostly near the surface. While trees may grow slowly at first, they can grow at accelerating rates for long periods as their roots become increasingly extensive.

What may seem like poor sites initially become much better. If the trees of crowded groups differ enough in height, the leading trees will forge ahead even without thinning as their root systems become deeper.

Eastern Hardwoods

Some of the strongest efforts to apply the selection system have involved complicated mixtures of species of the sort that often grow where site factors do not greatly limit species diversity. The basic idea appears to have been the effort, subconscious or otherwise, to equate complexity with the uneven-aged condition, and stem diameter with age. Chapter 17 deals with such stands and additional ways of interpreting complexity of stand structure.

While most of this discussion is postponed to that chapter, it is well to recount some of the ideas about managing complicated mixtures by the selection system. Many have developed with respect to the vast and variable deciduous forest of eastern North America (Fig. 15-13); only the tropical rain forest is more complicated. Different modes of approach can be distinguished, and their success with the attainment of different objectives has varied from place to place.

Most regeneration actually starts as advance growth under partial shade. The spatial pattern of removal cutting determines stand structure

FIGURE 15-13 Application of the group-selection system in a stand of northern hardwoods, Menominee Indian land in Wisconsin. (*Photograph by American Forestry Association.*)

more by releasing advance growth than by causing new regeneration. If openings are small enough to keep direct sunlight off the ground most of each day, it is possible to reduce the amount of intolerant pioneer vegetation that appears, if that is a desired objective. The regeneration usually develops in the form of the stratified mixtures of species described in Chapter 17. It is, of course, possible to adjust the composition by early cleaning operations if there is reason not to let the trees sort themselves out unaided.

In the use of the selection system in eastern hardwood stands, much attention has been given to the development of the kinds of J-shaped diameter-distribution curves that were discussed earlier in this chapter (Leak, Solomon, and Filip, 1969; Trimble, 1970). Most of the kinds of diameter distributions that are created and maintained seem to be skewed in favor of a large representation of trees of the sizes that grow well in terms of board-foot volume. In this way they are much like the distributions of growing stock described earlier in connection with small holdings in Central Europe. This is perhaps appropriate because most of the eastern hardwood forest is in small holdings. This approach may sacrifice provision for regeneration to the desire for high financial returns over the next few decades. It will take time to tell, but there is also some time in which to make readjustments if policies of owners and society about sustained yield require them.

The worst sins committed in hardwood forests in the name of the selection system can be avoided by keeping proper track of distribution *of species* by diameter classes. The chronic problem with selection cutting in all forests, especially complicated ones, is that it may degenerate into high-grading. In mixed stands, the problem arises mainly when tree size is mistaken for an index of tree age. If, as is often the case, the only yellow-poplars in a stand are large and one blithely cuts all of then without leaving space for new seedlings, the species will be eradicated. If the diameter distribution of the remaining trees seems "correct" but the smaller trees consist of low-value species such as American beech, red maple, hop-hornbeam, or hackberry, this hardly perpetuates yellow-poplar.

Worse yet is the case in which each generation high-grades a stand and memories are too short to see that some species are disappearing and that the average diameter and quality of the trees are dwindling. Sometimes foresters observe the result of this process and, without knowing the sad history, ascribe it to poor site conditions or think of it as a condition that always existed. While one cause of high-grading is ignorance, another may be lack of markets for small trees or unwillingness to spend money on killing them. In a pure, even-aged stand, high-grading eliminates only the best individuals of one species; in complex mixtures, however, it is very likely to eliminate species, especially in such diversified assemblages as the tropical moist forests.

Sometimes the effects of high-grading can be remedied by switching to cutting undesirable trees to favor better ones and gradually developing good, uneven-aged stands. Some of the most successful applications of the selection method have started with such changes. Ironically, the most expeditious and successful solution in many eastern hardwood stands has been to start over again with the kind of heavy removal cutting that is often called "clearcutting." This usually depends on favorable markets for firewood or other small-tree products. If there is enough desirable regeneration from advance growth, sprouts, and new seedlings, a new, complex stand of good quality arises. When, later in life, this even-aged stand develops a wide range of diameter classes and species, there will be various options about imposing irregularities of the sort that may lead either to the true uneven-aged condition or back to the even-aged one.

The selection method can be misapplied, but it can also be the most versatile kind of silviculture. It can work well if stands are highly accessible and easily logged. It is the opinion of this writer that efforts to make stands into self-contained, sustained-yield units usually interfere with securing the more important benefits of the selection system. If the concept is not hobbled with that restriction, it can be very useful, especially in multiple-use forestry and in the management of small holdings. If stands are already uneven-aged and partial cutting is feasible in them, it is often well to perpetuate the condition.

BIBLIOGRAPHY

Adams, D. M., and A. R. Ek. 1974. Optimizing the management of uneven-aged stands. *Can. J. For. Res.* **4**:274–287.

Alexander, R. R., and C. B. Edminster. 1977. Uneven-aged management of old-growth spruce–fir forests: cutting methods and stand structure goals for the initial entry. *USFS Res. Paper* RM-186. 12 pp.

Anderson, H. W. 1960. Research in management of snowpack watersheds for water yield control. *J. For.* **58**:282–284.

Boyce, S. G., and H. A. Knight. 1979. Prospective ingrowth of southern pine beyond 2000. *USFS Res. Paper* SE-200. 48 pp.

Crow, T. R., *et al.* 1981. Stocking and structure for maximum growth in sugar maple. *USFS Res. Paper* NC-199. 16 pp.

Davis, K. P. 1966. *Forest management: regulation and valuation*. 2nd ed. McGraw–Hill, New York. 517 pp.

Geiger, R. 1965. *The climate near the ground*. 3rd ed. Harvard Univ. Press, Cambridge, Mass. 611 pp.

Gingrich, S. F. 1978. Growth and yield. *In* Uneven-aged silviculture and management in the United States. *USFS Gen. Tech. Rept.* WO-24. Pp. 115–124.

Gordon, D. T. 1970. Natural regeneration of white and red fir: influence of several factors. *USFS Res. Paper* PSW-58. 32 pp.

Hallin W. E. 1959. The application of unit area control in the management of ponderosa–Jeffrey pine at Black Mountain Experimental Forest. *USDA Tech. Bull.* 1191. 96 pp.

Isaac, L. A. 1956. Place of partial cutting in old-growth stands of the Douglas-fir region. *Pac. Northwest For. & Range Exp. Sta., Res. Paper* 16. 48 pp.

Knuchel, H. 1953. *Planning and control in the managed forest.* Transl. by M. L. Anderson. Oliver & Boyd, Edinburgh. 360 pp.

Leaf, C. F. 1975. Watershed management in the Rocky Mountain Subalpine Zone: the status of our knowledge. *USFS Res. Paper* RM-137. 31 pp.

Leak, W. B., D. S. Solomon, and S. M. Filip. 1969. A silvicultural guide for northern hardwoods in the Northeast. *USFS Res. Paper* NE-143. 34 pp.

Munger, T. T. 1941. They discuss the maturity selection system. *J. For.* **39**:297–303.

Paine, D. P. 1981. *Aerial photography and image interpretation for resource management.* Wiley, New York. 571 pp.

Pearson, G. A. 1950. Management of ponderosa pine in the Southwest. *USDA, Agr. Monogr.* 6. 218 pp.

Putnam, J. A., G. M. Furnival, and J. S. McKnight. 1960. Management and inventory of southern hardwoods. *USDA, Agr. Hbk.* 181. 102 pp.

Reynolds, R. R., J. B. Baker, and T. T. Ku. 1984. Four decades of selection management on the Crossett farm forestry forties. *Ark. AES Bull.* 872. 43 pp.

Roach, B. A. 1974. What is selection cutting and how do you make it work—What is group selection and where can it be used? *State Univ. of N. Y., Coll. of Env. Sci. and Forestry, Appl. For. Inst. Misc. Rept.* 5. 9 pp.

Schlesinger, R. C. 1976. Sixteen years of selection silviculture in upland hardwood stands. *USFS Res. Paper* NC-125. 6 pp.

Schnur, G. L. 1937. Yield, stand, and volume tables for upland oak forests. *USDA Tech. Bull.* 560. 88 pp.

Scott, V. E. 1978. Characteristics of ponderosa pine snags used by cavity-nesting birds in Arizona. *J. For.* **76**:26–28.

Trimble, G. R., Jr. 1970. Twenty years of intensive uneven-aged management: effect on growth, yield, and species composition in two hardwood stands in West Virginia. *USFS Res. Paper* NE-154. 12 pp.

Tubbs, C. H., and R. R. Oberg. 1978. *How to calculate size-class distribution for all-age forests.* USFS, North Central Forest Experiment Station, St. Paul, Minn. 5 pp.

U.S. Forest Service. 1978. Uneven-aged silviculture and management in the United States. *USFS Gen. Tech. Rept.* WO-24. 234 pp.

Walker, L. C., and K. G. Watterton. 1972. Silviculture of southern bottomland hardwoods. *S. F. Austin State Univ. Sch. For. Bull.* 25. 78 pp.

Williston, H. L. 1978. Uneven-aged management in the loblolly–shortleaf pine type. *South. J. Appl. For.* **2**:78–82.

C H A P T E R   1 6

Methods Based on
Vegetative Reproduction

Silviculture that depends on vegetative sprouting is not only remarkably successful but is also the most ancient kind of deliberate forest regeneration. A **low forest** is one originating vegetatively from natural sprouts or layered branches as constrasted with a **high forest**, which develops from seeds or planted seedlings. The ancient Germanic origin of these words refers to the fact that most sprout forests in Europe were grown on short rotations and the trees were not tall. It could not be known at the time that the coast redwood, the tallest of all plants, commonly reproduces by sprouting (Fig. 16-1). The term **coppice** denotes a stand arising primarily from sprouts; the means of regenerating such forests is called the **coppice** or **sprout method**.

In the simple coppice method all standing trees are cut at the end of each rotation and a perfectly even-aged stand springs up almost immediately. The coppice-selection method is not used very much but would be basically similar to the selection method except for the origin of the reproduction. A **compound** or **composite coppice** is a stand in which certain selected trees or **standards** are carried on a much longer rotation than a simple coppice beneath them. The technique of handling them is the **coppice-with-standards** or **compound coppice method**. Sometimes vegetative regeneration is accomplished by relying on **layering**, which is the natural induction of roots on the stems of foliated branches buried in organic matter. **Pollarding** refers to the practice of cutting the tops off trees and thus inducing them to sprout at points high above the ground. The planting of rooted cuttings or other asexually regenerated material is not construed as a low-forest method of regeneration.

VEGETATIVE REPRODUCTION

The ability to reproduce vegetatively is of tremendous survival value for many species of perennial plants. It is the normal regeneration mode

FIGURE 16-1 Sprout and seedling regeneration of coast redwood, 7 years old, Six Rivers National Forest, California. The taller stems that are in clusters (around huge invisible stumps) are the sprouts. (*Photograph by U. S. Forest Service.*)

for most shrub species and many grasses and forest plants that have perennial roots and annual shoots, as well as a few trees. For such species, sexual regeneration from seed seems like a safety factor retained to preserve capacity for genetic change and adaptation to new conditions. For many other species, mostly angiosperms, sprouting is just an additional mode of regeneration. For others, mostly conifers, it occurs never or infrequently.

Vegetative regeneration produces good grass for grazing and much growth of shrubs for browsing animals and other purposes. It is a traditional mode of fuelwood production throughout much of the world. It is reputed to be a poor means of timber production, but this opinion may derive from cases in which the technique has been overused. The sprouts of desirable arborescent species often produce stems that are not as straight or well-formed as plants of the same species grown from seeds.

As far as most woody plants are concerned, vegetative sprouting is a natural adaptation to fires that are hot enough to kill the above-ground parts. Only by this means could a species maintain itself in the face of catastrophes that occur so often that no individuals reach seed-bearing age. Almost every forest region has areas where fires have occurred so often that sprouting species predominate.

Sprouting is a response to sudden death of the parent tree caused by agencies such as fire or cutting. It may also occur as a result of disease, injury, or serious physiological disorder. In fact, the appearance of sprouts of any form near the base or on the bole of a tree is frequently the first sign of unhealthiness. The process is triggered by interruption of the flow of auxins that are produced in the top of the tree and normally inhibit the development of basal buds.

The shoots that arise from portions of an older tree grow much more rapidly in youth than seedlings of equal age. This tendency is to some extent the result of inherent physiological differences that are not clearly understood. It is probable, however, that much of the unusual rapidity of growth of young sprouts can be accounted for by the inheritance of extensive root systems and supplies of carbohydrate from the parent trees.

Cuttings made during the dormant period lead to much more vigorous sprouting than those made during late spring and summer. This is because the reserves of carbohydrates in the roots are at a maximum during the winter and at a minimum immediately after the formation of new leaves and shoots (Kramer and Kozlowski, 1979). If cuttings are made late in the growing season, sprouts sometimes appear almost immediately, but they rarely harden before the first frost. Therefore, coppice cuttings are best done in late fall or winter.

If full advantage is to be taken of vegetative reproduction in renewing a stand, all the trees of the species to be reproduced should be cut in one operation to stimulate sprouting to the utmost. Trees of the same species in a single stand are often interconnected by root grafts. Therefore, if a few scattered trees are left behind they may draw on the food supply of the common root system or contribute hormonal inhibitors of sprouting and thus reduce the vigor of the new sprouts. The reduction is often greater than could be caused by the shade of the residual trees.

Although trees of vegetative origin grow faster during early life than those developing from seed, they do not retain this advantage on long rotations. As a general rule, vegetative reproduction may be expected to attain larger size on a short coppice rotation than trees grown from seed would in the same time. In fact, that is the primary advantage of coppicing stands on short rotations. If the rotations are extended just enough to produce saw logs there appear to be no consistent differences in size that can be related to the mode of origin (Leffelman and Hawley, 1925). McIntyre (1936), however, found that oak sprouts grown beyond a century in age on good sites tended to be smaller than trees originating from seed. He found no evidence that oak sprouts were inherently shorter lived than oaks grown from acorns.

Vegetative regeneration has a bad reputation because of evidence that repeated coppicing from the same root-stocks with short rotations has caused progressive declines in the quality and vigor of successive crops. Almost all observations of this have been made in Europe in stands that have been coppiced for centuries. There it is regarded as necessary to renew coppice forests from seed occasionally to restore the quality of production. The declines do not always take place and when they do the causes are not altogether clear.

Some degradation has probably come from depletion of the inorganic nutrient capital of the soil as a result of the nearly complete utilization of the forest production that is associated with coppicing. Declining productivity could result from understocking of the site if there were a progressive decrease in the numbers of basic sprout clumps. It is theoretically possible that loss of vigor might result from a gradual increase in the infection of root-stocks by fungi or obscure viruses.

The basic problem, at least with stump-sprouts, seems to be that both the form and the vigor of individual sprouts suffer from competition among the very crowded stems that can arise from some sprout clumps that have been repeatedly coppiced. If repeated vegetative reproduction was inherently deleterious, deterioration would have been observed in certain varieties of domesticated plants that have been perpetuated entirely from rooted cuttings for centuries.

The conditions required for successful coppice management sometimes place serious limitations on the size and spacing of the trees that can be grown. In this connection, however, it is important to distinguish between the different kinds of vegetative regeneration.

Stump-Sprouts

The sprouts that arise from the root collars of stumps represent the most important type of vegetative reproduction. These shoots almost invariably develop from **dormant buds** that were originally formed on the leading shoot of the seedling and grew outward with the cambium but without previously developing into branches. The pith of dormant buds can be traced all the way back to the pith of the original stem. If this connection is broken during the growth of the tree the bud becomes incapable of developing into a new shoot. If the bark over the dormant bud becomes too thick, it may be impossible for the bud to break through and develop into a sprout. Both the thickness of the bark and the possibility of interruption of the bud trace increase with age. Therefore, the sprouting capacity of the tree tends to decline with age.

Dormant buds can give rise to new shoots at any level on the bole, although only those that arise at or very close to the ground line have much potentiality for developing into good trees. Trees that develop from sprouts that originate too far above ground are often crooked, forked, or subject to breakage or decay. Shoots that develop even higher on the living boles of trees are epicormic branches, discussed in Chapter 3, that defeat natural pruning.

The risk that rot will spread from the old stumps to the heartwood of the new shoots varies with species. It does not appear to happen in such diffuse-porous species as the maples in which any infected interior of the trees is "compartmentalized" or sealed off from the sapwood by gum formation (Shigo, 1985). This does not appear to help with another diffuse-porous species, yellow-poplar. With the oaks the incidence of such rot increases with the diameter of the parent stump and the distance of the sprout above ground (Fig. 16-2) (Roth and Hepting, 1943). However, American oak forests have many large, sound trees that grew so rapidly when young that they must have been of sprout origin. There seems to be no problem with coast redwood, which produces sound sprouts from even the most gigantic stumps.

Stump sprouts usually cluster in dense rings around sprouting stumps. Their numbers usually dwindle rapidly because of severe competition between them.

If coppicing is repeated frequently over hundreds of years it can lead to the formation of sprouting masses of tissues, sometimes more than a meter in diameter, that produce densely spaced shoots. The shoots are so spindly that they become crooked or curved, probably because they are not nourished well enough to stand erect.

The sprouts of conifers come from dormant buds that form at the bases of either primary or secondary needles. The American conifers that can be usefully regenerated by sprouting are coast redwood, bald-cypress, and three pines, shortleaf, pitch, and pond.

FIGURE 16-2 Two desirable stump sprouts of low origin on an oak stump with one large and two undesirable ones on top of the stump. (*Photograph by U. S. Forest Service.*)

Seedling Sprouts

Sprouts that come from small stumps usually have the same anatomical and physiological origin as stump sprouts but also so many attributes of seedlings that they have this compound name. Seedling sprouts are defined as coming from stumps less than 2 inches in diameter. This generally means that the stumps may have only sapwood and may quickly be covered over with callus tissue; both factors greatly reduce the risk of heart-rot spreading from stump to sprout. They are often superior to seedlings as sources of regeneration because the new shoots grow more rapidly as the result of having established root systems to nourish them.

The stems tend to be straighter simply because the faster a tree grows through a certain zone of height, the less time there is for the operation of agencies that create deformities.

While almost any sprouting species can produce seedling sprouts, they are especially important with many large-seeded species such as the oaks, hickories, chestnut, and walnuts. The seedlings of these species are seldom numerous but, once established, they are very persistent and may survive for many years in the shade of older trees. There is an ample supply of sprouts at the ground line. The seedling tops continually grow up and get killed back but the carrot-like roots often survive and grow larger. When some permanent release takes place these kinds of plant grow rapidly; they often are a chief source of regeneration for the species involved. The mechanism enables a kind of long-term storage of advance growth, especially of species that produce small numbers of large seeds.

On the other hand, this particular source of regeneration is really much more important with the so-called high-forest systems than with the coppice method. Simple coppice rotations would often be too short to be conducive to the kind of seed production that would, over some years, allow this kind of regeneration to develop.

Lignotubers

Many of the Australian eucalypts, which are planted widely in the warmer parts of the world, have such an affinity for fire that they have some special modes of sprout regeneration (Hillis and Brown, 1984). The one that is largely limited to this genus is the **lignotuber**. These are paired globular masses of tissue, full of buds, that form in the axils of the opposite leaves of small seedlings. These tissues grow much larger but usually become buried at the base of the tree stems. If the stems do not get more than about 20 cm in diameter, there is prolific sprouting from the lignotubers when the tops of the trees are cut, killed, or damaged.

Many species of eucalypts also have typical dormant buds that form in the leaf axils and produce stump sprouts or epicormic branches. However, some of the most important members of the genus do not sprout but respond to infrequent catastrophic fires by mass seeding. It is significant that another nonsprouting eucalypt is *Eucalyptus deglupta*; it occurs naturally only in New Guinea and the southern Philippines and in rain forests where there has not been enough fire to call forth adaptations to it.

Sprouts from Adventitious Buds

Sprouts can also arise from adventitious buds that sometimes develop in the callus tissue that forms over wounds or in the root cambium of a limited number of tree species. Sometimes if sprouts from dormant buds are cut, new sprouts can arise from adventitious buds that form anew in the callus tissue of the wounds.

The most important role of adventitious buds as constructive sources of forest regeneration involves several hardwood species that produce **root-suckers** or **root-sprouts** along the roots of trees that are killed or even slightly damaged. This is so nearly the normal mode of regeneration in the aspen poplars that true seedlings are rare in many localities (Schier, 1975). It is common with sweet-gum, black-gum, American beech, black locust, and sassafras. Trees that arise from root-suckers can be very much better than those from stump-sprouts. The shoots are much more likely to be free of rot as well as evenly spaced and straight because they are not clustered around any stumps.

In a few species, including American beech, adventitious buds produce **stool-shoots**; these arise between the bark and wood and develop into peculiar, usually short-lived rings of sprouts around the tops of stumps.

Layering

This peculiar kind of vegetative reproduction arises from living, low-hanging branches that have been partially buried in moist organic matter. Layering is a process of silvicultural significance mainly in peat bogs where sphagnum moss tends to overgrow the lower branches of trees in open stands. Black spruce and northern white-cedar are the only American species that have been extensively regenerated by natural layering (Johnston, 1977).

COPPICE METHOD

In this method stump-sprouts or root-suckers are relied upon as the main source of regeneration, although some seedlings and seedling sprouts may form part of the new crop. The stand is cut clear so that nothing is left to reduce the vigor of sprouting (Fig. 16-3). A clear distinction should be drawn between species that sprout from stumps and those that form root-suckers.

Most species that can be handled by the coppice method reproduce vegetatively from stump-sprouts. The characteristics of this kind of reproduction place certain restrictions on the details of application. In the first

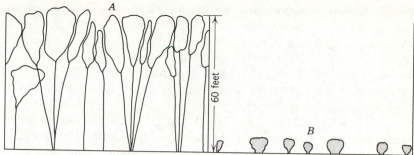

FIGURE 16-3 A stand reproduced by the simple coppice method, shown before and after the reproduction cutting. *A:* Stand ready to cut clear. *B:* Same area one year later with tall, wide-spaced clumps of sprouts growing from the stumps of the felled trees.

place, the rotations must be shorter than those ordinarily contemplated for the production of good sawlogs. Ordinarily the objectives are the production of fuelwood, pulpwood, and animal browse.

The spacing of a coppice stand derived from stump-sprouts is controlled by the arrangement of the parent trees. If the rotation is extended sufficiently to produce saw timber, many of the basic sprout clumps will die of suppression. If dependence is then placed on reproduction by the simple coppice method, the new sprout clumps will grow up so far apart that production suffers, and the stems become crooked and branchy before the stand closes. Best results are obtained when the rotation is so short that the final spacing of sprout clamps is acceptable as an initial spacing for a new coppice stand.

The vigor of sprouting from stumps also declines rapidly with increasing age and diameter. The period of satisfactory sprouting is usually coincident with that of most rapid growth and ordinarily ends before the trees become effective seed bearers. The age at which sprouting can be expected varies widely with species. The longest rotation likely to produce satisfactory results varies between 30 and 40 years. Some species will sprout well when much older, but the problem of spacing then becomes the limiting factor.

The incidence and extent of decay in stump sprouts also increase with age, although the rate of deterioration varies greatly with species and climate. Some species are affected very slowly and to an insignificant extent; in others, decay spreads quickly and may be so injurious as to preclude use of the method.

Attention must be given to preparing the stumps suitably. Saws are better than axes or power shears for the cutting because they are less likely to loosen the bark of the stumps. It is reputed to reduce incidence of rot if the cut surfaces are slanted so that water runs off them. Normally it is best to cut the stumps as low as possible so that the new sprouts will not have to grow on the side of a decaying and unstable stump (Fig. 16-2). Sometimes the deliberate use of slash burning to kill the sides of the stumps is useful in restricting the sprouting to buds at or below the ground line. However, these steps, which are useful for oaks, are not advantageous for the eucalypts because the best sprouts of those are ones that come from undamaged stumps 10 cm tall (Jacobs and Métro, 1981).

If repeated crops are to be secured from stump-sprouts, the rotations are ordinarily less than 35 years and sometimes only several years long. The logical rotation length is that which will produce the maximum mean annual increment of fuelwood and other small products. The form of the stems is usually so poor that these are the only products intentionally grown under the coppice method. The rotations can be extended to produce saw timber of acceptable quality but this is most often done, intentionally or otherwise, in the process of converting coppice stands to high forests derived from seed. However, even if the coppice method is to be continued, it is still desirable to obtain some reproduction from planting or natural seeding to replace decadent root-stocks and improve the spacing. This kind of restocking can sometimes be encouraged by thinning the coppice stand late in the rotation. Another important objective of any thinning is the reduction of the number of sprouts in a clump; the faster this number can be reduced to one, the better is its growth.

Application of the coppice method to the management of species that regenerate from root-suckers is even simpler than creating a new crop from stump-sprouts. Sprouting is not confined to the stumps of the original stand so that spacing is not a problem nor is any care necessary in preparing the stumps. There is no evidence that the ability to produce root-suckers declines with age. The maximum length of the rotation is determined largely by the rate at which the losses from decay increase. It so happens that the aspens, the American species most commonly regenerated from root-suckers, have an unusually short pathological rotation, although the fungi responsible do not invade the sprouts through the old roots of the parent tree.

Application of the Simple Coppice Method

The chief role of the coppice method is in growing fuelwood, pulpwood, and other products of small trees from sprouting species of angiosperms. For centuries it has been the most efficient way of growing crops of fuelwood. Since this has always been the most important human use of wood, it may be presumed that the coppice system has always been the most common kind of silviculture. In times and places in which fossil fuel is cheap and available, the method has usually been out of favor for a variety of conditional reasons.

Provided that the species is one that sprouts, there is no regeneration method with a higher certainty of success. There is no delay except until the start of the next growing season; the growing space of the soil remains almost continuously and completely occupied. The new stems grow astonishingly fast in height so that the above-ground growing space is quickly reclaimed. The rotations are generally confined to the short periods during which accumulation of total dry matter is most rapid. The optimum rotation length is usually regarded as that which maximizes the mean annual increment of dry matter. Although the trees are seldom impressive in size or appearance, the return on the small investment in growing stock is very high and it may be obtained in a short period.

The method has always been very useful in situations in which the continuing existence of a human community depends on the present and future supply of fuelwood. This was true for many centuries in much of Europe and for about two centuries in parts of the Atlantic seaboard of the United States. The availability of cheap coal and, later, oil caused the situation to change in the first half of the present century. In both places the modern concern about the coppice method has mostly taken the form of replacing coppice forests with high forests. There is, however, one noteworthy difference. In Europe the coppicing had been repeated often and for centuries, sometimes for dozens of times (Rackham, 1980). The typical root-stock had proliferated into a large, gnarled aggregation of tissue that would give rise to dense clusters of sprouts after each of the frequent coppice cuttings. The effects of extreme competition between the sprouts, coupled with those of phototropic bending, often made the individual sprouts into slender, stunted, curving shoots. The grotesque sprout clumps yielded firewood and products useful in the labor-intensive society of a former time.

American foresters were for some years convinced that what they saw in the so-called Sprout Hardwood Region (Fig. 16-4) of the Boston–Washington Megalopolis was the same sort of degraded coppice forest about which they had read in the European literature. However, coppicing in the American cases had usually not been repeated more than two or three times before the collapse of the fuelwood market caused it to cease. Early in the twentieth century it was possible to see many potential deformities in the asymmetries of multiple-stemmed sprout clumps in the aging stands once coppiced for firewood. Now that most of these trees have grown to sawlog size and the old sprout clumps have thinned themselves down to single stems, it takes close examination to detect what once caused concern. The difference is probably one of degree between round stumps ringed with as many as a dozen sprouts and the gnarled callus growths, common in Europe, that produced congestions of hundreds of sprouts.

It may be that the best American case of what caused European concern about excessively repetitive coppicing has come not from cutting of hardwoods but from fires in stands of pitch pine, a sprouting conifer. In parts of southern New Jersey there are small areas called the Pine Plains that are covered with dense clusters of crooked, stunted pine sprouts. The cause of the stunting is controversial. The most plausible is that these areas have been burned so long and often that the pines have come to be the same kind of shoots from gnarled callus tissue that gave hardwood coppice a bad name in Europe.

FIGURE 16-4 Simple coppice stands of mixed hardwoods in Connecticut. The stand in the background is 30 years old. That in the foreground is part of the same stand that was cut for fuelwood during the previous winter; the new sprouts have not yet finished the first year's growth in height. (*Photograph by Yale University School of Forestry and Environmental Studies.*)

Association with Close Utilization of Resources

Part of the bad reputation of the coppice system is of guilt by association with other forms of cruel, but not unusual, punishment of soil and vegetation. When people are reduced to using small sticks of wood for fuel they are also likely to be driven to other deleterious practices. Among these are overgrazing, excessively close utilization of foliage or twigs, and litter removal. Regardless of what defects it may have, it is better for people to use the coppice method to grow fuelwood than to ruin the soil by scraping up litter and animal dung to use for cooking their food.

Pollarding is a special case of coppice cutting that is often associated with situations in which there is pressing demand for both wood and fodder for domestic animals. Instead of having the tops of the stumps and the sprouting level close to the ground, it is kept far enough above the ground to be protected from the animals. At the start of the rotation, the whole tree, usually one of pole size, is decapitated to produce a cluster of sprouts. These are usually thinned out during the short rotation between decapitations. The cutting is normally done when the animals are in most need of the leaves and twigs as fodder. Shade trees are sometimes pollarded where it is desirable to keep their crowns low.

The coppice method is associated with the kind of close utilization and frequent cutting that is most likely to remove chemical nutrients from the site and thus cause their depletion. This is true whether twigs are being removed for fuel by labor-intensive means or whether they are chipped up for pulpwood under circumstances in which the labor needed for manual delimbing is too costly. The losses can be reduced to some extent by measures such as harvesting trees when they are leafless or delimbing or debarking by the stump and thus leaving the twigs, leaves, and branches on the site. If the depletion becomes too great, it may be necessary to apply fertilizer just as is the case in the essentially depletive practice of agronomy.

Another difficulty historically associated with the coppice method is that of changing from short rotations to long ones if it becomes desirable to replace the coppice method with high-forest methods. If, as was once the case in southern New England, large areas were coppiced every 30 years and it becomes necessary to switch to saw-log management with 80-year rotations, the situation becomes very awkward. During the 50-year transition period it is possible to have large areas of forests but with few trees large enough to harvest. Circumstances of this kind were part of the reason why European foresters have cursed the coppice method for so long. Foresters who have inherited situations of this sort are likely to take dismal views about short rotations and proposals for departure from the objective of sustained yield.

The reputation that the coppice method has for being highly productive of dry matter for fuel or pulp is the result of very prompt reclamation of the growing space and the very close utilization often associated with it. The fact that some of the stems increase in dimensions rapidly may be useful for some purposes but is not necessarily an indicator of high site productivity. If the rapid growth of the individual trees is associated with wasteful respiration, it may even be an indicator of a low rate of production per unit of land area.

Modern Use

One virtue of the coppice method is that it thoroughly preserves the genetic attributes of any trees grown by it. This attribute is often used in the culture of various poplars not only by coppicing but also, at least with the cottonwood poplars, by the planting of cuttings that easily take root. If a genetic form of chestnut could be developed to replace the ill-fated American chestnut, it would be logical to grow it by the coppice method to perpetuate whatever planted genetic combination was necessary. Fortunately this species sprouted vigorously even from large stumps and grew rapidly into stems suitable for sawlogs and even for rot-resistant telephone poles.

It would be theoretically possible to use the coppice method to grow almost any American hardwood and the few sprouting conifers listed earlier. This might be done in cases representing special or particular kinds of circumstances. This would be the logical system to follow if one wanted to obtain domestic fuel for a single household entirely from a woodlot of several hectares. A paper company might use it as a means of providing a supply of hardwood pulpwood that was always readily available, regardless of mud conditions, from lands close to the mill. Coppicing is a very useful way of producing browse and low cover for some wildlife species, especially those herbivores that are such important game species. It is not, however, any universal silvicultural panacea of wildlife management; many small mammals and birds require tall trees and these also can be important cover for the large ones. One important and traditional use of the coppice system is for the production of posts and small poles from such species as black locust.

As far as most sprouting species of North American trees are concerned, there is the benefit that sprout regeneration can be an important supplement to seedling regeneration with almost any kind of high-forest management (Wendel, 1975). This is seldom true with clearcutting and planting because volunteer sprout regeneration can be the most troublesome kind of competition in most sorts of forest plantations. Nevertheless, the sprouts of many hardwood species, coast redwood, and even the sprouting pines often develop as well as or even better than associated individuals of seed origin in naturally regenerated stands. Sprout regeneration is not inherently bad.

From the aesthetic standpoint, the coppice method can be even less desirable than the clearcutting method; the cutover areas have the same appearance and the harvests are repeated more often than is the case with true clearcutting. On the other hand, the new vegetation covers the land more quickly. Nevertheless the stands are generally low in stature, monotonously regular, often full of poorly formed trees, and sometimes little different from brushfields.

Coppice stands are almost always dealt with as being even-aged but there is no basic reason why patches of sprouts of different age should not be intermingled in uneven-aged stands by the **coppice selection method.** There might be problems with phototropic effects and suppression of sprouts from root-grafted stumps along the edges of the age groups. If the object is to include production of some large trees then the coppice-with-standards method is more appropriate.

Aspen Culture

The coppice system functions at its best in the management of species that can regenerate, as stands of rather uniformly spaced sprouts, from root-suckers. By far the most important North American example of this is the culture of the quaking and bigtooth aspen poplars, especially in the

FIGURE 16-5 A well-stocked stand of 4800 aspen root-suckers per acre, which originated after a clearcutting made during the dormant season 11 years previously. Chippewa National Forest, Minnesota. (*Photograph by U. S. Forest Service.*)

Great Lakes Region (Graham *et al.*, 1963; Brinkman and Roe, 1975) (Fig. 16-5). Vast aspen forests replaced virgin stands of pine and hardwoods after they were devastated by heavy cutting and fires near the beginning of the present century.

A large and diversified wood industry has come to subsist on coppice management of aspen. This works best on the better soils where the trees can grow to middle age and good diameters before the heart-rotting fungi harm them. Soils that are excessively wet or dry are best converted back to conifers. Aspen regeneration from root suckers depends mainly on being sure to cut all the trees down, preferably during the dormant season. As with other sprouting species, the regeneration step is simple and almost magically automatic.

COPPICE-WITH-STANDARDS METHODS

At various times and places, such as Europe before the Age of Coal, it has been important to combine the production of large amounts of fuelwood with that of some larger trees. This is done by reserving a few of the better trees, called **standards**, at the time each crop of coppice material is cut. The standards are kept to grow while successive crops of coppice sprouts grow up around them; new standards are recruited from each coppice crop to replace the standards removed in thinnings or final-harvest cuttings. This is not the same as the coppice selection method because the standards are allowed to grow to much larger size. However, the compound coppice or coppice-with-standards method produces a kind of stand which has an uneven-aged selection stand superimposed on an absolutely even-aged coppice stand. Above the coppice are several different stories of standards, each story representing a class with an age that is a multiple of the length of the coppice rotation (Fig. 16-6).

The coppice-with-standards method is most easily understood by considering the steps by which a simple coppice is converted into a compound coppice stand. When the coppice reaches the end of the rotation all the trees except for certain carefully selected standards are removed. The individuals reserved should be dominants of good form and species, preferably of seedling origin. The sprouts that arise after the cutting form a distinct story beneath and between the standards.

At the end of the second coppice rotation another age class of standards is selected from the best trees in the coppice and the remainder are cut. Some of the standards left at the end of the first coppice rotation may be harvested, but the better ones are left. This process is continued through as many coppice rotations as desired; the number of age classes increases by one for each cutting of the coppice until the last of the first standards chosen are harvested. The number of standards in a given age class is gradually decreased through successive coppice rotations. If the stand is to be a sustained-yield unit, the distribution of diameter classes of standards should conform to the J-shaped curve that is characteristic of a balanced, uneven-aged stand.

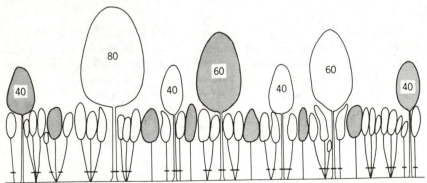

FIGURE 16-6 A compound coppice, with three age classes of standards, at the end of a 20-year coppice rotation. All trees of the coppice stratum will be cut except those (*shaded crowns*) chosen for new standards. The 80-year-old standard has reached the end of its assigned rotation and will be cut. The cutting of three younger standards is an operation analogous to thinning by which the number of standards in a particular age class is gradually reduced with advancing age, just as in a balanced, uneven-aged stand.

When the old standards are finally harvested, their stumps may be too large to sprout and they may have to be replaced by planting or natural seeding from the remaining standards.

Cleanings are sometimes necessary to bring young prospective standards of seedling origin up through the competition of the faster-growing stump-sprouts. The growth of well-established seedlings can also be accelerated by cutting them back to the ground and converting them into seedling sprouts. Plants of this type have the rapid initial growth of true sprouts, combined with the good form and rot resistance of trees of seedling origin. If the seedling sprouts spring up in pairs or clusters, all but the best one should be eliminated.

The standards and the coppice need not be of the same species. Intolerant species with light foliage are usually favored as standards, but trees capable of thriving under partial shade are desirable in the coppice. The species of the coppice must sprout readily and produce wood that is merchantable in small sizes. The standards do not have to be of species that sprout; therefore, a conifer can be grown over a hardwood coppice, being introduced by planting if necessary.

The standards ordinarily are sufficiently isolated to develop deep crowns so their rather short boles grow rapidly in diameter. It is necessary that the coppice rotations be long enough that the sprouts will grow up and cause adequate natural pruning of the standards. This effect is also necessary to eliminate the epicormic branches that may develop on the naturally pruned boles as a result of the extreme exposure after each cutting.

Epicormic branching is a problem in many species of hardwoods and a few of conifers. It is by no means confined to the coppice-with-standards method but may be encountered after any kind of partial cutting. Artificial pruning is not a useful solution because resprouting is usually prompt, so repeated operations would be necessary to keep the stems clear. The physiological causes for the sprouting of dormant buds into epicormic branches are obscure. However, dominant trees with full, vigorous crowns are less likely to develop new branches after exposure than those that have unhealthy crowns or are from lower crown classes. The best way to avoid epicormic branching is to develop good crowns by thinnings and to maintain crop trees as dominants throughout their lives.

Individual trees with stems of high quality can be grown rapidly in open stands without risk of creating large vacancies in the growing space. The invasion of undesirable vegetation is prevented by the rapid development of the young coppice, which occupies the soil continuously. Trees of given diameter can be grown on substantially shorter rotations than in high forests. The compound coppice does not represent as heavy an investment in growing stock as a normally distributed high forest. Therefore, the net return on the investment is comparatively high, especially if fast-grown trees of large diameter command good prices.

FIGURE 16-7 A 30-year-old coppice of beech root-suckers beneath a fine red oak standard about 60 years old on a farm woodlot in southern Connecticut. The coppice, which has been cut periodically for fuelwood, has been used to control the form and pruning of the standard. (*Photograph by Yale University School of Forestry and Environmental Studies.*)

The danger of logging damage to the residual stand is very low. The irregular appearance and the presence of fine individual trees make a compound coppice almost as pleasing aesthetically as a selection forest.

The main problem with this method is that it is difficult to practice unless there is a good market for the fuelwood or other small products that come from the coppice. This is why the method fell into disfavor in Europe and has been practiced only in rare cases in America (Fig. 16-7). The problems are more with the coppice than with the standards. While it is often regarded as an archaic technique, use can be found for it in cases in which demands for fuelwood, pulpwood, or animal browse impinge on the use of forests to grow larger timber, especially on small holdings.

REPRODUCTION BY NATURAL LAYERING

In the peat swamps of the northern forest, regeneration of black spruce and northern white-cedar by layering is an important supplement to that from natural seeding. Although these species bear seed in abundance, a high proportion of the seedlings are suppressed by the relatively fast-growing sphagnum moss, which is the main constituent of the ground cover. The seedlings have the best chance of survival on scattered, elevated hummocks and on decaying wood, a seed-bed that should not be abundant in the managed forest. As time progresses the peat grows up and engulfs the bases of the trees, covering any low-hanging branches that remain alive. The adventitious buds of buried portions of the branches develop into roots while the living ends of the branches turn upward and grow into separate trees. Young shoots of this kind are sturdier and more vigorous than seedlings, but their form is not as satisfactory.

Reproduction of black spruce by layering is most readily achieved in peat bogs by the use of a modification of single-tree selection cutting (Johnston, 1977). If continued reproduction from layers is to occur the uneven-aged form must be maintained so that the living crowns of the trees will remain in contact with the ground long enough for layers to become firmly rooted. It is significant that the residual stands are surprisingly resistant to wind because the peat is so resilient that the force of the wind tends to be dissipated in agitating the soil itself rather than in damage to the trees. However, black spruce growing on the firmer soils of uplands lacks wind-firmness and rarely layers; therefore, it is best reproduced by clearcutting after advance regeneration has become established or by planting.

In northern white-cedar, reproduction from layers is fully as important as that from seed. Although normal layering from low-hanging branches is the most common form of vegetative reproduction, it is also possible for new, upright stems to develop from the branches of windthrown trees. Appreciable amounts of reproduction may also rise from the rooting of small branches that become partially buried after being severed from the parent tree by browsing animals or during logging. The species is important not only for the production of posts and poles but also as a favored winter food of deer.

CONVERSION FROM COPPICE TO HIGH FOREST

Although the coppice methods really have an important place in silviculture, much of the concern of foresters with coppice stands is that of replacing them with stands of seed origin. The very same sprouting habit that makes coppice stands easy to regenerate also makes them hard to eradicate. The degenerate coppice stands of Europe have been a classic problem.

The most serious North American problem of this kind has to do with stands of sprouting woody plants that are not normally thought of as coppice stands. These are the extensive brushfields of shrubs or poor hardwood species that have developed after heavy cutting and repeated fires on soils that are best adapted to the growing of nonsprouting species of conifers. Many of these sites are ones on which the conifers grow well and the broad-leaved plants grow badly but the conifer seed source has been eradicated. Fortunately the development of herbicides and mechanical methods of site preparation has made it much easier and cheaper than it once was to eliminate the sprouting species and restore the better ones by planting. In these situations prescribed burning by itself is seldom helpful, although it can be useful in conjunction with the chemical and mechanical methods.

In the case of formerly coppiced hardwood stands in the eastern United States, there has been reason for replacement chiefly on soils where conifers would grow better. If the soils are good enough for the hardwoods to grow well, it has generally been sufficient to let the stands grow and thin them while the crop trees grow to sawlog size. It is not desirable to thin the stands so hard that they become a simple coppice with one age class of standards unless there is some deliberate purpose in doing so. Shelterwood cuttings can be used for ultimate regeneration of the stands, although both seedlings and sprouts can be anticipated in the ultimate regeneration. However, some of the regeneration will also come from stump-sprouts. As is the case with vegetative regeneration in general, the question of whether this is good depends on the characteristics of the sprouts and not upon some ancient dicta about the relative merits of sprouts and seedlings.

BIBLIOGRAPHY

Brinkman, K. A., and E. I. Roe. 1975. Quaking aspen: silvics and management in the Lake States. *USDA, Agr. Hbk.* 486. 52 pp.

Graham, S. A., R. P. Harrison, Jr., and C. E. Westell, Jr. 1963. *Aspens: Phoenix trees of the Great Lake Region.* Univ. Mich. Press, Ann Arbor. 272 pp.

Hillis, W. E., and A. G. Brown (Eds.) 1954. *Eucalypts for wood production.* Academic, Orlando, Fla. 427 pp.

Jacobs, M. R., and A. Métro. 1981. Eucalypts for planting. *FAO Forestry Series* 11. 677 pp.

Johnston, W. F. 1977. Manager's handbook for black spruce in the North Central States. *USFS Gen. Tech. Rept.* NC-34. 18 pp.

Kemperman, T. A., and B. V. Barnes. 1976. Clone size in American aspens. *Can. J. Bot.* **54:**2603–2607.

Kennedy, H. E., Jr. 1975. Influence of cutting cycle and spacing on coppice syca-more yield. *USFS Research Note SO*-193. 3 pp.

Kramer, P. J., and T. T. Kozlowski. 1979. *Physiology of woody plants.* Academic, New York. 811 pp.

Leffelman, L. J., and R. C. Hawley. 1925. Studies of Connecticut hardwoods. The treatment of advance growth arising as a result of thinnings and shelterwood cuttings. *Yale Univ. School of Forestry Bull.* 15. 68 pp.

McIntyre, A. C. 1936. Sprout groups and their relation to the oak forests of Penn-sylvania. *J. For.* **34**:1054–1058.

North Central Forest Experiment Station. 1972. Aspen: symposium proceedings. *USFS Gen. Tech. Rept.* NC-1. 154 pp.

Panel on Firewood Crops. 1980, 1983. *Firewood crops, shrub and tree species for energy production.* 2 vols. National Academy Press, Washington, D.C. 319 pp.

Rackham, O. 1980. *Ancient woodland, its history, vegetation and use in England.* Edward Arnold, London. 402 pp.

Roth, E. R., and G. H. Hepting. 1943. Origin and development of oak stump sprouts as affecting their likelihood to decay. *J. For.* **41**:27–36.

Schier, G. A. 1978. Vegetative propagation of Rocky Mountain aspen. *USFS Gen. Tech. Rept.* INT-44. 13 pp.

Shigo, A. L. 1985. Compartmentalization of decay in trees. *Sci. Amer.* **252**(4):96–103.

Wendel, G. W. 1975. Stump sprout growth and quality of several Appalachian hardwood species after clearcutting. *USFS Res. Paper* NE-239. 9 pp.

CHAPTER 17

Silviculture of Stratified Mixtures of Species

Structural Concepts

Silvicultural practices are based on understanding of both the structural arrangements of stands and the developmental processes that cause such structures to exist and change over time. The traditional basis is the concept of the even-aged aggregation of trees in which all of the crowns are in a single canopy stratum. In such aggregations the trees differ enough in height growth that the Kraft crown classification (Fig. 2-3) of trees into dominant, codominant, intermediate, and overtopped exists to categorize the comparatively small differences in height that develop between them. The concept of the single-canopied stand was developed to describe pure stands. However, it seldom fits the development and structure of mixed stands and even impedes attempts to deal with them.

The concept of the stratified mixture, on which this chapter is based, is yet another device to systematize understanding of stand development. This structure comes about from processes in which different species grow in height at differing rates so that their crowns tend to become dispersed into different horizontal strata; this can happen even when the trees are of essentially the same age.

The species of even-aged stratified mixtures are ordinarily arranged with intolerants in the upper strata with species of increasing tolerance in each successively lower stratum (Fig. 17-1). There may be more than one species in a stratum. The order of vertical arrangement of species is not necessarily the same at all stages or ages. Not only do some species drop out but some may even exchange their positions.

Some intolerants race quickly ahead but soon slow down and die; other intolerants grow rapidly and steadily to endure to old ages. The more shade-tolerant species almost always start slowly and accelerate later. The very shade-tolerant ones may endure as practically dormant seedlings or saplings and shoot for the sky only when some tree above is eliminated.

FIGURE 17-1 A simple kind of even-aged stratified mixture, consisting mainly of aspen over balsam fir, but with one white spruce and one white pine, remnants of an earlier stand, projecting above the aspen stratum. Superior National Forest, Minnesota. (*Photograph by U. S. Forest Service.*)

Some species grow rather rapidly at first and then slow down such that they lapse into the understory while another that had lagged forges ahead.

The storied structure can have the following advantages: (1) the overstory species get plenty of room in which to grow and (2) the understory species can usually make effective use of the leftover growing space that might otherwise fill up with undesirable vegetation. The latter point is especially important on well-watered sites, which are very difficult to keep under silvicultural control (Fig. 17-2). There are many valuable species of low or moderately low shade tolerance that are simply not capable of occupying good sites fully enough to exclude other species from the lower strata. It is much better to fill the lower strata with useful vegetation than to try to keep them vacant by continual attempts to eliminate undesirable plants from them. In other words, this concept provides a way of living with complexity rather than trying to fight it.

So many of the constituent species start as suppressed advance growth that age is best reckoned from the year when some destructive event eliminated all or most of the overtopping trees and allowed a new age class of mixed species to commence accelerated height growth.

FIGURE 17-2   A stratified mixture, consisting of a 60-year-old upper story of various hardwoods above a somewhat younger lower story of eastern hemlock, in Connecticut, two years after the third thinning of the hardwood stratum. The straight yellow poplars tend to project as emergents above the less erect red oaks and other hardwoods. If the hemlocks were as old as the hardwoods, they would extend up to the bases of the hardwood crowns. (*Photograph by Yale University School of Forestry and Environmental Studies.*)

It is, in other words, possible for one species or group of species to grow for long periods under others and sometimes to grow quite well. In general, there can be as many strata as there are groups of species that differ from one another in height growth and shade tolerance. The stratification is not always perfect or readily apparent. Even when the strata are well differentiated a few individuals of a species that goes with one stratum may have had opportunity to grow up into some gap in a higher stratum, or have been tardy and left behind in a lower one.

Furthermore, there can also be real differences in ages of trees in mixed stands, which can alter the stratification that would exist if all were of the same age. Differences in height can exist, even within species, between trees that started at the same time but with some from new seedlings and others from either released advance growth or sprouts.

Sometimes mixed stands are thought of, and managed, in the context of the single-canopied structure. If there are only two or three species and they do indeed grow at the same rate, there is no reason why the stands cannot be treated almost as if they were pure stands. Ordinarily, however, one species tends to suppress its associates; very small differences in height growth become greater as the leaders forge ahead and the laggards suffer. This can happen in mixtures of loblolly and shortleaf pine; loblolly gets ahead unless something happens to allow the shortleaf to take advantage of its superior adaptability to water shortages. Mixtures of beech, spruce, and fir, all tolerant species, in central Europe must be thinned constantly to keep the beech or spruce from exterminating the fir; in fact, trees that stray out of the single canopy stratum are usually either eliminated or promoted into it by releasing operations.

Mixtures of Pure Groups

There is another concept that accommodates the idea of single-canopied structure to the tendency of different species to grow at different rates. This is that mixed stands are mosaics composed of little pure stands, each arising from a small, pure patch of young trees. Since each patch develops independently (except at the edges), it is possible for each species to grow at its own rate. A process can be postulated in which each patch might be reduced to a single tree in a mixture that became stratified in some late stage of development.

Patchwise mixtures of this kind are most likely to arise from deliberate planting. Patchy variations in microenvironmental conditions might cause them to start from natural seeding, but the randomness imparted by seed dispersal and other factors usually causes species to be intermingled. Clumps of natural root-suckers are commonly pure but they are not necessarily mixed with pure patches of other species.

Interpretation of Diameter Distributions

In Chapter 15 it was pointed out that using diameter distributions to regulate harvests in all-aged stands was really interpreting the stand as a series of little, even-aged, *pure* stands of differing age. As was shown in Fig. 15-6, the J-shaped diameter-distribution curves were composed of the bell-shaped curves for series of different age classes. This kind of J-shaped curve is, therefore, a manifestation of the concept of the single-canopied structure of each age class.

A potentially dangerous misinterpretation arises because even-aged stratified mixtures of species can have J-shaped curves of diameter distribution (Wilson, 1953) (see Fig. 1-3). This can happen because the species that have grown at differing rates in height also have correspondingly different diameters. Each species can be represented by its own bell-shaped diameter-distribution curve. The overall diameter distribution is especially likely to conform to a J-shaped curve if no one bothers to take account of the seedling and sapling classes, which are often absent.

Observation of the existence of the J-shaped diameter distribution can lure one into attempting to manage an even-aged stratified mixture under the selection system just as if it were really uneven-aged. It is well to bear in mind that a real uneven-aged stand, pure or mixed, will have a crown canopy that is very irregular on top. If an even-aged stratified mixture is managed on the pretense that it is all-aged, the species of the uppermost stratum are likely to be eradicated. This may not be bad if the consequences are recognized and if the trees of the lower strata develop acceptably after their release. If the subordinates are crooked, feeble, or rotten, the process is simply one of high-grading.

This kind of misapplication in mixed stands has sometimes given the selection system undeserved guilt by association. In Appalachian hardwoods, for example, one cannot cut all the large red oaks and pretend that understory red maples are going to turn into oaks through some magic conferred by a J-shaped diameter distribution.

Tropical Origin of Concept

Stratified mixtures were first recognized in the wet evergreen and moist deciduous tropical forests, which are practically incomprehensible without this means of analyzing their structure (Richards, 1952). The alphabetical designation of the strata, with *A* for the uppermost, is described in Chapter 1 (Fig. 1-3).

The fact that stratified mixtures can be even-aged has been learned in the temperate zone where trees have annual rings. From these it has been possible to reconstruct the patterns of height growth that lead to whatever structural arrangement exists at any developmental stage (Oliver, 1981). The myriad of such patterns that must exist in moist tropical forests is almost unknowable until ways are invented to determine ages of the trees.

Relation to Non-restrictive Sites

Stratified mixtures with two or more tree strata are most common on sites where soil moisture conditions are favorable enough to allow many species to grow. This is why the concept was first recognized in the humid tropics and fits those temperate forests in which soil moisture does not become deficient during the growing season. If the soil factors are not seriously limiting, the vegetation collectively uses more of the photosynthetically functional light. The vertically distributed foliage in a stratified mixture probably represents a sequence of sun- and shade-leaves more fully adapted to do this than the rather similar array of leaves within a single species.

While the biological potential of such sites to produce wood is very great, the stands can easily get out of silvicultural control. In the early stages particularly they can be the confused tangles for which the term "jungle" was invented. Much of the gross production may be diverted into species of low economic value so that the very sites that are fundamentally most productive may be regarded as poor from the economic standpoint. This view is often applied to moist tropical forests or the bottomland hardwood forest of the southeastern United States. In the latter region, the adjacent dry uplands may be deemed more productive only because it is easier to control hardwoods and grow pines.

It is ironic that the most successful silviculture is conducted not on the very best sites, but where the soil factors are just restrictive enough that species composition can be kept under control. On the best sites, it may be easier and cheaper to learn to understand and live with the problems than try to row upstream in the rapids. The concept of the stratified mixture is one way of systematizing and harnessing the productivity of complex forests.

Stratification Process

Orderly stratification of species is most likely to occur after some disturbance, such as a heavy cutting or blow-down, that was complete enough to cause many species to start growth simultaneously. The presence of advance-growth species is often an essential condition. The developmental pattern may be obscure in the early stages; chaos may seem to reign. Quickly growing, weedy, pioneer vegetation may seem to overwhelm everything and hide the tree seedlings that have started beneath it.

Even where the tendency for the species to segregate into different strata is strong, the pattern will be obscured if the initiating disturbance is not complete and scattered older trees have remained. Such trees are often shade-tolerant ones that were too small to cut, but they can be of almost any origin. These older residual trees usually occupy higher canopy strata than individuals of the species would in a truly even-aged stand. Within a species and single age class, trees of sprout origin may occupy a higher stratum than those originating from new seed. The presence of these kinds of aberrant trees is not necessarily undesirable but it can mask the processes leading to development of stratified mixtures. It has also been an important reason why the concept of the uneven-aged stand has been inflicted on mixed stands without regard for the actual age-class structure.

While it is technically possible to convert complex forests to pure stands it can be costly if natural tendencies resist the conversion. There can also be a variety of unfortunate consequences of an ecological nature. Simple silviculture has important virtues, but the forest should not be torn down and replaced with simple stands just because they are easy to understand.

North American Stratified Mixtures

The most outstanding example is the eastern deciduous forest, which, with its admixture of conifers, is the most complex forest formation outside the tropics (Braun, 1950). The most complicated mixtures of conifers on earth are the coastal forests of the Pacific Northwest with their outliers in the Northern Rocky Mountains; those of the west slope of the Sierra Nevada are only slightly simpler.

Mixed forests in these regions sometimes have definite emergents; distinction can often be made between species of the upper, middle, and lower stories. A species that behaves as an emergent or upper-stratum species in one part of its range may elsewhere occupy the *B* or *C* stratum in mixture with different associates. Douglas-fir is an upper-stratum species in relation to hemlock or true firs on the West Coast but there are places in the dry interior where it is a lower-stratum species with respect to ponderosa pine and may not be able to start in the open. Some valuable intolerants, such as yellow-poplar, black cherry, and black walnut in eastern hardwood forests, can scarcely exist in mixture except as emergents. Very tolerant species, such as sugar maple, beech, and hemlock, almost always form the lower strata of those stratified mixtures in which they occur.

One of the well-differentiated stratified mixtures is that of eastern white pine, various hardwoods, and eastern hemlock occasionally found in the northeastern United States and adjacent regions (Fig. 17-3). In old stands, the scattered white pines are the emergents simply because they continue growing in height after the hardwoods have stopped. The mixed hardwoods form stratum *B*; the very tolerant hemlock forms stratum *C*, so that the two layers of conifers are separated by one of hardwoods. If the stands are viewed casually from a long distance to the side, the few pines may appear to form a nearly pure stand; from above, as nearly pure hardwoods; from underneath, as nearly pure hemlock with some hardwoods and scarce pines.

Of course, many stratified mixtures are less complicated. Two-storied mixtures of fast-growing pioneers, such as aspen poplars, red alder, or the cecropias of the American tropics, over more tolerant species are easily recognized for what they are and treated accordingly.

Integration with Agriculture

The concept of the stratified mixture is also applied in **agro-silviculture** or **agro-forestry** in which the production of crops of trees is combined with that of agricultural plants or forage for domestic animals. These practices are especially important for subsistence agriculture in the humid tropics on soils where stratified mixtures of trees occur naturally. Farmers have even more difficulty controlling the vegetation of such sites than foresters do; the development of stratified mixtures is one way they have of doing this.

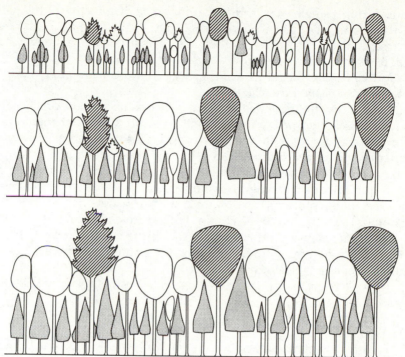

FIGURE 17-3   Stages in the natural development of an untreated stratified mixture in an even-aged stand of the eastern hemlock-hardwood-white pine type. The upper sketch shows the stand at 40 years with the hemlock (*gray crowns*) in the lower stratum beneath an undifferentiated upper stratum. By the seventieth year (*middle sketch*) the emergents (*hatched crowns*) have ascended above the rest of the main canopy, except for the white pine which has only started to emerge. The lower sketch shows the stand as it would look after 120 years with the ultimate degree of stratification developed.

The stratified mixtures of agro-silviculture are not limited to soils with plenty of water. Most species of trees require lots of water over extended periods but this is not true of a number of species of annuals, grasses, or other short plants. Many of these are active only during brief periods when there is plenty of water but are dormant the rest of the year. The water that goes to produce them may be too transitory to be useful to the trees or the trees can be thinned out to divert water to the other vegetation. The combination of a sparse cover of trees with grass or similar plants, as in natural savannah vegetation, is common in semiarid forests.

Management and Regeneration of Stratified Mixtures

The constituent trees of a stratified mixture can be of many different modes of orgin and initial age. If the stands arise after fires or the kinds of clearcutting that simulate them, all of the trees are likely to germinate or sprout during a single year. If the origin is some disturbance in the nature of a windstorm or shelterwood cutting, there is likely to be a mixture of new trees and advance growth of distinctly greater age. Occasionally additional trees may germinate some years after the initiating disturbance but usually there is no vacant growing space for newcomers after the initial trees have established themselves.

Many of the stratified mixtures of the eastern deciduous forest seem to regenerate well after heavy removal cuttings, provided that there is adequate advance growth of those desired species that must start as such. The cuttings are often called clearcuttings, but are better thought of as forms of shelterwood cutting because of the source of some of the important regeneration. It is rather surprising that complicated stands can be reestablished with such simple procedures. It is not clear how widely this approach will work but it is not limited to the eastern United States. The conifer mixtures that sometimes develop on clearcut areas in the Pacific Northwest often form stratified mixtures with Douglas-fir above and red-cedar and hemlock or other tolerants beneath.

Relation between Development and Treatment

In many instances, especially with hardwoods, it is often hard to know when or whether to intervene with cleanings in the early stages of stand development. Usually there is not much experience to guide one. Often it is best to leave the stands alone and let the trees fight it out among themselves; some remarkably fine stands have developed from what may seem like neglect. One advantageous kind of intervention is the removal of certain trees or removal of vines and other plants that are obviously detrimental and likely to endure.

The key to guiding the development of such stands lies in knowing how the different species interact with one another in their height growth. Those that race ahead and soon die must be harvested early or allowed to go to waste. It is indeed possible that one might harvest successive overstory layers as each matured, but provision could have to be made for the fact that the seed source of each species might also be successively eradicated. The advisability of working down to the lower strata and leaving them to continue development usually depends on the potential quality and prospective growth of the trees that are released.

There are cases in which species can exchange vertical positions because of changes in rate of height growth. Eastern red oak, for example, grows at a steady, moderate rate that cannot be accelerated; after several decades associates such as red maple and black birch that had previously gone ahead slow down and lapse into the lower strata (Oliver, 1978). Some species, such as the white pines and spruces, may linger below the top of the canopy and ultimately rise above it simply because they survive or continue growing in height longer than their associates.

If the uppermost species are the most desirable, as is usually the case, stratified mixtures have some useful developmental attributes not found in pure stands. If these species can be depended upon to get on top, it is not necessary to have anywhere near as many of them in the initial regeneration as is the case with pure stands. The rest of the growing space is filled with trees of the species that lag or fall behind. These trees of the subordinate strata are usually tall enough to provide the crowding necessary for the natural pruning and other kinds of training of the taller trees but without restricting the horizontal expansion of the upper parts of the crowns. The congestion that is common in the crown canopies of pure stands is thus avoided.

If the uppermost trees ascend increasingly farther above the lower strata, the freedom of their crowns to expand becomes greater so that one may get the effect of a thinning without having to conduct one. These trees grow much like the deep-crowned standards of the coppice-with-standards system. However, the trees of the lower strata do not have to be cut repeatedly and are more effective as trainers and more salable than true coppice trees would be.

Upper-stratum trees that develop in this manner may have such large crowns that they can continue good diameter growth to advanced ages and they often have fat boles of high quality, even in untreated stands. When one sees fine, wide-crowned cherry-bark oaks in stratified mixtures of bottomland hardwoods in the southern United States it is prudent to resist the temptation to assume that a pure plantation of the species would be better yet. The result of that is apt to be a congested stand of slender oaks. Whenever one finds a large, fine tree in the woods and ponders the ways of growing similar ones, it is sobering to measure the great width of the crown and consider the implications of that for stand management. It is very difficult to thin single-canopied stands hard enough to produce such results without major waste of growing space.

When such stands are thinned it is necessary to examine the relationships between tree crowns carefully and in three dimensions. Although one should scrutinize the tree crowns while marking trees for thinning in single-canopied stands, more reliance can be placed on the diameters and spacings of the tree stems as indicators of crown relationships. With stratified mixtures, it is necessary to note that the crowns of two trees may touch above those of several trees that have stems that stand between them. Much craning of the neck is required.

It often helps to look upon each stratum as a separate stand and thin it as if it were (Fig. 17-4). Sometimes one can discern, within each stratum and if it is viewed separately, the same crown classes that are recognized in pure stands. If a lower stratum is ultimately to be released and made into the main top-canopy stand, it is desirable to anticipate this and thin it down to an appropriate spacing in advance.

The question of whether one or more of the lower strata can be released and made into successive main-canopy stands depends on whether

FIGURE 17-4   A stand in southern Maine, similar to that of Figure 17-3, which has just been thinned with removals from each of the strata and sufficient opening of the lowermost to allow establishment of new advance growth. (*Photograph by U. S. Forest Service.*)

appropriate logging can be done and the qualities of the subordinate trees. Sometimes they have grown in height such that the tops of their crowns have battered against the bottoms of the crowns above. This can leave the crowns too seriously deformed, but usually not. Stem malformations from phototropic leaning are more common. The length of time that the subordinate trees linger in the lower strata increases the period during which they can suffer accidents or pest attacks. Negatively geotropic species such as conifers or sugar maple, sweet-gum, and hickories tend to remain straight and are thus the most promising kind to release.

It is not always possible to make a "new" stand by removing the top layer and releasing the next. If it is, however, one can get two stands for the prices of money, difficulty, and lost production that are normally paid for regenerating one. The countervailing items to consider are mainly those of more complicated logging and the question of whether the production by the released stand is worth as much as that of an entirely new one.

Regeneration

It should not be presumed that the uppermost stratum is always the one to remove first. The trees of any stratum can be removed whenever they are economically mature or when this will facilitate regeneration or accomplish some other purpose. Reasons could be found in different cases to remove the strata in almost any sequence, including starting in the middle. When eastern hemlocks grow under hardwoods they can actually become larger in D.B.H. than the trees above and, furthermore, attain optimum size for sawlogs without ever reaching the upper stratum. Since such hemlocks also inhibit regeneration, it helps to take them out before the hardwood stratum above them.

Since advance growth is usually important for the regeneration of at least some of the desirable species of stratified mixtures, regeneration cuttings commonly take the form of some sort of shelterwood cutting. However, precautions are necessary to maintain sources of seed of the desired species if reliance is placed on natural regeneration. When a given stratum is removed it would be easy to eliminate a species rather permanently.

Aside from artificial regeneration, there are two basic ways of trying to prevent such loss. The simplest conceptually is to follow the uneven-aged group-selection system as shown in Fig. 17-5. With this approach there is a means of keeping the seed source of all desired species in proximity to any group that was to be regenerated, provided that the objective is kept in mind during the mangement of the stand. If this kind of combination of the complexities of both uneven-aged stands and stratified mixtures is feasible, it provides a moderately orderly way of administering the situation.

The other general approach is not as orderly but may often be more pragmatic. It is simply to manipulate the stands such that seed-producing trees of the desired but endangered species are kept scattered within the stands. This would often mean avoiding the complete removal of any one

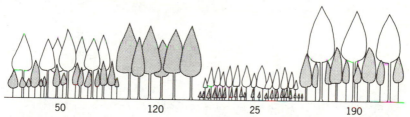

| 50 | 120 | 25 | 190 |

FIGURE 17-5 Part of an uneven-aged stand being managed by the group-selection system and composed of a stratified mixture of two conifers arranged in even-aged groups. The trees with the *light crowns* are of an intolerant species that grows comparatively rapidly when young and attains greater size than the more tolerant (*dark-crowned*) species of the mixture. The numbers indicate the ages of the groups. The upper stratum is ready for the first thinning at 25 years of age and, at 50 years, most of the crowns of this stratum have been freed around most of their perimeter. This stratum is harvested soon after 90 years; this releases the lower stratum which has been left untreated in the meantime to fill any gaps. The remaining stratum is managed for the rest of the rotation as a simple, single-canopied, even-aged group. After 120 years, it is ready for uniform shelterwood cuttings designed to replace the mixture. The trees reserved in these cuttings provide the seed of the tolerant species; that of the intolerant species comes from adjacent groups. There is no silvicultural reason why either stage of the rotation could not be prolonged or terminated earlier; the timing indicated here is purely an example.

species or stratum in any single operation. One would, in other words, follow a kind of irregular shelterwood system (Fig. 17-6). Advantage can usually be taken of the fact that some trees are likely to need to be removed long before or long after most of the others of the same species. If this is done and if there is also deliberate effort to develop potential seed-trees or to store advance growth of the endangered species, ways can ordinarily be found to perpetuate them. With this approach, the stands are likely to become increasingly irregular in height.

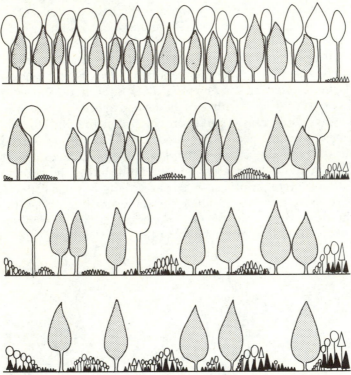

FIGURE 17-6 Reproduction of an even-aged stratified mixture of eastern hemlock (*dark crowns*), hardwoods (*rounded white crowns*), and white pine (*pointed white crowns*) by the irregular shelterwood method. The upper sketch shows a 75-year-old stand prior to treatment. Successive sketches, from top to bottom, show the stand at 15-year intervals, just before each cutting. They indicate how the upper stratum is opened up rapidly, releasing the hemlock but leaving some of the upper stratum for seed and extra increment. The hemlock stratum is opened more slowly and only to the extent necessary to allow regeneration of the intolerant species of the upper stratum. The bottom sketch shows the stand at 120 years, just before the last of the old hemlocks are removed; this leaves a new stratified mixture of somewhat broader range of age than the previous stand. One could accelerate the whole process and make the new stand more uniform by eliminating the hemlock stratum first.

Both approaches can be combined, sometimes with schemes in which final removal cuttings proceed in advancing strips. With either approach, the stands are likely to become more irregular with the passage of time. The orderly stratification of species is also likely to become less distinct, but it is more important that the stands be regenerated and treated intelligently than that they conform to some diagram in a textbook. If the irregularity complicates actual operations on the ground so much that they become too costly, then it may be desirable to maintain a simpler structure.

Multiple-Rotation Species

Some shade-tolerant species can lapse into long stages of postponed development such that it may take them twice or even thrice as long to grow to mature size as some more quickly growing associate. This is not necessarily bad and may even represent a common natural growth habit of some species. Marquis (1981b) has shown that small pole-sized sugar maples and beeches, left as being too small to cut during very heavy removal cuttings, in Allegheny hardwood stands developed remarkably well as components of otherwise even-aged stands that came up around them.

As shown in Figs. 17-7 and 17-8, such trees, if of good form, can become valuable components of what might otherwise seem like even-aged stands. The irregularity that these hold-over trees imposed on the strata of black cherry that grew up around them caused broader differentiation of crown classes in the cherries and, therefore, much better diameter growth of the potential crop trees among them. The cherries also grew up rapidly enough to prevent the whippy hold-over trees from becoming large-crowned wolf-trees. It has often been presumed that such seemingly stunted, small residuals are of no value and ought to be eliminated as a routine matter. As is usually the case, it is well to look for the exceptions that seem to lie behind each general rule.

The same phenomenon often occurs with shade-tolerant conifers such as certain spruces, true firs, and hemlocks. For example, in mixtures of red spruce and balsam fir the spruces initially grow more slowly than contemporaneous firs (Fig. 17-7). Some of the spruces, when less than about a meter in height, stall in height growth. Branch elongation continues such that these saplings develop "umbrella" tops or the form of a candelabra. They survive in this quasi-dormant condition and are able to resume height growth when released by the cutting or death of the short-lived balsam firs. Large trees with this history of release can be identified by dense clusters of dead branches that once formed the umbrella tops. More normal advance-growth saplings with deep, conical crowns actually respond to release more quickly, but the peculiar umbrella spruces are striking evidence of the reality of the phenomenon.

The red spruces may even go into two stages of arrest such that they complete their development during three balsam fir rotations. If they grow well after their ultimate release, their total age at that time is of no conse-

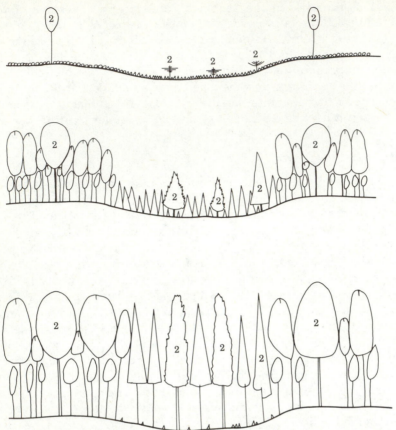

FIGURE 17-7   Three stages in the development of an unthinned mixed
stand, on two different soils, in which trees held over from the previous
rotation become major components of the next stand. The trees marked 2 are
taking two rotations to make full development. Those with rounded crowns
and vertical marks at the tips are the fast-growing black cherries of the case
described by Marquis (1981b), while those without such marks are shade-
tolerant maples or beeches. The conifers on the poorly drained site are red
spruces (rough, pointed crowns) and balsam firs (smooth, conical crowns).
The trees marked 2, which were subordinates in the previous stand become
the largest trees of the new stand and also prevent it from developing into
excessively uniform thickets. Top sketch shows initial condition.

quence and does not add to the length of the "rotation." Since their size
gives the effect of having planted a very tall seedling, any rotation reckoned from the time of release is actually shortened.

The important point about these examples is that they are cases in
which one species of a mixture may mature while two or more rotations of
another one grow. This may be a common state of affairs in nature. There
is evidence (Veirs, 1982) that coast redwoods may endure for more than a
dozen centuries while Douglas-firs come and go around them every century or two. The question of whether such developmental patterns are

FIGURE 17-8 Two stages, 43 years apart, in the development of a stratified mixture of black cherry (rough-barked in lower picture) and sugar maple (smoother bark) in northwestern Pennsylvania. The small pole-sized trees left after the heavy cutting shown in the first picture are of the kind shown in the upland hardwood portion of Figure 17-7; they are obscured by younger trees in the later picture. (*Photographs by U. S. Forest Service.*)

503

useful in managed stands usually depends on the feasibility of harvesting timber without destroying residual trees. Just as there is good reason to distinguish between single-canopied stands and stratified mixtures, it also helps to note that the two-aged stand can be an important and useful condition intermediate between even- and uneven-aged.

Moist Tropical Forests

While the concept of the stratified mixture was originated in these complex tropical forests, neither it nor any other mode of interpretation has led to any universally satisfactory silvicultural procedure. With such complexity it is logical to suppose that a large number of procedures, many of them yet unimagined, will be developed.

Much of the more theoretical kind of silvicultural and ecological thought has always gravitated toward the perpetuation of the very complicated "primary" or climax forests of evergreen broad-leaved species on those sites that are biologically the most favorable. It must be recognized that there is no silvicultural magic that will enable such forests to be harvested and regenerated without altering them in some degree. This has not happened with any other kind of forest. There are all sorts of reasons, including the guidance of silvicultural practice, to safeguard significant areas of old, natural forests as areas of vegetation that are simply preserved and not treated.

There have been many efforts to achieve acceptable natural regeneration by partial cutting in these forests but so far success has been achieved only in somewhat exceptional cases. The longest experience and closest approach to success has been in southeast Asia (Whitmore, 1975), where the "tropical shelterwood system," in which one or more lower strata are removed to make way for regeneration, has sometimes been successful. It is advantageous that the dipterocarps, which include some of the most valuable species of the region, are heavy-seeded and adapted to start as advance growth. When this approach is followed on other continents, the vacant lower strata are sometimes claimed by vines and other undesirable vegetation.

In any part of the tropics, including those of high rainfall, there can be soils, such as those of poorly aerated swamps, conducive to stands of pure or simple composition. Silvicultural management of these has often been comparatively uncomplicated.

It is probable that the silvicultural problems would be much easier to solve if uses were found for more of the species. In the past, as in many other parts of the world, there has been preoccupation with a few species highly valued for export or some other specific purpose. Although some species are always more useful than others, a weed may cease to be a weed if a use is found for it. Furthermore, if there is need to eliminate a plant to guide some silvicultural development, it is usually more economical to use it than to kill it. This problem is easily solved where fuelwood is in short supply, but there are often exasperating difficulties of transporting low-value wood from remote forests.

It is often stated that the production of primary forests in the humid tropics is very low even though there are other statements that are quite the opposite. The most plausible explanation of these contradictory views is that if only the production by a few scattered trees of cabinet wood is counted it would be small even if the total biological production were large. However, the situation is also complicated by the lack of annual rings and difficulty of knowing ages of trees. Stand development may take place so fast that the net production by a 50-year-old stand of primary or climax vegetation in Panama may be as small as that of a 350-year-old stand in Oregon.

Attention sometimes turns to managing for the "secondary" or early successional species that are often rather easily regenerated. These may or may not be associated with or followed by regeneration of species of later successional status. The combinations and permutations of what may happen or be done are both numerous and mostly yet to be learned.

It is important to note that most tropical regions are actually rather dry and that deciduous forests, savannahs, and grasslands are much more common than moist forests and certainly more so than true rain forests. Most people tend to live in the drier regions. Production of fuelwood is really more important to the population than old-style exploitation of distant humid forests for export species. Under drier conditions, the silvicultural practices are not different in principle from those applicable in most of the temperate zone. Coppice management for fuelwood may be more important but pure stands are probably just as common.

Even in the humid tropics, the main forestry problems do not turn on knotty questions of whether to preserve, manage, or destroy remote stands of primary vegetation. They are, as has commonly been the case with forestry, how to rehabilitate forests and lands that have been damaged by hard or unwise use. Much of this has involved various kinds of planting.

The simplest schemes involve establishment of pure plantations of fast-growing native or exotic species, such as *Gmelina arborea* or *Eucalyptus deglupta*, which are both from southeast Asia. The pines, such as *Pinus caribaea*, are especially useful not only for providing long-fibered softwood but also for coping with the tall grasses that often usurp lands degraded by agricultural use. A more complicated planting technique is that of **enrichment**, which usually involves putting plants of the most valuable species in lanes or openings cut in the stands. This works only if followed by frequent cleanings.

Stratified Mixtures in Agro-forestry

Given the shortages of agricultural land in the settled parts of the tropics, much attention has turned to the ancient practices of agro-forestry or agro-silviculture (Wiersum, 1981). Farmers in the humid tropics have long grown stratified mixtures as a way of coping with weeds and for both utilizing and controlling solar radiation. Often the mixtures are only of herbaceous plants, but strata of shrubs and trees are sometimes included. The growing of coffee shrubs beneath taller trees is one of the simplest cases. The most complicated are typified by the "forest gardens" of Indonesia. These are small plots with overstories of timber trees over middle strata of citrus or other fruit trees with food plants in the lowest strata. The tall trees may also shade the farmhouse. The gardens are managed by adroit removals, releases, and replantings in the different strata.

The **taungya** method is a common way of integrating agriculture and reforestation. Just before land is taken out of cultivation it is planted with both trees and plants. The weeding of the crop also keeps the tree seedlings free and the initial clearing provided the site preparation. The food plants use space that the slower-growing trees would not claim during the early stages. This method is often associated with shifting cultivation in which farmers keep shifting around, clearing land, farming it for a time, and then abandoning it to clear some more when the soil starts to deteriorate.

One kind of agro-forestry that can be conducted not only in humid areas but also in dry ones involves combinations of tall, woody plants and low grass and browse for animals. In dry areas this may merely imitate the savannahs, mixtures of scattered trees and grass, that naturally characterize the driest sites on which trees can grow. Forage production can be enchanced by growing leguminous trees over grass to increase the supply of soil nitrogen. Sometimes the trees are pollarded in such cases to provide branchwood and fodder when there is no edible grass.

Planting of Mixed Stands

Stratified mixtures can be created by planting. The most successful mixtures are composed of faster-growing intolerant species above slower-starting tolerants (Fig. 17-9). If the trees of the uppermost species are not too dense and numerous, they grow more rapidly in diameter than they would if crowded into the single canopy of a pure plantation; the lower-stratum species can provide much the same training effect of a pure stand. The upper-stratum species should not comprise more than a quarter of the total number of trees.

If there are species that differ sufficiently in rate of growth and tolerance it is possible to create stands with several different strata, although it is rare that much advantage can be gained by having more than three species. Properly selected species can be planted simultaneously at random or according to a variety of patterns and the stratified mixture will develop automatically. Sometimes the fastest growing can be planted several years after those destined for the lower strata.

Pioneer species, such as the larches, hard pines, and poplars, are the

FIGURE 17-9 A Connecticut forester in a stratified mixture of conifers that he planted 33 years earlier, with equal numbers of 4 species distributed at random. The species, in order of decreasing present height, are European larch, red pine, white pine (mostly removed in thinnings), and Norway spruce. The larches, such as the one beside the forester, form a topmost story and each has had plenty of space for growth of large crowns and diameters. (*Photograph by Yale University School of Forestry and Environmental Studies.*)

most satisfactory overstory species because they grow rapidly in height and cast light shade. The best species for the lower strata are tolerant ones, typified by spruces, hemlocks, and others characteristic of late successional stages. Species of intermediate tolerance, such as Douglas-fir and soft pines, can be used for any story depending on the associated species.

It is seldom that two species will grow at the same rate and thus develop into single-canopied mixtures. If they do it is usually only within some narrow range of site conditions. The one with the more rapid juvenile height growth usually goes to the top and the other either lapses into the understory or dies of suppression. One may be a leader on poor sites but a follower on good sites. Even small differences in initial height growth lead to this development by way of increasing differences.

The resulting self-thinning effect enhances the diameter growth of the ascendant species, but the other becomes a rather costly filler and trainer if it does not grow to useful size. If mixtures of closely similar species are desired it is necessary to separate them into groups and strips wide enough to allow space for mature individuals of the particular species. If horizontal mixtures of this sort are planted, it may be well to arrange then in ways that will expedite future operations or fit variations in site and existing vegetation.

Relative Merits of Pure and Mixed Stands

The question of whether mixed stands are superior to pure stands depends on the circumstances and not on sweeping generalizations. It takes more care and skill to treat and manage them. Their yields and patterns of development are less predictable than with pure stands. They may be safer, more productive, less costly to maintain, and more attractive, but not always.

Mixtures of species are usually less damaged by biotic pests than pure stands for much the same reasons that uneven-aged stands can be safer than even-aged. The physical separation of susceptible plants inhibits the spread of many pests, but probably not highly mobile ones such as spore-dispersed fungi. It may be noted that no real physical separation exists between species in a stratified mixture in which there is an essentially pure stratum above or below ground. If the food supply of some pest does not include all of the species of a mixed stand then it is diluted in ways that may inhibit buildup of the pest population. It is likely that mixtures of species are, in general, more secure against pests than would be the case with uneven-aged pure stands; a given pest is more likely to attack many age classes of a single species than to attack different species.

Nevertheless, there are important instances in which pure stands are more resistant to certain pests than stands of the same species mixed with highly susceptible ones. Pure stands of spruce are more resistant to the misnamed spruce budworm than spruce–fir stands that have overmature balsam firs on which the insect can thrive. It may be noted, however, that mixtures of hardwoods with spruce and fir are even more resistant. The fact that the American chestnut grew in complex mixtures did not save it from the chestnut blight fungus; however, if it were still present in eastern forests, there might be less difficulty with the gypsy moth, to which chestnut is unpalatable. The heteroecious stem rusts of conifers, or any other organism with alternate hosts, could not exist in absolutely pure stands; they are instead adaptations to the association of different species. The resistance of any kind of stand to its biotic enemies depends on the susceptibility to attack and vulnerability to damage of its individual components and not on doctrinaire generalizations.

The relative resistance of mixed or pure stands to damage from fire and from mechanical agencies such as wind or frozen precipitation depends on the physical characteristics of the individual trees or stands and the physics of action of the agencies. If the mixed condition is going to increase resistance to such damage, it must be by some effect that changes some physical condition.

Regardless of the agency of damage, however, a mixed stand of both resistant and nonresistant species is bound to be more secure against injury than a pure stand of a nonresistant species, but less so than a pure stand of a resistant species. Sometimes the main value of mixed stands in this respect is the rather dreary consolation that the risk of losing whole stands all at once is reduced.

The question of whether mixed stands are more productive than pure depends on (1) the units of volume or value on which comparison is based, (2) relative vulnerability to losses, and (3) relative degree of utilization of the growth factors of the site.

In terms of total dry-matter production, single-canopied mixtures, with species mixed in the horizontal dimension, are almost sure to give less than pure stands of the most productive member. The productive potential of the fastest growing species is normally diluted by its intermingling with the less productive. Exceptions would exist if the mixed stand suffered less loss or had some beneficial characteristic, such as improved nitrogen fixation or soil porosity, that increased growth.

Stratified mixtures, however, are likely to show greater total productivity than pure stands of the intolerant overstory species. The tolerant species of the lower strata are, by their very nature, more efficient in photosynthesis than the overstory species. Furthermore, they must to some extent use solar energy not absorbed by the overstory. They also use some of that part of total production that might exist in any lesser vegetation that would otherwise develop beneath a "pure" stand. It is, however, possible that a stratified mixture might be less productive than a pure stand of an evergreen, tolerant species adapted to grow as an understory. Evidence about these matters is very scanty. The situation may differ depending on whether soil factors or light are most limiting or whether production is measured in terms of weight or volume of wood.

Foresters are usually more concerned about merchantable yield than total production, even though one cannot be managed well without understanding the other. One highly important consideration is the rate of diameter growth, which in turn affects rotation length. This often means that the typical overstory species may have some valuable attributes even though it is fundamentally less productive than some associate normally in the understory. The wood of one species is often inherently more valuable than that of another without respect to its ecological status. If the stratified mixture has virtue with respect to yield, it is often because of the opportunity to combine the quick growth of the individual trees of the overstory species with the slow-starting but high and long-sustained production that the understory species achieves after its release.

Mixed stands probably utilize the soil more effectively than pure stands, although the extent to which they do is not clear. There is no evidence that the root systems of different species become stratified like the crowns, but some penetrate more deeply than others and thus extract water and nutrients from deeper levels. The nutrients ultimately return to the forest floor in fallen leaves and thus become available to the whole community. Some species also extract more nutrients than others. The addition of a species effective in capturing nutrients would thus improve upon the nutrient status of a pure stand of a species that was not so effective.

It has sometimes been postulated that mixed stands may be more effective than pure stands in preventing nutrients from leaching out of the forest system. Most evidence suggests that if one component of the vegetation does not fully use some available nutrients some other will appear or expand to do so. Some nutrients leach into streams but there is so far nothing showing that the rate is faster with pure stands.

If stand composition is such that litter decomposes very slowly, much of the nutrient capital of the site may be bound in unavailable form in thick layers of humus. This is most likely to occur under very dry, wet, or cold conditions. This problem is best handled by cuttings or site treatments to speed decomposition. Sometimes, however, the decomposition of coniferous litter is enhanced by an admixture of hardwood litter, which supports a more abundant and versatile population of decomposing and disintegrating organisms.

No manipulation of vegetation can create any increase in that part of the total nutrient capital of a site that is of geological origin. Fixation of atmospheric nitrogen can be improved but that is about all. The fact that poor soils often have pure stands whereas mixed stands are associated with good ones is an effect rather than a cause of the condition of the soil.

The choice between pure and mixed stands is often one between pure conifers and mixtures of conifers and hardwoods. Evergreen conifers generally produce more substance than deciduous hardwoods and a higher proportion of it is deposited in the usable bole. Furthermore, the growth of conifers is diminished less by unfavorable site conditions than that of hardwoods. Consequently, a pure coniferous stand plagued with some ills might still be superior to a conifer–hardwood mixture. Therefore, the conversion of hardwood stands on poor and mediocre sites to pure conifers is usually regarded as desirable. It is on some of the best sites, where hardwoods (and most other species) grow well and where the vegetation is hard to control, that growing quasi-natural stratified mixtures may be the best solution.

CONCLUSION

This book has presented some of the thoughts and ideas that foresters have developed about the art and science of silviculture. The main purpose has been to help those who are starting to learn about it and to stimulate their thoughts. It is hoped that this is only the beginning of the process. It should not be presumed that success can be achieved by copying some solution from the book. Good silviculture depends on foresters who think analytically about what is known, and what they themselves can learn, about forests and how they can be managed to serve society.

BIBLIOGRAPHY

Braun, E. L. 1950. *Deciduous forests of eastern North America.* 1974 reprint. Macmillan, New York. 596 pp.

Kellison, R. C., D. J. Frederick, and W. E. Gardner. 1981. A guide for regenerating and managing natural stands of southern hardwoods. *N. C. State Univ., Sch. Nat. Resources Bull.* 463. 24 pp.

Longman, K. A., and J. Jénik. 1974. *Tropical forest and its environment.* Longman, New York. 196 pp.

Marquis, D. A. 1981a. Removal or retention of unmerchantable saplings in Allegheny hardwoods: effect on regeneration after clearcutting. *J. For.* **79**:280–283.

Marquis, D. A. 1981b. Survival, growth, and quality of residual trees following clearcutting in Allegheny hardwood forests. *USFS Res. Paper* NE-277. 9 pp.

Means, J. E. (Ed.) 1982. *Forest succession and stand development research in the Northwest.* Oreg. State Univ. For. Res. Lab., Corvallis. 171 pp.

Oliver, C. D. 1978. The development of red oak in mixed stands in central New England. *Yale Univ. Sch. For. and Env. Studies Bull.* 91. 63 pp.

Oliver, C. D. 1981. Forest development in North America following major disturbances. *For. Ecol. and Mgmt.* **3**:169–182.

Richards, P. W. 1952. *The tropical rain forest, an ecological study.* Cambridge Univ. Press, London. 450 pp.

Siedel, K. W., and P. H. Cochran. 1981. Silviculture of mixed conifer forests in eastern Oregon and Washington. *USFS Gen. Tech. Rept.* PNW-121. 70 pp.

Veirs, S. D., Jr. 1982. Coast redwood forest: stand dynamics, successional status, and the role of fire. *In:* Means, J. E. (Ed.) *Forest succession and stand development research in the Northwest.* Oreg. State Univ. For. Res. Lab., Corvallis. Pp. 119–141.

Whitmore, T. C. 1975. *Tropical rain forests of the Far East.* Oxford Univ. Press, London. 282 pp.

Wiersum, K. F. 1981. *Viewpoints on agroforestry.* Agricultural University Wageningen, Netherlands. 185 pp.

Wilson, R. W. 1953. How second-growth northern hardwoods develop after thinning. *USFS, Northeastern For. Exp. Sta., Sta. Paper* 62. 12 pp.

A P P E N D I X I

Common and Scientific Names of Trees and Shrubs Mentioned in the Text

Common Name	Scientific Name*
Alder, red	*Alnus rubra* Bong.
Ash, mountain (eucalypt)	*Eucalyptus regnans* F. Muell.
Ash, white	*Fraxinus americana* L.
Aspen, bigtooth	*Populus grandidentata* Michx.
quaking	*Populus tremuloides* Michx.
Bald-cypress	*Taxodium distichum* (L.) Rich.
Basswood, American	*Tilia americana* L.
Beech, American	*Fagus grandifolia* Ehrh.
European	*Fagus sylvatica* L.
Birch, black	*Betula lenta* L.
gray	*Betula populifolia* Marsh.
paper	*Betula papyrifera* Marsh.
yellow	*Betula alleghaniensis* Britton
Blackberry	*Rubus* spp.
Black-gum	*Nyssa sylvatica* Marsh.
Cecropia	*Cecropia* spp.
Cedar, Spanish	*Cedrela mexicana* Roem.
Cherry, black	*Prunus serotina* Ehrh.
choke	*Prunus virginiana* L.
pin	*Prunus pensylvanica* L. f.
Chestnut, American	*Castanea dentata* (Marsh.) Borkh.
Cottonwood, eastern	*Populus deltoides* Bartr. (ex Marsh)
Currant and gooseberry	*Ribes* spp.
Dogwood, flowering	*Cornus florida* L.
Douglas-fir	*Pseudotsuga menziesii* (Mirb.) Franco
Eucalypts	*Eucalyptus* spp.

*Scientific names of North American trees based on: E. L. Little, Jr., *Checklist of United States trees*. USDA, Agr. Hbk. 541, 1979, 375 pp.

512

Fir, balsam	*Abies balsamea* (L.) Mill.
European silver	*Abies alba* (Mill.)
grand	*Abies grandis* (Dougl.) Lindl.
noble	*Abies procera* Rehd.
Pacific silver	*Abies amabilis* (Dougl.) Forbes
subalpine	*Abies lasciocarpa* (Hook.) Nutt.
white	*Abies concolor* (Gord. & Glend.) Lindl. (ex Hildebr.)
Hackberry	*Celtis* spp.
Hemlock, eastern	*Tsuga canadensis* (L.) Carr.
mountain	*Tsuga mertensiana* (Bong.) Carr.
western	*Tsuga heterophylla* (Raf.) Sarg.
Hickory	*Carya* spp.
Hinoki	*Chamaecyparis obtusa* Endl.
Hop-hornbeam, eastern	*Ostrya virginiana* (Mill.) K. Koch.
Incense-cedar, California	*Libocedrus decurrens* Torr.
Larch, European	*Larix decidua* L.
Japanese	*Larix leptolepis* (Sieb. & Zucc.) Gord.
western	*Larix occidentalis* Nutt.
Locust, black	*Robinia pseudoacacia* L.
Madrone, Pacific	*Arbutus menziesii* Pursh.
Mahogany, Honduran	*Swietenia macrophylla* King.
Manzanita	*Arctostaphylos columbiana* Piper
Maple, big-leaf	*Acer macrophyllum* Pursh
red or soft	*Acer rubrum* L.
sugar	*Acer saccharum* Marsh.
vine	*Acer circinatum* Pursh
Oak, bear	*Quercus ilicifolia* Wangenh.
black	*Quercus velutina* Lam.
bur	*Quercus macrocarpa* Michx.
cherry-bark	*Quercus falcata* var. *pagodaefolia* Ell.
northern red	*Quercus rubra* L.
pin	*Quercus palustris* Muenchh.
post	*Quercus stellata* Wangenh.
scarlet	*Quercus coccinea* Muenchh.
southern red	*Quercus falcata* Michx.
white	*Quercus alba* L.
Palmetto, saw	*Serenoa repens* (Bartr.) Small
Pine, Aleppo	*Pinus halepensis* Mill.
bishop	*Pinus muricata* D. Don
Caribbean	*Pinus caribaea* Mor.
eastern white	*Pinus strobus* L.
jack	*Pinus banksiana* Lamb.
jelecote (Mexican)	*Pinus patula* Schl. & Cham.
loblolly	*Pinus taeda* L.
lodgepole	*Pinus contorta* Dougl.

longleaf	*Pinus palustris* Mill.
Monterey	*Pinus radiata* D. Don.
pitch	*Pinus rigida* Mill.
pond	*Pinus serotina* Michx.
ponderosa	*Pinus ponderosa* Laws.
red or Norway	*Pinus resinosa* Ait.
sand	*Pinus clausa* (Chapm. ex Engelm.) Vasey ex Sarg.
Scotch	*Pinus sylvestris* L.
shortleaf	*Pinus echinata* Mill.
slash	*Pinus elliottii* Engelm.
sugar	*Pinus lambertiana* Dougl.
Virginia	*Pinus virginiana* Mill.
western white	*Pinus monticola* Dougl.
white-bark	*Pinus albicaulis* Engelm.
Poplar	*Populus* spp.
Raspberry	*Rubus* spp.
Red-cedar, eastern	*Juniperus virginiana* L.
western	*Thuja plicata* Donn
Redwood, coast	*Sequoia sempervirens* (D. Don.) Endl.
Sassafras	*Sassafras albidum* (Nutt.) Nees
Sequoia, big-tree	*Sequoiadendron giganteum* (Lindl.) Buchholz
Spruce, black	*Picea mariana* (Mill.) B.S.P.
Engelmann	*Picea engelmannii* Parry
Norway	*Picea abies* (L.) Karst.
red	*Picea rubens* Sarg.
Sitka	*Picea sitchensis* (Bong.) Carr.
white	*Picea glauca* (Moench.) Voss.
Sugi	*Cryptomeria japonica* D. Don
Sweet-gum	*Liquidambar styraciflua* L.
Walnut, black	*Juglans nigra* L.
White-cedar, Atlantic	*Chamaecyparis thyoides* (L.) B.S.P.
northern	*Thuja occidentalis* L.
Willow	*Salix* spp.
Yellow-poplar	*Liriodendron tulipifera* L.

APPENDIX II

Abbreviations in Reference Lists

Standard abbreviations are used in the names of serials and periodicals with the exception of those listed below. The abbreviations for some regional experiment stations of the U. S. Forest Service of recent years are prefixes to the serial numbers of the publications.

AES   Agricultural Experiment Station
FAO   Food and Agriculture Organization of the United Nations
INT   Intermountain Forest and Range Experiment Station
NE   Northeastern Forest Experiment Station
NC   North Central Forest Experiment Station
PNW   Pacific Northwest Forest and Range Experiment Station
PSW   Pacific Southwest Forest and Range Experiment Station
RM   Rocky Mountain Forest and Ranger Experiment Station
SE   Southeastern Forest Experiment Station
SO   Southern Forest Experiment Station
USDA   U.S. Department of Agriculture
USFS   Forest Service, U.S. Department of Agriculture

SUBJECT INDEX